FRONTISPIECE   (11) THE KING'S MANOR. E. front. *c.* 1480 and later.

ROYAL COMMISSION ON HISTORICAL MONUMENTS

ENGLAND

An Inventory of the
Historical Monuments in the

CITY OF
YORK

Volume IV
Outside the City Walls
East of the Ouse

LONDON · HER MAJESTY'S STATIONERY OFFICE

MCMLXXV

ISBN 0 11 700335 2★ (Red binding)
ISBN 0 11 700719 6★ (Grey binding)

TABLE OF CONTENTS

Map showing the position of Monuments

St. Mary's Abbey, plan

The King's Manor, plan of ground floor

The King's Manor, plan of first floor

Fig. 1. Floor-tile from St. Mary's Abbey.

LIST OF ILLUSTRATIONS

(The prefixed numerals in brackets refer to the Monument numbers in the text.)

ix

Embrasure in Postern Tower, Bootham.

Fig. 2.

PREFACE

THIS fourth Inventory of Monuments in the City of York includes St. Mary's Abbey and the King's Manor, several areas of mediaeval settlement just outside the walls, and other areas which have become part of the City as 19th-century development spread over them. The walls enclosing St. Mary's Abbey were described and illustrated in the earlier Inventory volume, *York* II: *The Defences* (1972), but the relevant entry is repeated here to give a complete view of the whole abbey complex.

The description of St. Mary's Abbey has benefited from the researches of Dr. C. A. R. Radford and the late Professor F. Wormald, who have thrown much light on the uses of the various parts of the abbey, and of Professor G. Zarnecki, who has contributed to the study of the sculpture. The account of the King's Manor owes much to the researches of Sir John Summerson and Mr. H. M. Colvin.

The Yorkshire Museum contains a collection of sculpture from various parts of York, many of the pieces having come from buildings which no longer exist. It is not the Commission's general policy to list the contents of museums, but the sculpture here preserved is discussed in so far as it relates to the buildings of York, past and present. Similarly a brief indication of the extent of the topographical drawings in the City Art Gallery has been included, being relevant to any study of the buildings of York.

In the domestic field the destruction caused by the siege of York in 1644 and later rebuilding have left no substantial remains from a period earlier than the middle of the 17th century, and early documentary records of the kind that contributed much to the interest of *York* III (1972) do not exist for the extramural area now described. Here the early 19th-century material occupies the larger part of the Inventory, and a number of the houses of that period which provide good examples of contemporary design and workmanship have been treated more fully than the early 19th-century houses in our previous volumes; the descriptions of the numerous smaller houses have been rigorously compressed. In the Sectional Preface to our third volume on the City of York the design of houses was discussed at some length and many comments then made are equally applicable to the houses in the area now covered. Descriptions of all the buildings that have been recorded by the Commission's staff in the area have been included, even where the buildings have since been pulled down. Demolitions have resulted in the disappearance of some whole streets, but the map in the Inventory which shows the position of monuments is from an edition purposely chosen to show all the monuments recorded still standing. The most important change since the date of the map has been the bridging of the Ouse at Clifton.

In accordance with the Commission's practice no monument has been included which has not been inspected, and the written descriptions are amplified by drawn plans and elevations. Plans and elevations of houses are reproduced at a uniform scale of twenty-four feet to one inch; where a full key to the dating conventions used is not given, black represents original work and white or dotted represents alterations and additions. Some of the plans and elevations have been redrawn from the architects' original drawings. The photographs have all been taken by the Commission's photographers and include reproductions of some drawings which throw light on the earlier appearance of the buildings described.

I would draw attention to the fact that the record cards for the City of York may be consulted by accredited persons who give notice of their intention to the Secretary of the Commission. Copies of photographs may be purchased from the National Monuments Record.

ADEANE

LIST OF COMMISSIONERS

The Right Honourable The Lord Adeane, P.C., G.C.B., G.C.V.O. (*Chairman*)
Her Majesty's Lieutenant in the City of York and the West Riding (*ex officio*)
Henry Clifford Darby, Esq., O.B.E.
Courtenay Arthur Ralegh Radford, Esq.
Sir John Newenham Summerson, C.B.E.
Howard Montagu Colvin, Esq., C.B.E.
William Abel Pantin, Esq.
Arnold Joseph Taylor, Esq., C.B.E.
William Francis Grimes, Esq., C.B.E.
Maurice Willmore Barley, Esq.
Sheppard Sunderland Frere, Esq.
Richard John Copland Atkinson, Esq.
Sir John Betjeman, C.B.E.
John Nowell Linton Myres, Esq.
Harold McCarter Taylor, Esq., C.B.E.
George Zarnecki, Esq., C.B.E.
John Kenneth Sinclair St Joseph, Esq., O.B.E.

Secretary
Robert William McDowall, Esq., O.B.E.

THE ROYAL WARRANT

WHITEHALL,
2ND OCTOBER, 1963

The QUEEN has been pleased to issue a Commission under Her Majesty's Royal Sign Manual to the following effect:

ELIZABETH R.

ELIZABETH THE SECOND, by the Grace of God of the United Kingdom of Great Britain and Northern Ireland and of Our other Realms and Territories, QUEEN, Head of the Commonwealth, Defender of the Faith,

To

Our Right Trusty and Entirely-beloved Cousin and Counsellor Robert Arthur James, Marquess of Salisbury, Knight of Our Most Noble Order of the Garter;

Our Trusty and Well-beloved:

Sir Albert Edward Richardson, Knight Commander of the Royal Victorian Order;

Sir John Newenham Summerson, Knight, Commander of Our Most Excellent Order of the British Empire;

Nikolaus Pevsner, Esquire, Commander of Our Most Excellent Order of the British Empire;

Christopher Edward Clive Hussey, Esquire, Commander of Our Most Excellent Order of the British Empire;

Ian Archibald Richmond, Esquire, Commander of Our Most Excellent Order of the British Empire;

Henry Clifford Darby, Esquire, Officer of Our Most Excellent Order of the British Empire;

Donald Benjamin Harden, Esquire, Officer of Our Most Excellent Order of the British Empire;

John Grahame Douglas Clark, Esquire;

Howard Montagu Colvin, Esquire;

Vivian Hunter Galbraith, Esquire;

William Abel Pantin, Esquire;

Stuart Piggott, Esquire;

Courtenay Arthur Ralegh Radford, Esquire;

Arnold Joseph Taylor, Esquire;

Francis Wormald, Esquire.

GREETING!

Whereas We have deemed it expedient that the Commissioners appointed to the Royal Commission on the Ancient and Historical Monuments and Constructions of England shall serve for such periods as We by the hand of Our First Lord of the Treasury may specify and that the said Commissioners shall, if The National Buildings Record is liquidated, assume the control and management of such part of The National Buildings Record's collection as does not solely relate to Our Principality of Wales and to Monmouthshire, and that a new Commission should issue for these purposes:

Now Know Ye that We have revoked and determined, and do by these Presents revoke and determine, all the Warrants whereby Commissioners were appointed on the twenty-ninth day of March one thousand nine hundred and forty six and on any subsequent date:

And We do by these Presents authorize and appoint you, the said Robert Arthur James, Marquess of Salisbury (Chairman), Sir Albert Edward Richardson, Sir John Newenham Summerson, Nikolaus Pevsner, Christopher Edward Clive Hussey, Ian Archibald Richmond, Henry Clifford Darby, Donald Benjamin Harden, John Grahame Douglas Clark, Howard Montagu Colvin, Vivian Hunter Galbraith, William Abel Pantin, Stuart Piggott, Courtenay Arthur Ralegh Radford, Arnold Joseph Taylor and Francis Wormald to be Our Commissioners for such periods as We may specify in respect of each of you to make an inventory of the Ancient and Historical Monuments and Constructions connected with or illustrative of the contemporary culture, civilisation and conditions of life of the people in England, excluding Monmouthshire, from the earliest times to the year 1714, and such further Monuments and Constructions subsequent to that year as may seem in your discretion to be worthy of mention therein, and to specify those which seem most worthy of preservation.

And Whereas We have deemed it expedient that Our Lieutenants of Counties in England should be appointed ex-officio Members of the said Commission for the purposes of that part of the Commission's inquiry which relates to ancient and historical monuments and constructions within their respective counties:

Now Know Ye that We do by these Presents authorize and appoint Our Lieutenant for the time being of each and every County in England, other than Our County of Monmouth, to be a Member of the said Commission for the purposes of that part of the Commission's inquiry which relates to ancient and historical monuments and constructions within the area of his jurisdiction as Our Lieutenant of such County:

And for the better enabling you to carry out the purposes of this Our Commission, We do by these Presents authorize you to call in the aid and co-operation of owners of ancient monuments, inviting them to assist you in furthering the objects of the Commission; and to invite the possessors of such papers as you may deem it desirable to inspect to produce them before you:

And We do further authorize and empower you to confer with the Council of The National Buildings Record from time to time as may seem expedient to you in order that your deliberations may be assisted by the reports and records in the possession of the Council: and to make such arrangements for the furtherance of objectives of common interest to yourselves and the Council as may be mutually agreeable:

And We do further authorize and empower you to assume the general control and management (whether as Administering Trustees under a Scheme established under the Charities Act 1960 or otherwise) of that part of the collection of The National Buildings Record which does not solely relate to our Principality of Wales or to Monmouthshire and (subject, in relation to the said part of that collection, to the provisions of any such Scheme as may be established affecting the same) to make such arrangements for the continuance and furtherance of the work of The National Buildings Record as you may deem to be necessary both generally and for the creation of any wider record or collection containing or including architectural, archaeological and historical information concerning important sites and buildings through-out England:

And We do further give and grant unto you, or any three or more of you, full power to call before you such persons as you shall judge likely to afford you any information upon the subject of this Our Commission; and also to call for, have access to and examine all such books, documents, registers and records as may afford you the fullest information on the subject and to inquire of and concerning the premises by all other lawful ways and means whatsoever:

And We do by these Presents authorize and empower you, or any three or more of you, to visit and personally inspect such places as you may deem it expedient so to inspect for the more effectual carrying out of the purposes aforesaid:

And We do by these Presents will and ordain that this Our Commission shall continue in full force and virtue, and that you, Our said Commissioners, or any three or more of you, may from time to time proceed in the execution thereof, and of every matter and thing therein contained, although the same be not continued from time to time by adjournment:

And We do further ordain that you, or any three or more of you, have liberty to report your pro-ceedings under this Our Commission from time to time if you shall judge it expedient so to do:

And Our further Will and Pleasure is that you do, with as little delay as possible, report to Us, under your hands and seals, or under the hands and seals of any three or more of you, your opinion upon the matters herein submitted for your consideration.

Given at Our Court at Saint James's the Twenty-eighth day of September, 1963, in The Twelfth Year of Our Reign.
By Her Majesty's Command,

HENRY BROOKE

ROYAL COMMISSION ON THE ANCIENT AND HISTORICAL MONUMENTS AND CONSTRUCTIONS OF ENGLAND

REPORT to The Queen's Most Excellent Majesty

MAY IT PLEASE YOUR MAJESTY

We, the undersigned Commissioners appointed to make an Inventory of the Ancient and Historical Monuments and Constructions connected with or illustrative of the contemporary culture, civilisation and conditions of life of the people of England, excluding Monmouthshire, from the earliest times to the year 1714, and such further Monuments and Constructions subsequent to that year as may seem in our discretion to be worthy of mention therein, and to specify those which seem most worthy of preservation, do humbly submit to Your Majesty the following Report, being the thirtieth Report on the Work of the Commission since its first appointment.

2. We have pleasure in reporting the completion of our recording of the Monuments in the City of York lying outside the line of the City walls to the east of the river Ouse, an area containing 304 monuments.

3. Following our usual practice we have prepared a full, illustrated Inventory of the monuments in eastern extramural York, which will be issued as *York* IV. As in other recent Inventories, the Commissioners have adopted the terminal date of 1850 for the monuments included in the Inventory, but with a brief notice of one prominent later building.

4. The methods adopted in previous Inventories have in general been adhered to. This Inventory covers areas of extensive development of the first half of the 19th century, and the larger houses of this period make a substantial contribution to the record of domestic buildings; for the smaller houses many whole streets have been recorded as single monuments.

5. Our thanks are due to incumbents and churchwardens and to owners and occupiers who have allowed access by our staff to the Monuments in their charge or ownership, especially to the University of York and the Directors of the York Institute of Advanced Architectural Studies, Dr. P. J. Nuttgens and Mr. R. K. Macleod, and to the Curators of the Yorkshire Museum, Mr. G. F. Willmot and Mr. A. Butterworth, to the Yorkshire Philosophical Society and the Corporation of York. We are indebted to Mr. O. S. Tomlinson, York City Librarian, and Mr. L. M. Smith of the Reference Library, Mr. C. B. L. Barr of the York Minster Library, and Mrs. N. K. M. Gurney of the Borthwick Institute, to Mr. J. Ingamells of the York City Art Gallery and to Mr. J. M. Collinson, archivist at the Leeds City Library; also to the partners in the firm of Messrs. Brierley, Leckenby and Keighley, architects, who have allowed us to reproduce drawings from their archive, and to many other architects who have lent us drawings. Mr. R. J. Malden has assisted in the identification of ironwork.

6. We humbly recommend to Your Majesty's notice the following Monuments in the eastern extramural area of the City of York as 'most worthy of preservation':

EARTHWORKS

(1) LAMEL HILL, incorporating part of an Anglo-Saxon cemetery beneath a gun platform thrown up in 1644.

ECCLESIASTICAL

(4) ST. MARY'S ABBEY, the ruins of a great late 13th-century church, incorporating some remains of a church founded in the late 11th century, with the Hospitium and fragments of the claustral buildings dating from the 12th to the 15th centuries. The Gatehouse, walls and towers enclosing the abbey precinct were recommended to Your Majesty's notice in our twenty-eighth Report (*York* II).

(9) PARISH CHURCH OF ST. OLAVE, mostly rebuilt in the 15th and 18th centuries, incorporating part of the precinct wall of St. Mary's abbey and connected to the abbey chapel of St. Mary at the Gate.

SECULAR

(11) THE KING'S MANOR, formerly the house of the Abbot of St. Mary's, dating in part from the 13th century, rebuilt in the 15th century and enlarged to form a palace for the Council of the North in the 16th and 17th centuries.

(12) THE YORKSHIRE MUSEUM, built in 1827–9 to the design of William Wilkins, R.A.

(21) BOOTHAM PARK HOSPITAL, the front block completed in 1777 to the design of John Carr.

The monuments listed below in Bootham and Clifton include a small number which though not individually of especial distinction are part of a group of scenic value, giving character to one of the most pleasing approaches to the City.

(23) INGRAM'S HOSPITAL, BOOTHAM, a range of 17th-century almshouses.

(26) WANDESFORD HOUSE, almshouses opened in 1743.

(28) ST. PETER'S SCHOOL, buildings of 1838 in the Gothic style by John Harper.

(39) No. 33 BOOTHAM, a house built by Robert Clough, master builder, between 1753 and 1755.

(40) No. 35 BOOTHAM, a house built before 1688 and heightened in the 18th century.

(41) Nos. 39, 41, 43 BOOTHAM, part of a block of four houses of 1748.

(42) No. 47 BOOTHAM, a house designed by John Carr in 1753.

(43) No. 49 BOOTHAM, a house built in the late 17th century and heightened in c. 1738.

(44) No. 51 BOOTHAM, a large house designed by Peter Atkinson senior, completed in 1803.

(45) Nos. 53, 55 BOOTHAM, a pair of houses possibly designed by John Carr and built c. 1765.

(46) No. 57 BOOTHAM, a house built c. 1759 but remodelled c. 1830.

(47) No. 59 BOOTHAM, a house of the 18th century.

(48) No. 61 BOOTHAM, a late 18th-century house.

Monuments (41–48) all have street fronts of high architectural quality and many of them contain good fittings. They form the best group of domestic buildings in the City.

(50) Nos. 67, 69, 71 and 73 BOOTHAM, four houses of the early 19th century, with shallow bow windows, for their scenic contribution to the street as a whole.

(51) Nos. 75 and 77 BOOTHAM, houses of 1770, for their scenic value.

(56, 57) Nos. 54 and 56 BOOTHAM, two houses of c. 1840, both with interesting fronts and internal fittings and complementing the earlier group on the opposite side of the street.

(67) Nos. 2 and 4 CLIFTON, a pair of large houses of c. 1835.

(68) No. 8 CLIFTON, a house built between 1782 and 1784.

(69) THE WHITE HOUSE, No. 10 CLIFTON, a house of the early 18th century, heightened later in the century and enlarged c. 1830.

(70) Nos. 14, 16 CLIFTON, a symmetrical pair of houses of c. 1800.

(71) No. 18 CLIFTON, a house built after 1836.

(72) Nos. 26, 28, 30, 32 CLIFTON, a row of houses of c. 1841.

(73) No. 36 CLIFTON, a house built in the 18th century.

(74) No. 40 CLIFTON, a house of the early 19th century.

(75) Nos. 42, 44 CLIFTON, a house of the late 18th century divided into two.

(76) Nos. 64, 66 CLIFTON, a house of the late 17th century with curved gables, incorporating some earlier timber framing.

(81) ST. CATHERINE'S, No. 11 CLIFTON, a detached house of the second quarter of the 19th century, for its scenic value.

(82) BURTON COTTAGE, No. 27 CLIFTON, also a detached house of the second quarter of the 19th century, for its internal fittings as well as for its exterior.

(83) No. 29 CLIFTON, a house contemporary with (82).

(105) FISHERGATE HOUSE, a large house of 1837 by J. B. and W. Atkinson, for its internal planning.

(117) Nos. 3, 5 GILLYGATE, a pair of houses built in 1797 by the carver Thomas Wolstenholme, for the fittings by Wolstenholme.

(128) Nos. 26, 28 GILLYGATE, two houses built in 1769 by Robert Clough, with decorative plasterwork by Clough's son Robert.

(147) GARROW HILL HOUSE, HESLINGTON ROAD, built as a large private house in the early 19th century, now a nurses' hostel, for its internal fittings.

(278) MIDDLETON HOUSE, No. 38 MONKGATE, built c. 1700, later the house of the Rev. Charles Wellbeloved and for a time accommodating Manchester College, of which Wellbeloved was the principal.

(290) Nos. 13–18 NEW WALK TERRACE, six houses forming the best early 19th-century terrace in the City.

7. In compiling the foregoing list our criteria have been architectural or archaeological importance, rarity, not only in the national but in the local field, and the degree of loss to a city of national importance that the destruction of these monuments would involve; we have taken no account of such attendant circumstances as cost of maintenance, usefulness for present-day purposes, or problems of preservation.

8. We desire to express our acknowledgement of the good work accomplished by our executive staff in the preparation and production of this Inventory, in particular by the editor Mr. R. W. McDowall, O.B.E., M.A., F.S.A., and by Dr. E. A. Gee, F.S.A., F.R.HIST.S., and Messrs. H. G. Ramm, M.A., F.S.A., J. E. Williams, E.R.D., A.R.C.A., F.S.A., D. W. Black, B.A., F.S.A., the late J. Radley, M.A., F.S.A., I. R. Pattison, B.A., T. W. French, M.A., F.S.A., and Miss S. Spooner, B.A., also by our photographers Mr. C. J. Bassham and Mr. D. H. Evans, by our draughtsmen Mr. A. R. Whittaker and Mr. R. Meads, and by Mrs. J. Bryant who helped with the editorial work throughout.

9. We desire to add that our Secretary Mr. A. R. Dufty, C.B.E., F.S.A., A.R.I.B.A., has afforded constant and most valuable assistance to us Your Commissioners.

10. With deep regret we have to record the great loss to the Commission caused by the resignation for reasons of ill health of Lord Salisbury, who was our Chairman from 1957 to 1972 and who died on 23rd February 1972. Gratitude is due to him for his wise guidance, and the Commission staff join with us in paying tribute to his unfailing consideration and kindliness.

11. We thank Your Majesty for appointing Lord Adeane of Stamfordham in the County of Northumberland, P.C., G.C.B., G.C.V.O., Chairman of the Commission in place of the late Marquess of Salisbury.

12. Further Inventories of the Monuments in the City of York will be devoted to the Minster and to other buildings within the City walls, E. of the river Ouse.

Signed:

ADEANE (*Chairman*)	W. F. GRIMES
KENNETH HARGREAVES	M. W. BARLEY
H. C. DARBY	S. S. FRERE
C. A. RALEGH RADFORD	R. J. C. ATKINSON
JOHN SUMMERSON	JOHN BETJEMAN
H. M. COLVIN	J. N. L. MYRES
W. A. PANTIN	H. M. TAYLOR
A. J. TAYLOR	G. ZARNECKI

A. R. DUFTY (*Secretary*)

ABBREVIATIONS USED IN THE TEXT

APC	*Acts of Privy Council*
APS	Architectural Publication Society
Arch.	*Archaeologia*
Arch. J.	*Archaeological Journal*
ASC Text D	Anglo Saxon Chronicle in *English Historical Documents*, II, ed. D. Douglas (1953), 132–3
Benson, I	G. Benson, *York from its Origins to the end of the 11th century* (1911)
„ II	G. Benson, *Later Mediaeval York, 1100–1603* (1919)
„ III	G. Benson, *York from the Reformation to the year 1925* (1925)
BM	British Museum
Borthwick Inst.	Documents in the Borthwick Institute, York
Bowman	W. Bowman *et al.*, *Reliquiae Antiquae Eboracenses* (1855)
Brierley, Leckenby and Keighley	Original drawings by York architects in the archive of Messrs. Brierley, Leckenby and Keighley, Architects, York
Cave	H. Cave, *Antiquities of York* (1813)
Chron.	*Chronicle of St. Mary's Abbey*, Surtees Society, CXLVIII (1913)
CIM	*Calendar of Inquisitions Miscellaneous*
Colvin	H. M. Colvin, *Biographical Dictionary of English Architects 1660–1840* (1954)
CPR	*Calendar of Patent Rolls*
CSP Dom.	*Calendar of State Papers, Domestic*
CSP Dom. Add.	*Calendar of State Papers, Domestic, Addenda*
CSP Foreign	*Calendar of State Papers, Foreign*
Davies	R. Davies, *Walks through the City of York* (1880)
Drake	F. Drake, *Eboracum* (1736)
Dugdale *Mon.*	Sir William Dugdale, *Monasticon Anglicanum* (ed. 1846)
EETS	Early English Text Society
EHD	*English Historical Documents*, 9 vols. (1953–69), ed. D. Douglas
EPNS	English Place-Name Society
ERAS	East Riding Antiquarian Society
EYC	*Early Yorkshire Charters*, 12 vols.: Vols. I–III, ed. W. Farrer (1914–16); Vols. IV–XII, ed. C. T. Clay, Yorkshire Archaeological Society *Record Series*, Extra Series, I–X (1935–65)
Fallow and McCall	T. M. Fallow and H. B. McCall, *Yorkshire Church Plate*, Yorkshire Archaeological Society Extra Series, Vols. III and IV (1912)
Gardner	A. P. Gardner, *Medieval Sculpture* (1951)
Gunnis	R. Gunnis, *Dictionary of British Sculptors 1660–1851* (1953)
Halfpenny	J. Halfpenny, *Fragmenta Vetusta, or the Remains of Ancient Buildings in York* (1807)
Hargrove	W. Hargrove, *History and Description of the Ancient City of York*, 2 vols. (1818)
Hudson	W. Hudson (comp.), Miscellaneous notes, cuttings, prints, original drawings etc. in York City Library
Hutton	B. Hutton, *Clifton and its People* (YPS 1969)
Jourdain	M. Jourdain, *English Decoration and Furniture of the early Renaissance (1500–1650)* (1924)
Knight	C. B. Knight, *A History of the City of York* (1944)
L & P Hy VIII	*Letters and Papers of Henry VIII* (PRO)
Med. Arch.	*Medieval Archaeology*
New Guide	*A New Guide for Strangers and Residents in the City of York . . .* (1838), printed by W. and J. Hargrove
NG	National Grid Reference
NMR	National Monuments Record
Ord.	*Ordinale of St. Mary's Abbey*, Henry Bradshaw Society, LXXIII (1936); LXXV (1937); LXXXIV (1951)
OS	Ordnance Survey
PRO	Public Record Office
Raine	A. Raine, *Mediaeval York* (1955)
Sheahan and Whellan	J. J. Sheahan and T. Whellan, *History and Topography of the City of York, the Ainsty Wapentake and the East Riding of Yorkshire*, 2 vols. (1855–6)
Slingsby, *Diary*	*The Diary of Sir Henry Slingsby of Scriven, Bart.*, ed. D. Parsons (1836)
SS	Surtees Society
Stroud	Dorothy Stroud, *The Architecture of Sir John Soane* (1961)
TE	*Testamenta Eboracensia*, 6 vols., Surtees Society, IV (1836); XXX (1855); XLV (1865); LIII (1868); LXXIX (1884); CVI (1902)
Torre	J. Torre, *The Antiquities of York City, and the Civil Government thereof . . .* (1719)
VCH	The Victoria History of the Counties of England
Ward	*The History and Antiquities of the City of York*, 3 vols., printed by Ann Ward (1785)
Wellbeloved, *St. Mary's*	C. Wellbeloved, *Account of the ancient and present state of the Abbey of St. Mary, York, and of the discoveries made in the recent excavations* (1829)
Widdrington	Sir Thomas Widdrington, *Analecta Eboracensia*, 1660, ed. C. Caine (1897)

YAJ	*Yorkshire Archaeological Journal*	*YMH*	*A Hand-Book to the Antiquities in the Grounds and Museum of the Yorkshire Philosophical Society*
YAYAS	Yorkshire Architectural and York Archaeological Society		
YC	*York Courant*	YML	York Minster Library
YCA	York City Archives	*York* I	Royal Commission on Historical Monuments, *The City of York*: Volume I, *Eburacum* (1962)
YCh	*Yorkshire Chronicle*		
YCL	York City Library		
YCM	Castle Museum, York	*York* II	Royal Commission on Historical Monuments, *The City of York*: Volume II, *The Defences* (1972)
YCR	York Civic Records, 8 vols., Yorkshire Archaeological Society, *Record Series*, XCVIII, CIII, CVI, CVIII, CX, CXII, CXV, CXIX (1939–52)	*York* III	Royal Commission on Historical Monuments, *The City of York*: Volume III, *South-West of the Ouse* (1972)
YG	*Yorkshire Gazette*	YPS	Yorkshire Philosophical Society
YGS	York Georgian Society	YPSR	Yorkshire Philosophical Society *Report*
YM	Yorkshire Museum		

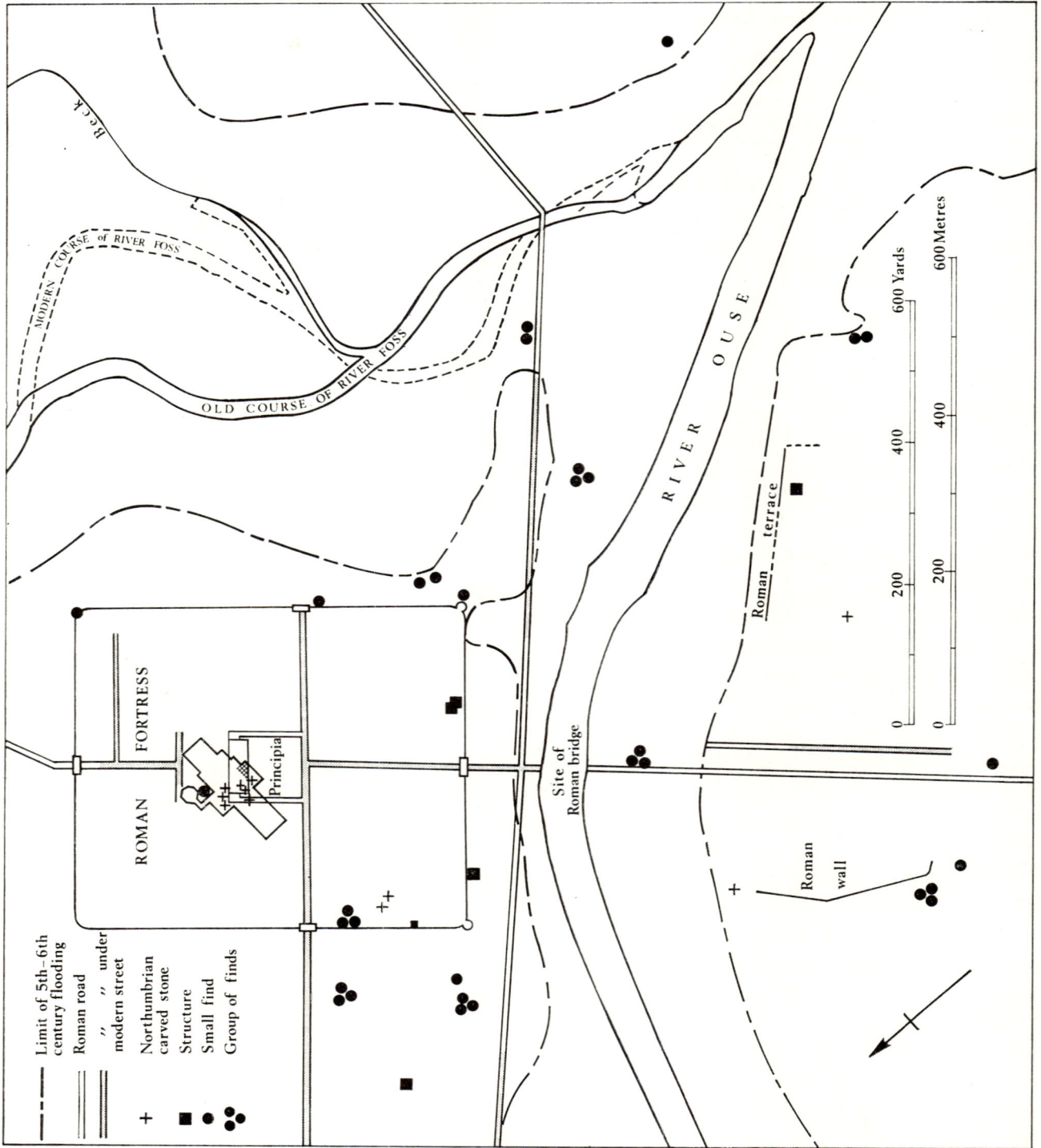

Fig. 3. York. 5th to 8th centuries. For Heworth Anglian Cemetery see map in end pocket.

CITY OF YORK
OUTSIDE THE CITY WALLS
EAST OF THE OUSE

GROWTH AND DEVELOPMENT OF THE CITY TO 1069

END OF ROMAN YORK

The buildings, walls and streets of the Roman fortress and city were still in the generations preceding the Norman Conquest among the most prominent features of the townscape, and determined the town plan of both mediaeval and modern York. It is convenient to begin with a summary description of this legacy, already more fully described in the early volumes (*York* I, Roman York; *York* II for subsequent history of the defences).

The Roman settlement comprised three parts:

(i) The *fortress*, a defended enclosure of 50 acres, built on an elevated spur between Ouse and Foss, including what is now the Minster and central area of the town on the N.E. side of the Ouse. Most of the internal buildings were low structures of stone, or occasionally stone and timber, but the central administrative buildings partly under the Minster were of monumental proportions. A subsidiary military enclosure was attached to the N.W. side on a site occupied in part by the choir of St. Mary's Abbey and the King's Manor.

(ii) The *canabae*, a civil settlement extending along the N.E. bank of the Ouse from St. Mary's Abbey to Castlegate, between the fortress and that river, but with the greatest concentration of buildings S. of the fortress, including stone temples, and baths in the Ousegate–Castlegate area. There were quays along the river Foss.

(iii) The *colonia* or main civil town, on the S.W. bank of the Ouse, defended by walls, rising up the then steep slopes of Micklegate Hill and Bishophill in a series of revetted terraces. The area corresponded approximately but not exactly with the mediaeval walled city on this side of the river. It was laid out in a regular grid of streets with the main road from Tadcaster as its major axis. There were monumental buildings notably in the Old Station Yard, where they had a slightly divergent alignment, and in the area of George Hudson Street.

Colonia and fortress were connected by a bridge opposite the Guildhall, which was the focal point of a road system of which the main constituents were the Tadcaster road approaching from the S.W., crossed immediately N.E. of the bridge by a N.W. to S.E. road which by-passed the fortress along its S.W. side. Along these and other roads outside the settlement area were cemeteries which included tombs of considerable size, and suburban houses.

Major alterations from this plan within mediaeval York are the following:

(1) The movement of the main Ouse crossing downstream from the Guildhall site to the present Ouse Bridge, and consequent realignment of roads.

(2) The removal of the distinction between the military and civil areas on the N.E. side of the river. This had three main consequences:

(a) The S.W. and S.E. defences of the fortress became superfluous and were partly removed.

(b) There was no longer any need to by-pass the fortress. The main crossing of the Foss was moved from the Castle to Foss Bridge with a consequent realignment of the S.E. approach road and the development of the Walmgate suburb.

(c) There was a parallel development on the N.W. where St. Mary's Abbey was allowed to block the road by-passing the fortress, and Bootham became the main road.

(3) The development within the fortress of the building of York Minster on its own E.–W. axis athwart the alignments of the Roman layout.

(4) The various changes in the Foss. After the Conquest the Castle mills and the *stagnum regis* prevented the continued use of that river for commercial purposes. The wharves sited on the Foss for the convenience of serving the legionary fortress and the later Danish wharves were replaced by others on the Ouse more convenient for the later town.

Development from the Roman town plan to the mediaeval was seriously affected by drainage and the water table. Certainly Roman level is in places below what is now the average summer level of the Ouse, and at Hungate pollen analysis showed an orchard type vegetation on the banks of the Foss in Roman times in an area later known as *in mariscis*. Archaeological evidence for disastrous flooding exists (see below) in the period between the end of Roman York and the 7th century. Fluctuation in water level may have affected York at other times—for example the development of Newgate in 1336 (Raine, 171–2) where the 12th-century name of Patrick Pool suggests there had previously been a wet area. The Ouse was tidal above York before the building of Naburn Lock in 1757 and was confined between stone walls revetting gardens in the later mediaeval period.

POST-ROMAN YORK

Little is known of York between the end of the Roman administration at the beginning of the 5th century and the emergence of Northumbrian York in the 8th century. There are two major pieces of evidence for this period. First there is considerable evidence for severe flooding between the late 4th and late 6th centuries (Ramm, 'The End of Roman York' in Butler (ed.) *Soldier and Civilian in Roman Yorkshire* (1971), 181–3), which at its greatest extent rendered uninhabitable that area of the town below the then 35 ft. contour (Roman levels are on average 10 feet below the present surface). Although the fortress and a great part of the *colonia* were unaffected, the economic results on a town already weakened by the break-up of Roman administration would be considerable. The silting of harbour facilities such as those at Hungate illustrates the loss for a period of York's status as an international port. On the other hand the site of the town at the N. extremity of a vast area affected by flooding (28 miles wide E. to W. and 40 miles long from York to Bawtry in the S.—*see* Radley and Simms, *Yorkshire Flooding* (1971), fig. 1, which illustrates the comparable area flooded in 1625) will have emphasised some of the underlying factors which influenced the choice of the site in the first instance as far as land communication was concerned, and ensured continuation of the settlement as a local market and strategic centre. Some evidence for the continuation of the Roman building traditions into the 5th century comes from Bishophill Senior and under the Minster.

The loss of the Roman bridge can be attributed to these floods with a consequent division of the town into two parts linked by the more tenuous communication by ford or ferry. When the flooding stopped, the 'heather and ling shewing that the ground hereabouts must at one time have been open moor' (described by James Raine (*YMH* (1891), 216) under Anglian and Viking remains in Clifford Street within the Roman *canabae*) illustrates the aspect of the formerly flooded areas and explains why the Roman town plan disappeared within them.

The second main piece of evidence derives from the cremation cemeteries on The Mount and Heworth Moor, where burned bones enclosed in pottery vessels of Germanic type date from the 5th century but include at least one at Heworth which is certainly a 4th-century type (Myres, *Anglo-Saxon Pottery and the Settlement of the English* (1969), 73 ff.). The distribution of similar pottery in burials outside Roman towns in Eastern England has led to the hypothesis that they belonged to mercenaries employed by Romano-Britons rather than to hostile invaders. Both sites (NGR 59375109 and 61055292) are alongside main roads in close proximity to the late 4th and possibly 5th-century Roman burials, a possible confirmation of a 4th-century date for the commencement of their use for cremations.[1] The siting of the two cemeteries on either side of the city is important in that it implies occupation of the Roman fortress and town at this date. Unlike contemporary or near-contemporary Roman or sub-Roman burials with their mixture of different styles of burial, these cemeteries are compact and uniform and imply that even if the newcomers came as friends, they were unassimilated and un-Romanised.

Germanic mercenaries and levies were no new thing in York at the end of the 4th century. Indeed in 367–9, the *Dux Britanniarum*, whose seat was at York, himself had a name of Frankish origin. Earlier, in 306, there was an Alemannic King Crocus present at York (*York* I, xxxiv). The German officers, at least, in the 4th century became thoroughly Romanised and lost contact with their homes (A. H. M. Jones, *Later Roman Empire* (1964), II, 622), but less is known about the rank and file. A break in Roman tradition is implied, as noted in *York* I (xxxiv), by the rifling of tombs, and the reuse of sarcophagi for new burials. In some cases this reuse is associated with gypsum burial, a 4th–5th century rite, deriving from Africa, and with Christian associations (Ramm, 'End of Roman York' in R. M. Butler (ed.), *Soldier and Civilian in Roman Yorkshire* (1971), 188–9). One burial, probably but not necessarily in gypsum, in a stone coffin found near Micklegate Bar (*YMH* (1891), 138), had preserved with it a piece of cloth, now in the National Museum at Edinburgh, made by a technique of Germanic origin (*YPSR* (1951), 22). But although we have some evidence of social revolution, material from 4th-century burials at York is usually unmistakably Roman. Earlier levies would appear to have been assimilated: the people buried in the Anglo-Frisian urns on The Mount and at Heworth, so far as their burial customs were concerned, were not.

Slight hints of other Germanic burials exist: a well-known 5th-century glass bowl comes from an unspecified site on The Mount (*YPSR* (1927), 7; *YAJ*, XXXIV (1958), 430; Harden (ed.), *Dark Age Britain* (1956), 142, Pl. XVI g); a possible urn from a separate cemetery on Heworth Moor is recorded by J. Raine

---

[1] On The Mount a 'Romano–Saxon' sherd (found in 1957, *YAJ*, XXXIX (1958), 434, fig. 4.1) and a male gypsum burial in the reused coffin of Aelia Severa with another tombstone for a lid (found in 1859 with other stone coffins, *York* I, 99, vii) come from the immediate vicinity of the cremations. Another area some 250–300 yds. to the W. was in use for burials on the basis of coin evidence throughout the 4th century, and included a gypsum burial in the reused coffin of Simplica Florentina (*York* I, 100–1, xviii, xx, xxi; *YAJ*, XXXIX (1958), 307–8; *YC*, 7 June 1838, for the gypsum). At Heworth a stone coffin (*York* I, 70, a(ii)) was found near the cremations and a large fragment of a 4th-century Crambeck face vase in the Yorkshire Museum is labelled 'Heworth, York, kept originally with material from the Anglian cemetery'.

(MS. notes, YCL 16, under date May 1879); and two presumably pagan Saxon cremations in urns are recorded from the junction of Parliament Street and Market Street (Hargrove, *New Guide* (1838), 52–3).

NORTHUMBRIAN YORK

Literary records reveal York's status in the 7th and 8th centuries as a cultural centre, the seat of an archbishopric, and the capital of the Saxon Kingdom of Northumbria, which stretched from the Humber to the Forth; the presence of a Frisian colony attested by Alcuin writing in the 8th century probably indicates a commercial centre too, but little is said of the topography or appearance of York at the time. Alcuin (*De Pontificibus*, 196. J. Raine (ed.), *Historians of the Church of York*, (Rolls Series), I, 355) refers to York 'still below its lofty walls' (*Euboricae celsis etiam sub moenibus urbis*) and as built first by the Romans 'high with walls and towers' (*hanc Romana manus muris et turribus altam fundavit primo*). The Roman walls and towers must have been an impressive survival in the 8th century and not only the walls and towers, but also the great gatehouses. In 685 St. Cuthbert was consecrated bishop in St. Peter's church at York and was given a grant of land within the fortress in terms that are now ambiguous but certainly imply the survival of a 'great west gate' whether it be the N.W. or the S.W. gate (*YAJ*, xxxix (1958), 436 ff.), and it is possible that the later Danish Koningsgarthr (Egil's Saga, *EHD*, I, 298–304) or Royal Palace used the S.E. gatehouse in King's Square. William of Malmesbury in the 12th century when the Roman walls were for the most part hidden beneath an earthen mound could still describe York as *urbs ampla et metropolis elegantiae Romanae praeferens inditium* (*Gest. Pont.* Prolog. iii), indicating the survival of other remains than the defences, and this is confirmed by the availability of Roman building material for reuse as late as the Norman period. *A fortiori* more survived in the 7th and 8th centuries, and when a 10th-century life of Archbishop Oswald describes York as a nobly built city then in decay through age, it is surely the Roman remains that inspire the comment.

Of Northumbrian buildings it is only the Minster that is described in any detail, although Alcuin says that St. Wilfrid adorned other churches with fine gifts and took care to increase congregations (*ecclesias alias donis ornaverit opimis . . . curam . . . gerebat multiplicare greges. De Pont.* (1227), *loc. cit.* 385) and he also refers to a church of St. Mary (*Christi genetricis in aula. De Pont.* (1605), *loc. cit.* 397). According to Bede, Edwin was baptised in 627 in a wooden oratory dedicated to St. Peter, around which he later built a square stone church which St. Oswald finished. This church fell into disuse and was repaired by St. Wilfrid *c.* 670. York suffered from fires in 741 and 761, the first of which burnt the Minster and the second devastated the city. Albert or Athelberht (Archbishop 767–80) first built a new altar over the place where Edwin had been baptised and provided it with fine ornaments. He then built a new basilica which he consecrated to God (*Sophiae sacraverat almae*) a few days before his death (Alcuin, *De Pont.* (1519), *loc. cit.* 394). Alcuin describes the church as a large one with arcades, numerous side chapels, galleries, splendid ceilings and windows. Archaeologically nothing is yet known of this building and its effect on York's topography is thus hard to assess. The excavations under the Minster (1966–9 and continuing) have not yet produced clear-cut evidence for the date at which the Minster assumed its present E.–W. alignment. Large parts of the cross-hall of the *principia* were standing and in use until the late 9th century, and other Roman buildings were repaired or adapted. Late Saxon (11th-century) graves under the N. half of the S. transept, and contemporary domestic structures under the E. end of the nave retain the Roman alignment. The place-names of York are Danish rather than Northumbrian. But if the name King's Square or Coney Garth reflects the palace of the Viking kings of York, then Coney Street may have reference to the palace of the earlier Northumbrian kings.

EVIDENCE OF BUILDINGS, 400–866

Although the remains of structures are few, significant deductions can be made from them. S.W. of the river at Bishophill Senior there is evidence of Roman building traditions continuing into the 5th century. The hypocaust system of a house built in the late 4th century fell into disuse and small buildings were built in the service yard over the furnaces. These buildings had good solid masonry built in a thoroughly Roman tradition. N.E. of the river we have evidence of structures of a different kind, of timber on rough stone footings, rectangular but with rounded angles, crowding the S.W. side of the Roman fortress with a complete disregard for its defensive value. In Museum Gardens such a building overlay the filled-in fortress ditch outside tower S.W.6; in St. Helen's Square, S.E. of the main fortress gate, rubbish pits have been dug into the berm between wall and ditch; in Davygate two separate fragments of building, 40 ft. apart, lay close inside the rampart behind the wall; and finally in Feasegate in front of the S. corner tower there were successively rubbish pits, an iron smelting hearth and a pit containing vegetable material and leather, of which the last is not later than the 10th century. In Museum Gardens the structure must date from a period when the fortress had gone out of use and, if a Danish rampart covered the fortress defences here as they did on the N.W. side, must precede this (i.e. date from the 5th–9th centuries). In Davygate beads of 9th–10th-century date came from just *above* the footings of the buildings. A stone tower, probably dating to the 7th–8th centuries, was built into a gap in the 4th-century fortress wall on the N.W. side (*York* II, *The Defences*, 112–15). The fortress walls even where they had become embedded in the expanding Northumbrian town still survived in an impressive condition as we know from Alcuin's poem and as indeed the surviving buried remains demonstrate. A burnt stone building on the N.W. side of Marygate in which was found a Saxon styca may indicate the spread of the town to the N.W. before the Danish occupation. Saxon sculpture of the 7th–8th centuries comes mainly from the Minster, but crosses from St. Leonard's Place, Bishophill Junior, and near the Old Station imply the existence of other churches. A hanging bowl from the Castle Yard (Haseloff, *Med. Arch.*, II (1958), 82, 83; Cramp, *Anglian and Viking York* (1967), 5–6), and a bronze bowl with drop handles from Clifford Street (*Arch.*, XCVII (1959), 60 fig. 1) might be part of the ecclesiastical equipment of a lost church or derive from 7th-century graves.

In the 7th–9th centuries the picture suggested is of a crowded town of timber houses, but with stone-built churches, expanding, on the N.E. bank of the Ouse at least, into areas kept clear in Roman times. Such a town would be liable to suffer from fire as indeed we are told it did, devastated (*vastata*) in 764 (Symeon Dunelmensis, *Opera Omnia* (Rolls Series), ii, 42. In 741 the Minster had been burnt, *ibid.*, ii, 38). It was a sufficiently peaceful city to disregard the surviving Roman defences and to allow civil building to encroach on them. The structural evidence then indicates the survival of Roman building into the 5th century, but the Northumbrian town of the 7th–9th centuries is mainly of timber and the surviving masonry fragment differs in style, structure and material from the Roman buildings. This change may have been assisted by natural disaster, as well as by the settlement of the English, whose coming in the late 4th or early 5th century, as we have seen, seems to have been as allies rather than invaders.

SMALL FINDS

The small finds, best illustrated by the map, are too few for firm deductions to be drawn from them. They occur on both sides of the river. In the S.W. they group in the Toft Green–Micklegate area, and N.E. of the river to the S. of the fortress and on the site of the later Earlsborough and St. Mary's Abbey. Some re-occupation of flooded land may have occurred before the Viking period: 9th-century finds occur within the flooded area in Tanner Row, and there is Anglian material as well as Viking from Clifford Street, including the bronze bowl perhaps of the 7th century (*Arch.*, XCVII (1959), 60 fig. 1). The absence of finds from within the fortress area is probably fortuitous. At least two coin hoards come from near Bootham Bar. One containing 10,000 coins was deposited in a pot buried into the earth rampart behind the Roman wall near the De Grey Rooms in St. Leonard's where it was found in 1842. Its latest coins were of Osberht (deposed 866) and Archbishop Wulfhere. The second hoard, found in 1879, came from the site of the Art Gallery or its vicinity and consisted of 400 coins up to mid 9th-century in date. During the removal of the city wall, mound, and upper stages of the Roman wall to make St. Leonard's Place in 1833–4, 47 Northumbrian stycas as well as 46 Roman coins were found. These came from a much higher level than the first hoard and presumably represent strays from another hoard or hoards disturbed in heaping up the mound of the subsequent city defences—the quantity is much greater than is normally found within the city mound. These hoards were perhaps deposited by citizens fleeing from the approach of the Danes who occupied the city in 866, or as a result of the fighting in 867, or the revolt against the Danes in 872. They indicate not only the insecurity of the time but the route taken by the fleeing citizens and the part of the city from which the wealthier citizens came. Yet another hoard was found near or on the site of the Old Railway Station in 1840, providing further evidence for settlement S.W. of the river at this period.

Outside the central area evidence is even more tenuous. S. of the city the area subject to flooding will have widened out to include Hob Moor, the Knavesmire, most of the Fulford area, Low Moor and Walmgate Stray. Settlement is to be looked for on the higher ground, such as Holgate Hill, Acomb, along the Hull road, and at Heworth. Acomb and Heworth have produced 4th-century Roman pottery, and the latter 4th-century burials just without the city boundary; both have English rather than Scandinavian names, and the name of Heworth belongs to an early phase of English settlement. The early cremation cemetery on Heworth Moor is to be associated geographically with the city rather than any subsidiary settlement at Heworth and the same is also true of the inhumation cemetery at Lamel Hill, to the S.E. of the city, with burials in wooden coffins laid regularly E.–W., but without grave goods, probably Christian and English, and belonging to a date when the rule of extramural burial was still followed.

VIKING AND LATE SAXON YORK, 876–1066

The Danes found a large and prosperous town, on both sides of the river, timber-built amongst the ruins of Roman fortress and *colonia*, with a cathedral symbolic of the leadership of the Church in the city's cultural life. The town did not entirely cover the area of Roman settlement and much of the formerly flooded areas must have been open land, dry but overlain with silt and covered with light vegetation, or still waterlogged and marshy, according to the drainage.

According to William of Malmesbury (*Gesta Regum*, II, 3) the city was burned in 866–7 (compare the evidence from Marygate above). Part of the Danish army settled in York, rebuilt the city (*civitatem reaedificavit*), and cultivated the neighbourhood (Surtees Soc., LI (1867), 144), or, as a later work (*loc. cit.*, 158) has it, rebuilt the walls of the city (*Eboracae civitatis maenia, una ex his restauravit*).

There is archaeological evidence for both forms of rebuilding. The Danes renewed three sides of the Roman fortress by covering the Roman wall and mound with an earth bank on which they erected a timber stockade, and, omitting the S.E. side, extended the enclosure to the river Foss where there was an open embankment to the river (*York* II, 8). This enclosure of 87 acres open-ended to a river compares with similar enclosures at Wareham, Dorset, of 80 acres, Hedeby in Schleswig-Holstein of 60 acres and at

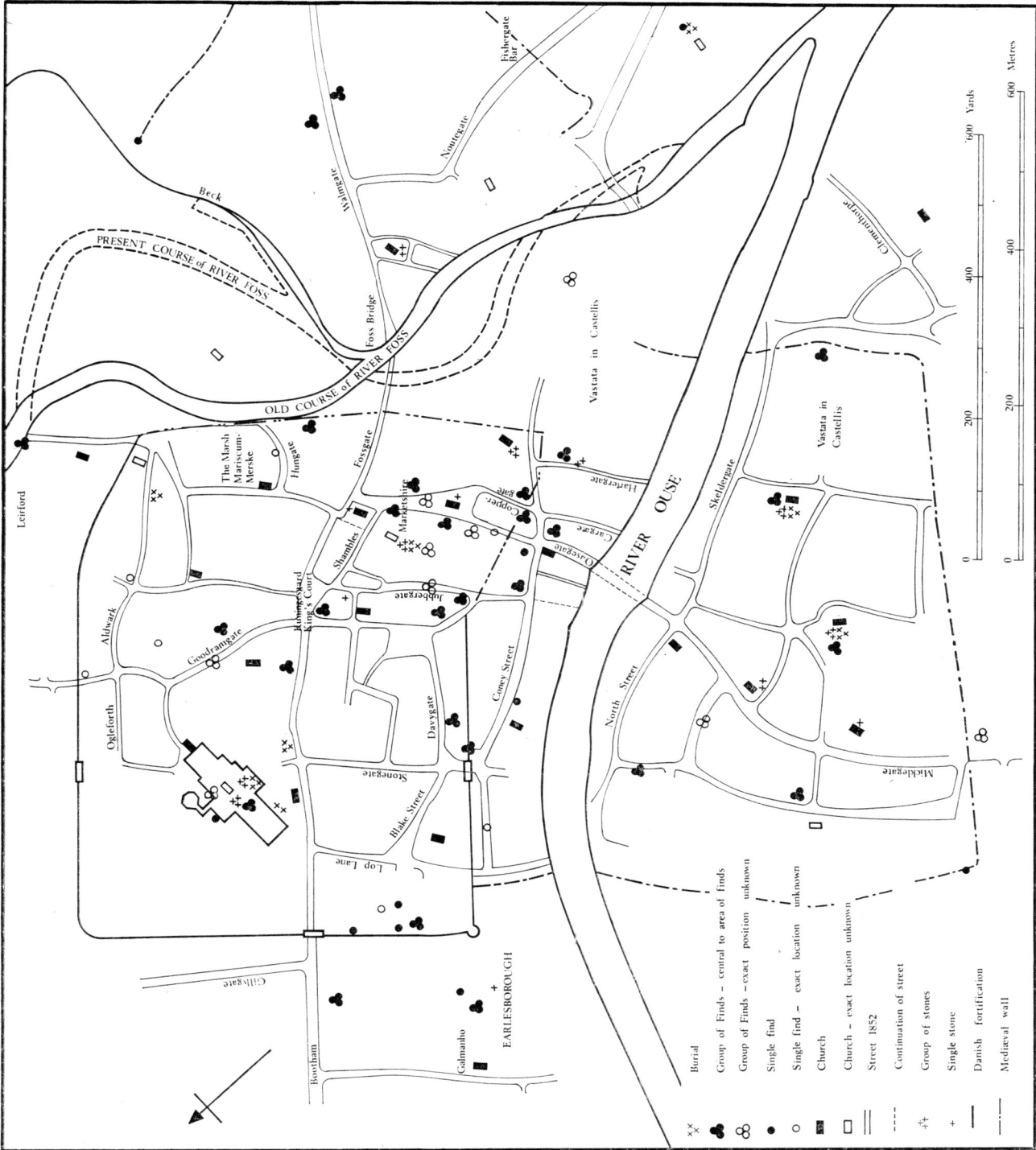

Fig. 4. York. 876–1066 A.D.

Burial

Group of Finds – central to area of finds

Group of Finds – exact position unknown

Single find

Single find – exact location unknown

Church

Church – exact location unknown

Street 1852

Continuation of street

Group of stones

Single stone

Danish fortification

Mediaeval wall

Birka in Sweden of 30 acres. But large as it was the side facing the Ouse was already abandoned by the later 10th or 11th century when tan pits were dug into the bank between Ousegate and Coppergate (YPSR (1902), 64), an industrial use of the site which had been replaced by stone houses in the 12th century (EYC, I, 183-5). Tenth-century occupation debris occurs outside the S.W. defences in Clifford Street and Coppergate.

A large part of the area added to the Roman fortress was on the area previously flooded. The Hungate excavations revealed evidence of systematic reclamation and drainage (Arch. J., CXVI (1961), 59-61). Within the enclosure the distribution of finds shows the weight of occupation to have been within the S.E. third of the Roman fortress and in the S.W. half of the added enclosure. The most prominent feature within the Danish enclosure must have been the royal palace sited, on place-name evidence (Smith, EPNS, York and East Riding, 285), in King's Square, possibly in the Roman S.E. gatehouse. The Egil's Saga (EHD, I, 299) described the house of Arinbjorn at York which had its court and a hall. In addition to the Minster and royal palace there were thus other large establishments, each presumably in its own enclosure. The picture in Egil's Saga is of a building of some size and sophistication.

Archaeological finds are more humdrum consisting, apart from occupation debris, of traces of timber buildings of 10th-11th century date from King's Square, Pavement, High Ousegate and Piccadilly (see Radley, 'Economic Aspects of Anglo-Danish York', in Med. Arch., XV, forthcoming). Radley suggests a commercial use for some of these buildings and there is evidence for crafts, particularly leather-working. The whole orientation of the new enclosure was to the river Foss; the weight of small finds illustrates this and the mediaeval street pattern confirms it. The later markets were in the enclosure: 14th-century documents refer to the Pavement area as 'marketshire' possibly implying one of the seven shires into which Domesday Book says late Saxon York was divided. The triangular street patterns of Ousegate-Coppergate-Pavement and King's Square-Colliergate-Shambles, the apex of the first towards Fossgate and of the latter towards King's Square, argue for roads able to converge over an open space to a natural focus rather than the grid of an artificially imposed pattern.

There need have been no great check on York's economic life as a result of the events of 866-7. York as a prosperous commercial centre had been to a considerable extent dependent on foreign merchants, particularly Frisians. Their position was decidedly insecure and they had to be ready to flee should enmity, aroused by envy of their wealth or by social prejudice, make this necessary. This is well illustrated by an episode described in Alcfrid's Life of St. Luidger (EHD, I, 725) who had to leave York: 'For when the citizens went out to fight against their enemies, it happened that in the strife the son of a certain noble of that province was killed by a Frisian merchant, and therefore the Frisians hastened to leave the land of the English, fearing the wrath of the kindred of the slain young man.' That was in the 8th century. The Danish army, part of which under Halfdan eventually settled in York, is said to have been led by Halfdan, Inguar and Ubba, and the last-named is described in the life of St. Cuthbert (loc. cit.) as dux Fresconum or General of the Frisians. York already had commercial links with N.W. Europe and these were maintained by merchants accustomed to the dangers of political instability and ready also to take advantage of it. York's commercial position may indeed have been enhanced by the conquest, merchants following the Danish army, just as in Egil's Saga Egil's companions and shipmates are left behind in York under safe-keeping to sell their wares before moving south to find him. In the late 10th-century life of St. Oswald (Raine (ed.), Historians of the Church of York, I, 154) York is said to be 'fantastically stocked and enriched with the treasures of merchants who come from all quarters particularly from the Danish people (ex Danorum gente)'.

PLATE I

(4) ST. MARY'S ABBEY. Statue, probably of St. John. *c.* 1200.

PLATE 2

(4) ST. MARY'S ABBEY CHURCH from S. Drawing by W. Lodge, c. 1678.

PLATE 3

SOUTH EAST VIEW OF THE REMAINS OF ST MARYS ABBEY
YORK

(4) ST. MARY'S ABBEY. Excavations of 1827–8. From *Vetusta Monumenta*, Vol. v (1829).

PLATE 4

From S.

From E.

(4) ST. MARY'S ABBEY CHURCH. 1270–94.

PLATE 5

From N.

Interior of W. end.

(4) ST. MARY'S ABBEY CHURCH. 1270–94.

PLATE 6

Part of W. front.

Nave aisle. Bay of N. wall.

(4) ST. MARY'S ABBEY CHURCH. 1270-94.

PLATE 7

Springing of vault.

Jamb of W. window and springing of vault.

(4) ST. MARY'S ABBEY CHURCH. N. nave aisle. 1270–94.

PLATE 8

Jamb of W. window of N. aisle.

Jamb of W. door.

(4) ST. MARY'S ABBEY CHURCH. 1270–94.

PLATE 9

N. nave aisle. Moulded capital. 1270–94.

S. transept. Buttress in cloister walk. 1270–94.

N. nave aisle. Carved capital. 1270–94.

N. transept. Buttress. 11th-century.

(4) ST. MARY'S ABBEY CHURCH.

PLATE 10

(4) ST. MARY'S ABBEY. Carved capitals, in Yorkshire Museum. Late 13th-century.

PLATE II

(4) ST. MARY'S ABBEY. Carved bosses, in Yorkshire Museum. c. 1300.

PLATE 12

Cloister. Fragments from arcading. *c.* 1300.

Carved bosses. *c.* 1300.

(4) ST. MARY'S ABBEY. Architectural fragments, in Yorkshire Museum.

PLATE 13

Pier at entrance to Chapter House,
partly reconstructed. *c.* 1200.

Warming House. Jamb of fireplace.
Late 14th-century.

Vestry passage. Vaulting shaft
against S. transept. *c.* 1280–90

S. cloister walk. Buttress and panelled walling, re-erected in Hospitium. Late 14th-century.

(4) ST. MARY'S ABBEY.

PLATE 14

(4) ST. MARY'S ABBEY. Supposed site of the chapel of St. Mary at the Gate. 15th-century.

PLATE 15

Hospitium. N.E. front. 14th-century and later.

Gateway S.E. of Hospitium, from S.W. *c.* 1500.

(4) ST. MARY'S ABBEY.

PLATE 16

a, b. Vestibule to Chapter House. Pier base. Late 14th-century.

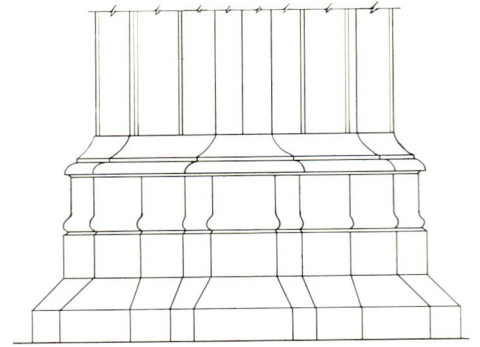

b. Plan and elevation.

c. Vestibule to Chapter House. Pier base at W. entrance. c. 1200.

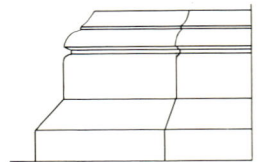

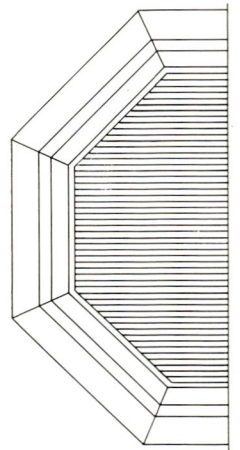

(4) ST. MARY'S ABBEY.

d. Inner Parlour. Base to vaulting shaft. Early 14th-century.

Inches
6   0   6   12   18   10   0   10   20   30   40   50   Centimetres

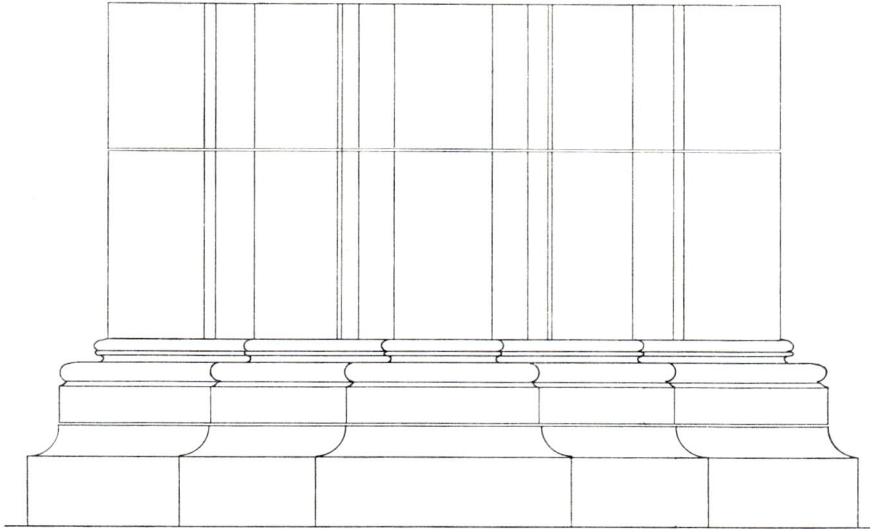

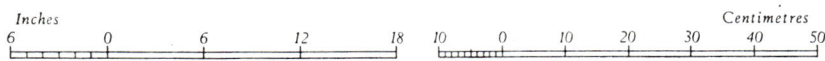

Drawings on Plate 16 are to the same scale

Fig. 5. (4) St. Mary's Abbey. Vestibule to Chapter House.
Pier base at W. entrance. *c.* 1200.

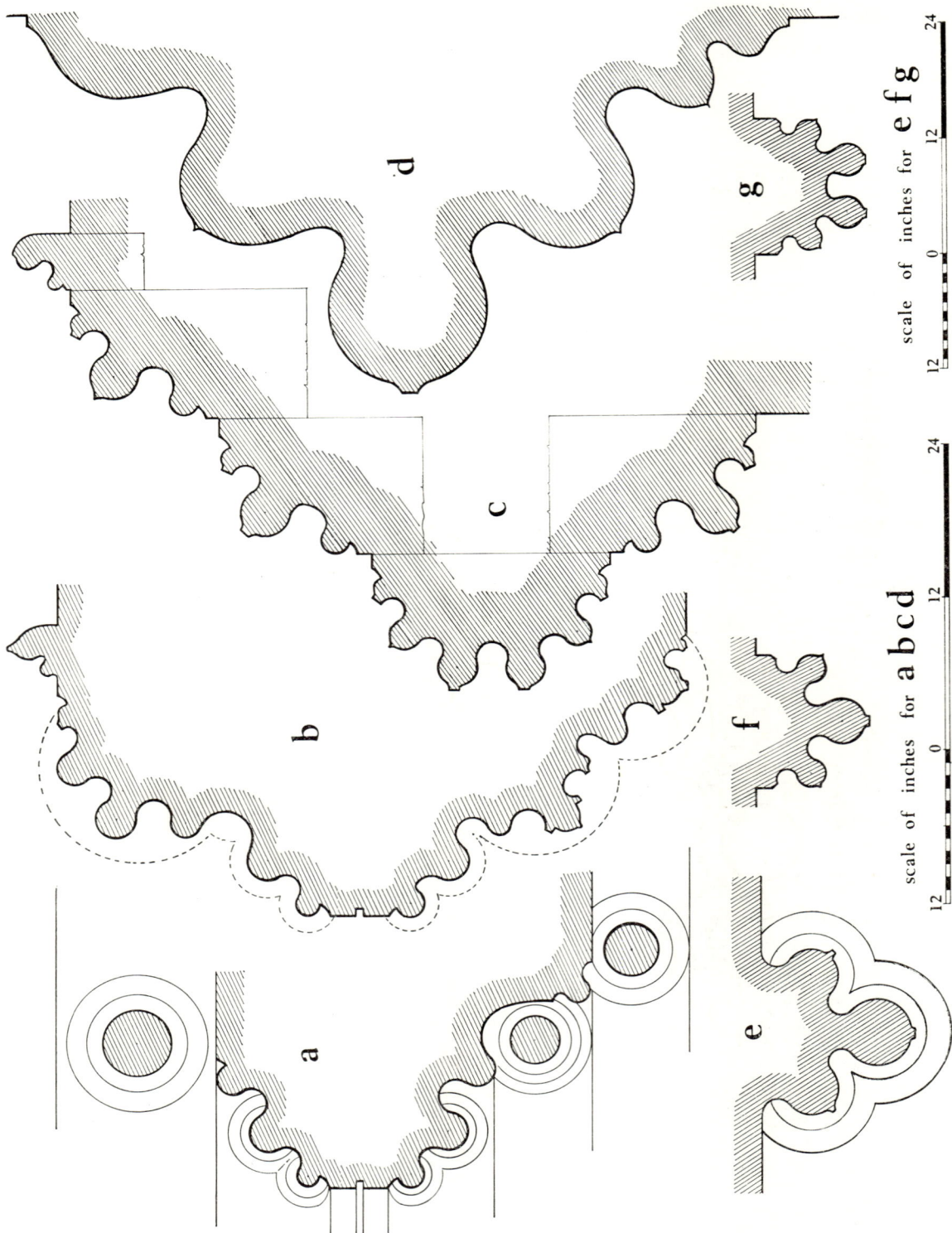

Fig. 6. (4) St. Mary's Abbey Church. Mouldings in stone.

scale of inches for **a b c d**

12   0   12   24

scale of inches for **e f g**

12   0   12   24

a. N. aisle, jamb of window.
b. „ „ arch over window.
c. Nave arcade, arch.
d. „ „ W. respond.
e. N. aisle, vaulting shaft.
f. „ „ transverse vaulting rib.
g. „ „ diagonal vaulting rib.

Within the added enclosure are two burial grounds—the one in Parliament Street is perhaps to be associated with a lost church of St. Swithin's and was probably never anything but Christian, the other in St. Saviourgate with rough wooden tree-trunk coffins may be pagan. One is reminded of the two cemeteries in the walls of Hedeby, the one Christian and the other, with wooden grave chambers, pagan. Of noble burials in 'hows' the only possible indication is the engraved bronze socket for an iron spearhead found at Howe Hill, between Holgate and Acomb (*Arch. J.*, VI (1849), 402).

The abandonment of the S.W. Danish defences in the 10th and 11th centuries did not greatly affect York's topography except to remove an artificial barrier to expansion. Nor did the destruction of the Danish *castrum* by Athelstan (*York* II, 8) affect greatly the Danish element in the city dependent on commercial rather than military settlement. The city is still weighted towards a central area S.E. of the old Roman fortress involving new crossings of the Ouse and the Foss at the present Ouse Bridge and in Fossgate. Whether we look at the distribution of archaeological small finds or Anglo-Danish sculpture, Scandinavian street names or of churches, for the greater number of which a pre-Conquest date can be argued on a variety of grounds, the concentration S.E. and S. of the fortress is marked.

Beyond the Foss the street named Fishergate, attested as early as 1070–80 A.D., implies urban development by that date and presumably before 1066 A.D. The six mediaeval churches along the route from Fishergate to Foss Bridge can all except one (St. George) be shown to have existed before 1100, three of them outside the later walls. Four of the remaining six churches S.E. of the Foss lie outside the later city walls and all except one along the line of the Roman road approaching the city from the E., but only one of these has evidence for an early date. The siting of the churches indicates the importance of Fishergate and the road from Fulford (site of a battle in 1066 A.D.) and the survival of the Roman line from the E. which met Fishergate at or near the church of St. George. Walmgate is a short cut already existing by 1070–80 A.D. but which had then not yet attracted much urban development. The later defences centring on Walmgate and defending a compact arc of the town within a large loop of the Foss, and the loss of the road linking Foss Bridge and Fishergate have obscured the shape of earlier development straggling along the E. bank of the Foss to the Ouse and filling out the old city boundary in the area, a riverine settlement presumably served by wharves and moorings along the bank and suggesting commercial expansion from the main trading area N. of the Foss.

N.W. of the Minster area was the Earlsborough, the fortified residence of the Earls of Northumbria in the 11th century, occupying a site important in pre-Danish times. The name Bootham, Scandinavian in origin, implies some kind of temporary development of huts or shanties and does not suggest an important suburb.

On the S.W. bank of the river the simpler parochial pattern suggests later and less congested development. Micklegate had already before the Norman Conquest been diverted towards the new river crossing (Ouse Bridge). North Street and Skeldergate, following the line of the river bank, indicate the importance of riverborne commerce in the life of the town. An important excavation by Mr. Wenham, adjacent to St. Mary's Church, Bishophill Junior, has located an area separated by a wall from the 10th-century churchyard where fish had been cleaned and gutted in quantity. Fish formed an important item of commerce and a staple food. Documentary evidence exists that it was imported, presumably dried, from the Hebrides and sold at a market under the control of the Archbishop.

Downstream development took a different form from that in the Fishergate area, and was not considered as an extension of the town. Satellite harbour settlements are indicated by the *-thorpe* names of

Clementhorpe, Bustardthorpe, Middlethorpe and Bishopthorpe, but the tolls from the ships which lay at Clementhorpe and beyond belonged by 1106 A.D. to the Archbishop. Satellite settlements on both sides of the river exist on the main E.–W. road line at Copmanthorpe, Dringthorpe and Layerthorpe and suggest that road traffic had its part, too, to play.

Outside the town a certain amount of arable was always included within the later city boundary, and the city had extensive rights of common. S.W. of the city the parishes of the city churches extend well beyond the city boundaries, and the -*thorpe* names suggest that many of the villages within them were satellite communities. E. of the Ouse some of the villages were also, in whole or part, within the city parishes, e.g. Naburn, Fulford, Heslington and Heworth. In Domesday there are 84 carucates which geld with York and are to be identified with the manors and vills listed immediately after the account of the city. These are all on the E. bank of the Ouse and their association may date back to the 9th century when a third of the Danish army settled in York and cultivated the land *in circuitu* (Surtees Soc., LI (1867), 144). The absence of Danish place-names could imply existing communities where the inhabitants remained to work for the Danes living in York.

By 1066 A.D. the town of York was a large and important commercial centre. The greater part of the old Roman fortress was occupied by the Minster and its precinct; adjacent to the N.W. was the Earlsborough, the fortified palace of the Earls of Northumbria. Roman buildings still survived but the business heart of the town had moved to the S.E. of the fortress on to land where Roman remains had been erased by the floods of the 5th and 6th centuries. Trade which had originally centred on the Foss had been transferred to the Ouse with wharves and riverine settlement stretching far downstream. Indeed, the town was far less compact than when it had retreated within the later mediaeval walls. None the less the basic plan of the mediaeval town is now recognisable.

SECTIONAL PREFACE

THE MEDIAEVAL SUBURBS

The present City of York includes considerable areas that have been absorbed since 1880 but in the mediaeval period the City boundaries already lay mostly well outside the City walls and enclosed extensive areas of pasture, gardens and orchards. Only between the river and Bootham was the City boundary defined by the City walls, leaving St. Mary's Abbey and the S.W. side of Bootham in the North Riding. Surviving stones marking the mediaeval boundaries are at the end of Burton Stone Lane, near Yearsley Bridge on the Huntington Road close to Rowntree's factory, and opposite the barracks in the Fulford Road (monuments 32–34).

St. Mary's Abbey stands on a site previously occupied by the palace of Earl Siward, who governed York and Northumbria from 1041 to 1055, and the church he built dedicated to St. Olaf, represented by the existing St. Olave's. On the southern side of the town and within the mediaeval boundary the church of St. Andrew, Fishergate, is recorded in Domesday Book, and All Saints, Fishergate, was in existence before the end of the 11th century when it was granted to Whitby Abbey. St. Helen's, Fishergate, was also in existence before 1100; it stood on a site crossed by the modern Winterscale Street. Early grants of land and buildings confirm the existence of dwellings by the middle of the 12th century in Bootham, Fishergate, Gillygate and Marygate. In Bootham building was confined to sites opposite the abbey walls or further to the N.W., and here in the late 12th century the Constable of Richmond had a lodging. There were also houses opposite the abbey walls in Marygate. The husgable roll of *c.* 1282 records 19 tofts in Bootham (YCA, c60). An agreement of 1354 between the City and the abbey confirmed that the abbey should maintain a ditch between the precinct wall and the street of Bootham provided that no buildings were erected there (*CPR 1354–8,* 84–6). At the end of the 13th century a complaint was made that the paving of Bootham was all broken up and the street was foul with the stench of pigsties and obstructed by loose pigs. At the City boundary close to the Burton Stone stood the Hospital and Chapel of St. Mary Magdalene, from which Burton Stone Lane derived its earlier name of Chapel Lane.

In Gillygate dwellings were probably confined to the S.E. side of the street but the church of St. Giles stood on the N.W. side. References to the street of St. Giles in the 12th century imply the existence of the church at that time, though the earliest direct reference to it is in a will of 1393. Beyond Gillygate was an open space called the Horsefair, extending beyond the junction of the Wigginton Road and Haxby Road. Adjoining the Horsefair were several religious houses: the Carmelites had their first house here before moving to Hungate in 1295; the Hospital of St. Mary, founded in 1314, stood at the S.W. corner of the Horsefair and two other hospitals dedicated to St. Anthony and St. Anne respectively probably lay along its N.W. side. St. Mary's Hospital later became St. Peter's School and was burnt down in 1644. St. Anthony's Hospital is mentioned in 1401, and was still 'lately founded' in 1420. Excavation by the York Archaeological Trust in 1972, on the site of Union Terrace, revealed a 12th-century building, enlarged *c.* 1200, which may have been a church, possibly used by the Carmelites, and other later buildings, interpreted as the remains of St. Mary's Hospital.

Lord Mayor's Walk was Goose Lane and undeveloped except perhaps in the immediate vicinity of St. Maurice's church. Between the Horsefair and Monkgate lay Paynelathes Crofts, later known as The Groves, an area of small enclosures belonging to St. Mary's Abbey.

Surviving remains of St. Maurice's church include an arch of mid 12th-century date and its existence argues for some suburban development in Monkgate by that time; the name Monkgate appears in the form Munccagate in 1080. In c. 1282 the husgable rolls record 50 tofts under the heading of Monkgate (YCA, c60) including two occupied by millers. Near Monk Bridge stood the hospital of St. Loy established in the 14th century in a house already standing; a little higher up the Foss were the mills of St. Mary's Abbey, perhaps opposite the N. end of Grove Terrace.

From Monkbridge the road to Heworth lies across heavy clayland which up to the middle of the 12th century was included in the Forest of Galtres. Heworth village itself lay outside the mediaeval boundaries of the City, although the ecclesiastical boundaries of the three City parishes of which Heworth formed part extended beyond the City boundary, to include the whole township.

At Layerthorpe there was a small settlement E. of the Foss probably in existence by the late 12th century, served by a small church dedicated to St. Mary. In the 15th century one of the great bells for the Minster was cast here.

Outside Walmgate Bar the City extended as far as Green Dykes Lane and included, in addition to the 12th-century church of St. Lawrence, churches dedicated to St. Edward, St. Michael and St. Nicholas. St. Edward's, on the N. side of Lawrence Street, is first mentioned in the 14th century; it was taken down in the reign of Edward VI. St. Michael's is supposed to have stood between St. Lawrence's and Walmgate Bar and was united with St. Lawrence's in 1365. St. Nicholas's was a larger building, of 12th-century date, from which the 12th-century doorways now at St. Margaret's and St. Denys's in Walmgate are both said to have come. The last remains of the church were removed in 1736; the surviving remains were sketched in 1718 (Evelyn Collection No. 539) and included a nave, large chancel, and substantial 13th-century W. tower. Associated with the church was the largest of York's four leper hospitals. No documents relating to private houses in Lawrence Street of a date earlier than the 14th century have been noticed.

Outside Fishergate Bar, on the site of the Cattle Market, stood the church of All Saints. This may have been a pre-Conquest foundation as already stated; its foundations were discovered in the early 19th century, and a rough plan published in the *New Guide* (p. 38) suggests an aisled church with an apsidal sanctuary. St. Andrew's church and priory lay to the W. of Fishergate; the pre-Conquest church was given to the Gilbertine Canons in 1202. A little further S. St. Helen's church stood to the E. of Fishergate; it was a small church built before 1100 when it was granted to Holy Trinity Priory. St. Helen's Hospital, Fishergate, must have stood nearby but its site is not known. Also outside Fishergate Bar was a chapel dedicated to St. Catherine. Houses and tofts in Fishergate figure in a number of 12th-century documents, but it is not clear how many of these lay outside the walls.

THE SUBURBS IN THE 17TH CENTURY AND LATER

All the suburbs described above were badly damaged or destroyed in the siege of 1644 but their appearance in the early 17th century was recorded by John Speed whose map of York was published in 1610. Ribbon development of continuous rows of houses is shown along the N.E. side of Bootham and on the S.W. side beyond the abbey walls, continuing a little way down Marygate with some further houses nearer the river. Houses in Gillygate are almost all on the S.E. side. A few houses stand between St. Maurice's church and

Lord Mayor's Walk, representing the mediaeval Newbiggin,[1] and both sides of Monkgate are built up. Houses continue a short distance into St. Maurice's Road and a few more appear in Jewbury. Layerthorpe is represented by less than a dozen houses; a road running S. leads to a larger house, presumably the house which gave its name to Hall Fields and Hall Fields Road. St. Mary's church is not shown. Lawrence Street is built up past St. Lawrence's church and then a few isolated houses lead to St. Nicholas's church. In the whole Fishergate area only one house is shown, possibly representing a remnant of All Saints' church, and three windmills. Archer's plan of c. 1680 indicates that houses had been built in Bootham, backing onto the abbey wall, and shows more houses in Marygate than Speed's plan. In Monkgate however the houses do not extend so far towards Monk Bridge as formerly. No new development is shown on the S. side of the City.

Not until after 1760 was there any significant increase in the population above the 17th-century level of about 12,000. The census of 1801 gives a population of over 16,000, an increase of one third but there was sufficient space within the walls and the existing suburban areas to accommodate this growth. During the first half of the 19th century the population was more than doubled. Baines's map of 1822 shows new development at Bootham Row and some extension of building in Marygate. New houses appear on the N. side of Lord Mayor's Walk and on the S. side adjoining Monkgate. Continuing eastwards, buildings are shown in St. Maurice's Road, then called Barker Hill, but little appears at Layerthorpe. Only the beginning of Lawrence Street is shown and some long buildings to the S. of Fishergate must represent the glassworks founded there before the end of the 18th century.

By 1851 the population of the City had risen to 36,000 and after 1820 the building of new houses proceeded rapidly, as is shown in the Inventory. The City boundaries were not extended until 1884 when Clifton, Heworth and an area around the Fulford Road were included, followed by further extension to the N.E. in 1893 and 1934.

ECCLESIASTICAL

ST. MARY'S ABBEY

Sources. Three mediaeval documents compiled by the monks of the abbey survive to the present day, and all three have been published. The Chronicle of St. Mary's Abbey, in the Bodleian Library (Surtees Society, vol. CXLVIII, 1934), includes an account of the founding of the abbey, and covers the years 1258 to 1325 except for a gap from 1284 to 1292 and part of 1293; it contains details of the great rebuilding of the church under Abbot Simon de Warwick at the end of the 13th century and of the building of the walls a little earlier, but little about the construction of the other monastic buildings. The Anonimalle Chronicle (V. H. Galbraith ed., Manchester 1927) covers the years 1333 to 1381 but has only four entries relating to the Abbey; these include an account of the fire in 1377. The Ordinale and Customary, in the library of St. John's College, Cambridge (Henry Bradshaw Society, vols. LXXIII, LXXV, LXXXIV) records the usage for the celebration of the Divine Office throughout the year. It was drawn up between 1390 and 1398 and contains many references to the church and other buildings of the abbey from which it is possible to deduce the uses made of the various parts of the abbey church and of the various buildings whose foundations have been uncovered by excavation.

Excavations carried out in the early 19th century are described and illustrated by the Rev. C. Wellbeloved

---

[1] Partially excavated in 1972 by the York Archaeological Trust.

in his *Account of the ancient and present state of the Abbey of St. Mary, York, and of the discoveries made in the recent excavations* (1829), which forms *Vetusta Monumenta*, vol. v.

The Abbey. The 11th-century archbishop of York, Thomas of Bayeux, objected to the foundation of the monastery so close to the Minster and laid claim to St. Olave's himself. The king however supported the monks and gave the archbishop another church in lieu. This was the first of a number of differences that arose between abbots and archbishops as the abbots rose to a position of great importance. The abbey became extremely wealthy and the abbot was one of the only two in the N. of England to have the privilege of *pontificalia* and a seat in Parliament. It was one of only four Benedictine Abbeys in the province of York, second only to Durham and much more important than Selby or Whitby.

The original foundation of the abbey was the result of a heroic and adventurous undertaking by three monks who came from Gloucestershire to Whitby to revive monasticism in the N. of England, but St. Mary's quickly became very traditional and settled in its ways with the result that in 1132 a party of monks from St. Mary's, dissatisfied with the discipline of the house, seceded to seek a stricter life; they established Fountains Abbey and were received into the Cistercian Order. St. Mary's was not exempt from visitation by the archbishop and the records of these visitations show that in spite of the great wealth of the abbey it was heavily in debt in the early 14th century. In the early 14th century Edward II, when involved in war with Scotland, moved the government from London to York and the Chancery was accommodated in the abbey.

Among the more important members of the community were John Gaytrick, who in the middle of the 14th century translated the Catechism into English verse, the so-called Layfolks Catechism issued by Archbishop Thoresbury (EETS, cxviii, ed. T. A. Simmons and H. E. Nelloth (1901)), and Thomas Spofford, abbot 1404–21, Bishop of Hereford 1422–48, who went to the Council of Constance and helped to organise the reform of German Benedictines.

Lists of monks total 33 in 1258, 48 in 1285 and 50 in 1539. There were eight dependent cells, including Wetherall and St. Bees in Cumberland.

The First Abbey Church. Of the abbey church begun in 1089 very little now remains visible but the general lines of its plan are known from excavation; it was a cruciform building with a short aisled presbytery, transepts with eastern chapels set in echelon, and an aisled nave. At the E. end were seven parallel apses. The choir aisles ended in apses finished square externally; the transept chapels completed the seven apses, progressively receding from the centre.

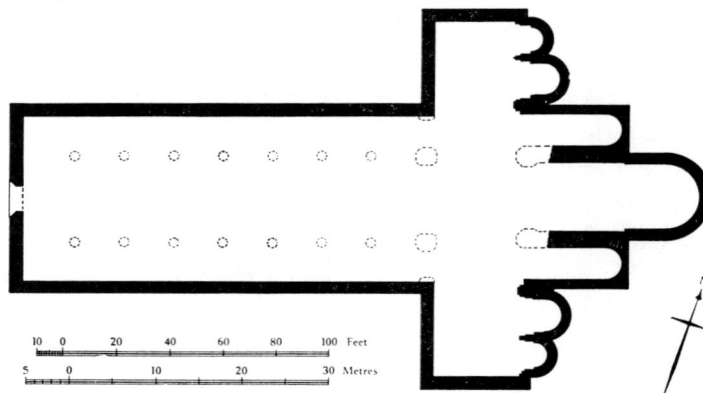

Fig. 7. (4) St. Mary's Abbey. Plan of Norman church.

The plan with parallel apses of varying projection occurs at about half of the known Benedictine abbeys of the late 11th century, the others having had an eastern ambulatory with radiating chapels. The parallel apse plan was first used in England at the Confessor's church at Westminster and then at Canterbury. It was not however confined to Benedictine churches as it was also used at the secular cathedrals of Lincoln and Old Sarum. These churches were begun between 1070 and 1080; all except St. Mary's had one transeptal apse each side. The only other English churches with two apses to each transept, making seven apses in all, were St. Albans, begun 1077–88, Binham Priory, a little later but before 1100, and Glastonbury after 1100. The short eastern arm of St. Mary's is similar to that of Lanfranc's church at Canterbury, but was only about half the length of the eastern arm of St. Albans which was of four bays giving a length of nearly 100 ft. Like St. Albans, St. Mary's seems to have had the presbytery separated from the aisles by solid walls. There were also solid walls between presbytery and aisles at Shaftesbury (RCHM *Dorset*, IV (1972), 60).

The fragments of the early church that remain *in situ* are mostly of dark-coloured gritstone, and fragments of a moulded cornice of the same stone are reused in the base of the W. wall of the present nave. The most likely source of this gritstone is at Bramley Fall on the river Aire near Leeds, whence the stone could easily be brought to York by water, but there is no evidence to suggest that any quantity of this stone was brought to York after the Roman period until the 17th century. The use of gritstone in any large quantity in mediaeval buildings in York is confined to the Saxon and early Norman periods and the stone is generally of Roman origin reused; gritstone at the abbey is certainly of Roman origin.

Surviving carved architectural fragments of the later 12th century, probably from the church and the chapter house, show that work of the highest quality was done here at that date, work which is remarkably like some of the work carried out in the Minster for Archbishop Roger after 1170 and also paralleled in Durham, in Selby Abbey and at Lincoln both in the Cathedral and in the domestic buildings. Particular features are the expansion of the chevron pattern by the alternation of straight sections between the angles of the chevron to give a chain link effect and the enclosing of roll mouldings within a trellis network of small rolls. Capitals carved with a formalised acanthus leaf are of excellent workmanship and form part of a series of capitals in the museum illustrating a wide variety of 11th and 12th-century caps.

The Second Church. The rebuilding of the abbey church in 1270–94 was carried out in a period which covers the finishing of the Angel Choir at Lincoln and the beginning of the new nave at York Minster, and probably also the building of the Chapter House at Southwell, which was closely connected with York. Stylistically it provides an interesting link between the 'Early English' of the 13th century, and the 'Decorated' of the 14th.

The Chapter House. The present vestibule occupies the full width of the E. range, and the Chapter House formed a rectangular projection E. of the range. The building of polygonal chapter houses outside the range was necessitated by their shape; for a rectangular one to be wholly outside the range was exceptional but this cannot have been the original arrangement. In the early 12th century the chapter house must have occupied the space of the vestibule and the remains of its E. wall can be seen under the entrance to the later chapter house. This entrance is of the late 12th century and must give the date of the building of the new chapter house, still rectangular but lying outside the range. The only remains of the new building found in the excavations of 1827–8 were 'the lowest portion of the foundations, built of grit-stone' (Wellbeloved) and later plans showing a triple E. window must be regarded as imaginative. An understanding of the character of the chapter house depends on consideration of a series of stone sculptures which are discussed

4

below; the plan suggests that it was covered by a barrel vault divided by transverse arches into five bays.

The Abbot's Lodging is described under The King's Manor, Monument (11).

Other Claustral Buildings. The remains of the claustral buildings are scanty. The plan of the principal buildings has been recovered by excavation and reading the records of excavation with the surviving mediaeval documents it is possible to arrive at a fairly clear picture of the layout. The surviving remains and the buildings that have disappeared are described together in the Inventory.

The Precinct Walls. The Gatehouse and the Precinct Walls are described in *York* II, *The Defences*, and the description given there is repeated in this Inventory. The wall with its towers forms the most extensive surviving monument of ecclesiastical defences in England.

Sculpture. 1. In the Yorkshire Museum are thirteen life-size figures or parts of figures which form one of the most important series of English sculptures that have survived from the Middle Ages (plates 1, 30–37). Seven of these figures were discovered in 1834 buried in the S. aisle of the abbey church under a layer of 13th-century window tracery set in mortar that was described as being the same as 16th-century mortar in the King's Manor. When found these seven figures, which include those in the best state of preservation, were painted in colour but little or no trace of colour now survives. Three figures were recovered from other parts of York where they had stood in exposed situations and had deteriorated badly. Two figures, purchased in 1954, came from Cawood.[1] The last figure is more fragmentary and its history is not known. The similarity of the last six to the first seven makes it likely that all the figures belonged to one sequence. Two of the figures represent Moses and St. John the Baptist; eight with bare feet and holding books are probably Apostles, but could also include Prophets. The others are too damaged to give any indication of their identity. All the figures show certain marked characteristics in treatment: the bodies are heavy, the heads are disproportionately large and the figures are apparently standing balanced on both feet although the drapery, falling straight over one leg and looped in folds over the other, is designed for figures with one leg straight and one flexed. The drapery is held up in one hand or looped over one forearm with a characteristic little swirl. The disposition of the drapery, originally invented to help in the representation of jointed limbs in varying positions, is here used on stiff immobile figures purely as a linear pattern without appreciation of its real purpose. The figures have vestigial columns emerging from their shoulders, but they are not true column-figures such as flank the doorways of many French cathedrals, where the figures are attached to full-height columns.

The figures are Romanesque in their proportions and in their heavy drapery falling to lively little pleats around the feet, but the Romanesque mannerisms are much less noticeable than in the comparable figures of the *Portico de la Gloria* at Santiago de Compostela of 1188. A new feeling of classicism in the York figures must place them later than the Compostela figures but they have not the simplicity and freedom from the Romanesque that was achieved at Laon *c.* 1190 and at Chartres, more strikingly, by *c.* 1210. The drapery looped over one leg and falling freely over the other has a longer history in other arts but first appears in French sculpture at Mantes *c.* 1175 and occurs commonly during the next thirty years, especially at Sens. One other detail affecting the date of the figures must be noticed: Moses is represented holding

[1] These two figures from Cawood must be two of the 'life-size figures of the four evangelists' described by E. Bogg (*The Old Kingdom of Elmet and the Ainsty of York* (1902), 233) as having been found buried adjacent to Cawood church and taken to The Grange, Wistowgate, Cawood. From The Grange they passed into the possession of Mrs. Wormald at Cawood Castle, who subsequently took them to the riverside garden of The Moorings in Cawood. Later, Mr. Samuel Wormald took them to Shoreham, Sussex, but at least one other incomplete figure had sunk too deeply into the riverside silt to be removed and has now disappeared. Bogg's statement makes it more doubtful if the figures did in fact come from York and opens the possibility of their having been made for the Archbishop's Palace at Cawood.

xliii

both the Tablet of the Law and the Brazen Serpent. This occurs in French sculpture only between 1170 and 1215. It seems clear therefore that the York figures cannot be much later than 1200. (See W. Sauerländer, 'Sens and York' in the *Journal of the British Archaeological Association*, XXII (1958), 54.)

It has been assumed that this series of figures flanked a great doorway to the abbey. The numerous French portals of the 12th century normally show the fore-runners of Christ, occasionally with St. Peter and St. Paul. Not until the end of the century was any French doorway flanked by the apostles. Indeed the earlier doorways did not provide space for them since they only had three or four figures each side. The number of figures surviving at York—and there is no evidence that the whole series has been recovered— make it unlikely that they flanked a doorway of c. 1200. Although some late 12th-century stonework is reused in the foundations of the later church, there is no evidence for any major building work on the abbey church at that time. Only if rebuilding necessitated by the fire of 1137 was not completed till c. 1200 could the erection of a portal of this date be explained, but Mr. G. F. Willmot's excavations at the W. end of the abbey church have failed to find any foundations of sufficient width to carry a portal or portals of the necessary size.

An alternative solution to the problem of the site from which these figures came is suggested by a group of earlier figures from the Chapter House at St. Etienne in Toulouse, also including eight apostles. These have long been displayed in an arrangement based on a figured portal but Linda Seidel of Harvard University has now shown that these figures, standing in pairs, must have formed the bay divisions in the walls of the chapter house, from which sprang transverse arches under a barrel vault (*The Art Bulletin*, L (1968), 33–41). The Toulouse figures are carved in pairs with columns between the heads supporting arched canopies, but the Camara Santa at Oviedo, rebuilt during the second half of the 12th century, also has a barrel vault with transverse arches springing from capitals and these are supported by true column-figures of the French type. Earlier, English figures supporting vaulting were erected in the Chapter House at Durham (c. 1135), which owed their inspiration to Southern France, but in the late 12th century it may not have been necessary to go so far as Southern France or Spain for examples of chapter houses decorated with figure sculpture, as numerous chapter houses in Northern France have been destroyed and the small chapter house at St. Georges de Boscherville shows that column-figures were not necessarily part of a portal. Another column-figure not forming part of a portal, from the cloister of Saint-Père, Chartres, is illustrated by L. Pressouyre in *Bulletin de la Société Nationale des Antiquaires de France* (Séance du 15 Mai 1968, Plate 10) with a drawing from the Gaignières Collection. The figure represents St. Benedict.

The surviving fragments of the entrance to St. Mary's Chapter House are elaborately moulded and decorated and may be dated stylistically to the late 12th century. It is difficult to suppose therefore that the Chapter House itself was not built in the last years of the 12th century, despite the large projection of the buttresses shown on Ridsdale Tate's plan of 1912; but the irregularity of those buttresses suggests that they may not have been original. A date at the extreme end of the century, c. 1200, is suggested by the character of the abbots in the second half of the century; Abbot Clement (1161–84) was described as 'rapax lupus omnia vastans' and his successor Robert was deposed by the archbishop at his visitation of 1197. Robert Longchamp who succeeded him, a man of wealth and power and brother of the Chancellor, is the more likely builder both of the Chapter House and the Gatehouse.

We are left then with a series of statues of c. 1200 some of which were buried close to the Chapter House at a time when the Chapter House, of much the same date, was pulled down to make way for new buildings for the Palace of the Council of the North. Most of the figures appear to be Apostles but two are certainly not. It appears therefore that the twelve Apostles can have marked the bay divisions of the Chapter House; Moses, St. John the Baptist, and other unidentified figures would then have marked the bay divisions of the Vestibule, replaced by moulded vaulting shafts in the late 14th century. The whole series would then have led from the precursors of Christ in the Vestibule, through the Apostles on the side walls of the Chapter House to the Christ in Majesty recorded as occupying an elevated position at the E. end of the Chapter House. This theory is supported by the shape of the backs of the figures, most of which are flat to stand against a straight wall but others are rounded or angled to stand in a corner where two walls meet. At Toulouse the eight Apostles formed the bay divisions and four other figures stood in the four corners of the Chapter House.

2. A group of large-scale fragments appear to come from a Coronation of the Virgin (Plate 40); the Christ figure is complete except for the right arm but the hand appears placing the crown on the Virgin's head. This is a late interpretation of a theme in which earlier artists had shown Christ's hand raised in blessing, and the crown held by angels. The middle part of the figure of Christ was found built into some late mediaeval repair work of the nave of the abbey, suggesting that the sculpture may have been broken as a result of the fire of 1377. No evidence has been found to suggest its original position, but it is possible that it formed a reredos to the altar of the Virgin in the nave, and was damaged in the storm of 1377.

OTHER MEDIAEVAL SCULPTURE

The Yorkshire Museum, built on the site of part of the cloister of St. Mary's Abbey, contains remains of the claustral buildings *in situ* and a collection of mediaeval architectural and sculptural stone carving. Most of the architectural fragments derive from the Abbey itself but among the decorative and figure sculptures there are more pieces from various other sources in York than there are examples *in situ* in the City outside the Minster. The principal items from St. Mary's Abbey (4) are listed on pages 23, 24; pieces associated with surviving buildings will be listed under those buildings; other pieces are listed here.

i. *Anglian Cross-head* (Plate 25b), probably found in York (YMC), of magnesian limestone, 6¼ in. by 6½ in. Listed as No. 12 by W. G. Collingwood in 'Anglian and Anglo-Danish Sculpture in York' (*YAJ* xx (1909), 178, 181, 185). On one side, within a medallion 3⅜ in. internal diameter, is a pentameter verse in fine lettering, which reads: 'Salve pro meritis, presbyter alme, tuis'. On the other side is a quatrefoil within a double circle.

ii. *Cresset* (Plate 25a), 9¼ in. high by 5¼ in. diameter at top (Collingwood, *ibid*, 200, 203, No. 22), of red gritstone. The shaft is plain up to a necking of two bands enclosing a two-strand twist. The bowl or cap has stylised leaf-forms growing from the necking in a chevron pattern, with a cable-moulded rim. This is not a reused baluster shaft, as the cable-mould forms a lip to the upper surface surrounding a hollowed-out bowl. Pre-Conquest.

iii. *Fragment*, of red sandstone, 9½ in. by 6 in. by 6 in., pattern on one side with interlaced double cable within a border of rectangular pellets; no provenance. Pre-Conquest.

iv. Top part of *Grave-slab* (Plate 25e), 23¾ in. by 17 in., 'found under the Mechanics' Institute in Clifford Street in July 1883'; the arms of the cross covered with tightly plaited interlace, the horizontal arms ending in animal masks; at the junctions of the arms are incised circles; the panels flanking the top of the cross each contain an animal with double outline, the hind leg extending as a double strap interlaced with the body; the panels are enclosed by a raised border treated, at the sides, as pilasters, one retaining a volute capital; pre-Conquest (Collingwood, *ibid*, 190, 193, No. 16).

v. *Grave-slab* (Plate 25c), of sandstone, 3 ft. by 1 ft., found in Parliament Street, with plain incised Latin cross with incised circles at the junctions of the arms; pre-Conquest (Collingwood, *ibid*, 162–3, No. 5a).

vi. *Grave-slab* (Plate 25f), of sandstone, broken, 2 ft. 9½ in. by 1 ft. 2 in., found in Parliament Street, with cross, cross-head with tapered arms in relief, cross-shaft incised only; the back similarly carved but unfinished; pre-Conquest (Collingwood, *ibid*, 162–3, No. 5b).

vii. Two *Capitals* (Plate 27e), carved with faces indicated by oval eyes and marked eyelids, found in a garden outside Bootham Bar; late 11th-century.

viii. Six *Corbels* (Plate 26), carved with grotesque or beasts' heads; 12th-century.

ix. *Corbel* (Plate 26b), of gritstone, carved with a complete animal, perhaps a wolf; 12th-century.

x. *Head*, damaged; late 12th-century.

xi. Two *Label-stops* (Plate 27f) carved with heads of a wolf and a bear respectively; late 12th-century.

xii. Fragments of an archway probably from All Saints', Pavement (Plate 28a–e): *Nook-shaft* richly carved with foliage; two *Capitals*, one with two-bodied lion with leaf-shaped tails, one with foliage and small leaf-tailed beasts; three *Voussoirs* with isolated roundels containing human figures and a bull; one *Voussoir*, from another order of the arch, with foliage on an undulating stem, and a roll-moulding; 12th-century. The nook-shaft is carved in a different style from the rest.

xiii. *Tympanum* (Plate 29a), carved with devils seizing the soul of a dead man as it emerges from his mouth, possibly the death of Dives, from the Hole in the Wall public house which stood on the N. side of Minster on the site of St. Sepulchre's chapel; late 12th-century.

xiv. Ten *Voussoirs* (Plate 28f) from an arch of at least three orders, one with beasts etc. in beaded roundels and other ornament, and a pattern of diagonal crosses on the soffit, a second order carved with formal foliage ornaments and a moulded soffit, a third order carved with beasts and a moulded soffit; all with roll mouldings; 12th-century.

xv. *Capital* to a nook-shaft (Plate 27b), with human head threatened by two dragons; 12th-century.

xvi. *Virgin and Child* (Plate 41b), headless torsos only, 15½ in. high; the Virgin wears a cloak held by a brooch secured with a cord, her hair falls in long plaits over her shoulders; late 12th-century.

xvii. Head of *Cross* carved with Christ in Majesty and, on reverse, Agnus Dei; 12th-century.

xviii. *Corbel* carved with human head; 13th-century.

xix. *Virgin and Child* (Plate 41a), both figures headless, 3 ft. high; the Virgin is seated and wears a simple dress with high belt, and cloak looped over left arm and draped across lap; the Child stands on her lap against her right arm; 14th-century.

xx. *Figure* of bishop (Plate 43a), head and legs missing, 20½ in. high, right arm broken, staff in left hand; 14th-century.

xxi. *Head* of bishop (Plate 42b), with low mitre; probably a label stop, 14th-century.

xxii. Small *Head* (Plate 42e), bearded, wearing a hood; found in Davygate; 15th-century.

xxiii. *Figure*, 14½ in. high, with name Maria Salome on base; c. 1500.

xxiv. *Figure* (Plate 42c), 27½ in. high, St. Margaret with dragon, perhaps from a tomb; the saint stands on dragon holding cross in right hand; early 16th-century.

xxv. *Angle Bracket*, 29 in. high, with half figure of angel, wings outspread, probably from the Carmelite Friary, Hungate; late mediaeval.

xxvi. *Head* of a man (Plate 43b), face badly damaged but carving of hair in good condition; late mediaeval.

xxvii. *Figure* of lady (Plate 43d), headless, 14½ in. high, with tight bodice above a full skirt; late mediaeval.

Monumental Effigies

xxviii. Mutilated recumbent effigy of a *Knight* in chainmail and surcoat, 4½ ft. long, formerly planted erect at the end of Clifton Village and known as Mother Shipton's stone; 14th-century.

xxix. Recumbent effigy of *Knight* with crossed legs, in chainmail and surcoat, 6½ ft. long, formerly used as a boundary stone for the parish of St. Margaret, Walmgate, on the E. side of Neutgate, now St. George's Street; the head rests on a pillow supported by angels; lion beneath feet; lizard-like creature by left leg biting shield; shield charged with *cross flory* and overall a *bendlet*, possibly for de Vescy (of the family of William de Vescy who gave the Carmelite Friary a site in Hungate in 1295); 14th-century.

Alabasters

xxx. *Virgin and Child* with bird (Plate 41c), headless fragment with traces of colour; found in the Ouse near St. Mary's Abbey; 14th-century.

xxxi. Six *Panels* from a sequence associated with St. William of York: (a) Birth of St. William in the presence of King Stephen; (b) Collapse of Ouse Bridge; (c) Edward I falling down a mountain, and the recovery of a fisherboy drowned in the Ouse; (d) Translation of St. William; (e) Virgin in Glory; (f) The Trinity with two donors; late 15th or early 16th-century. (G. F. Willmot, 'A Discovery at York' in *The Museums Journal* LVII no. 2 (May 1957)).

The bulk of pre-Conquest sculpture in York lies outside the scope of this volume; here, the most important item is the finely lettered crosshead fragment (i) (Plate 25b) which Collingwood ascribed to the early 8th century (*op. cit.* 178, 181, 185). The fragments of a grave-slab from St. Olave's church (Plate 25d) are of sufficiently high quality to suggest that they may have been from the tomb of a man of the importance of Earl Siward himself; they are similar to other slabs in York, one from St. Denys's church and others from the late Anglo-Danish cemetery under the S. transept of the Minster. The grave-slab (v) (Plate 25c) from Parliament Street is closely parallel to one from the same cemetery under the Minster.

Little is known of early Norman carving in York; capitals such as (vii) probably represent a local vernacular style much less sophisticated than contemporary quasi-Corinthian capitals in the Minster, which derive from Normandy. The collection includes a number of voussoirs from arches of 12th-century doorways which can be compared with the standing archways in various parts of the town, none of which is however in its original position. Those in the Museum are in better condition than most of those outside and the best picture of a complete doorway of this type is now to be derived from a drawing of the doorway of St. Margaret's church, Walmgate, made in 1791 and reproduced on Plate 44. This was an elaborate scheme with an archway in six orders. The outer order contains the Signs of the Zodiac and the Labours of the Months; to a similar series belong the voussoirs probably from All Saints' Pavement (xii). Two capitals from the same site are similar to capitals from St. Lawrence's and St. Maurice's. The foliage in a figure-of-eight pattern carved on other voussoirs resembles that on capitals in the crypt of the Minster.

Other Norman carvings include a number of corbels (viii–x), mostly of beasts' heads with open mouths; other heads probably formed the stops at the ends of hood-moulds.

Dragons appear in a number of carvings, either alone as at St. Lawrence's, where they terminate an order of the arch carved with foliage, and St. Margaret's, or in battle with men as on a capital probably from St. Mary's Abbey (Plate 29b, c).

The most ambitious figure composition is the damaged tympanum (xiii) from the cellar of a former public house which stood close to the N.W. tower of the Minster, found in 1817; the demons portrayed have their counterpart in a large scene with the mouth of Hell now in the Minster, found in the Deanery garden (John Bilson, 'On a Sculptured Representation of Hell Cauldron', in *YAJ*, XIX (1907, 435–45). The scene probably portrays the death of Dives and may be compared with the death of Lazarus portrayed at Lincoln where Dives and his companions are already in Hell (G. Zarnecki, *Romanesque Sculpture at Lincoln Cathedral*, n.d., pl. 11).

The great figures from St. Mary's Abbey, of *c.* 1200, were discussed above (p. xlii). Of the same date, are a Virgin and Child from Cawood (Plate 41d), now in a fragmentary state, and a series of voussoirs from the abbey, carved with scenes from the Gospel story (Plates 38, 39), of a type unusual in England but in the tradition of the greater French cathedral doorways.

With the coming of the 13th century a gradual softening of features is noticeable. Hair is stylised but the hard edges of eyes, eyelids and eyebrows become more naturalistic. This is first seen in the St. John the Evangelist (Plates 1, 30) and is continued in a fragment of a head from the abbey (Plate 43c) and a corbel head probably from the abbey (Plate 42a).

Sculpture in the monumental tradition of the figures from the abbey is found in the fragmentary group of the Coronation of the Virgin (Plate 40) where the eyes of Christ have the hard lids of earlier carving but the robes have the jagged crinkly outline of fully-fledged Gothic.

The roof bosses from the abbey include some fine naturalistic carving of *c.* 1300. Of later sculpture there is little anywhere in York outside the Minster and the shrines associated with St. William from the Minster, which are at present in the Yorkshire Museum, but the simple flowing lines of 14th-century work are shown in the headless statue of a bishop (Plate 43a) and an alabaster Virgin and Child (Plate 41c).

The fragments of panelling in St. Olave's church, carved with angels playing musical instruments, must derive from a stone screen and most probably come from the abbey (Plate 52).

CHURCHES

St. Olave's is the only complete church recorded; it is largely a building of the 18th century, retaining little mediaeval work undisturbed except for the tower, and it shows no trace of the church founded by Earl Siward in the 11th century. The chief interest lies in the way the church was integrated with the adjoining monastic structures, the parochial west tower being structurally one with the monastic chapel of St. Mary and the north wall of the church being built up on the base of the precinct wall of the abbey.

Of the other suburban churches only the tower of St. Lawrence's still stands; 12th-century doorways supposed to come from St. Nicholas's have been re-erected at St. Margaret's and St. Denys's, Walmgate, and from St. Maurice's at the church of St. James, Acomb Moor.

MONUMENTS

Some of the monuments at St. Olave's are to men of importance. William Thornton, joiner and architect, who died in 1721, showed considerable engineering ability in the restoration of the north front of Beverley Minster under Hawksmoor; he also worked under Colen Campbell on a house in Beverley, at Castle Howard under Vanbrugh, at Wentworth Castle for the Earl of Strafford, and at Beningbrough, N. of York, where he may have been the architect. The Wolstenholmes were of local importance in the field of carving and decoration (*see* p. lvi). William Etty, R.A., was very distinguished in his day as a painter of the nude and of history pictures; it appears that he was not connected with the Etty family of carpenters and builders who practised in York in the 18th century.

SECULAR

THE KING'S MANOR

The King's Manor, developed from the 13th-century lodging of the abbots of St. Mary's Abbey, is a building of great historical and architectural importance. A separate house for the abbot in the 13th century is to be expected at an abbey of the importance of St. Mary's. Abbot Samson at Bury St. Edmunds built himself a house c. 1200 and the abbot's house at Westminster and the prior's house at Ely, both rebuilt in the 14th century, include the remains of structures going back to the late 12th century. The surviving remains giving the most complete plan of an abbot's house of the 13th century are to be found at Battle where the house erected for Ralph of Coventry, 1235–61, lie on the W. side of the cloister (Sir Harold Brakspear, 'The Abbot's House at Battle', in *Archaeologia*, LXXXIII (1933), 139–66). The house at St. Mary's, as rebuilt in the late 15th century, stands partly on 13th-century foundations but the complete layout of the 13th-century house has not been preserved.

The rebuilding in the last years of the 15th century was carried out in brick, a material that had been used in York for the Merchant Adventurers' Hall in the middle of the previous century and in Hull c. 1320 for the building of Holy Trinity church, and then for the town walls at Hull and the gates at Beverley. Brickwork was not new to the area in the late 15th century but the use of terracotta for the windows is exceptional and is as early a use of this material for structural work as any that has been recorded in England; other examples are to be found in Norfolk and Essex. The rectory at Great Snoring may be the earliest of the southern examples, probably c. 1500 but not exactly dated; the work at East Barsham followed c. 1515 and at Layer Marney c. 1520. The great hall on the first floor with a flat ceiling hiding the roof timbers is remarkable, though not the earliest in York. The construction of the roof itself, with king-posts, is not in the local tradition of houses in the City but is akin to work being done in the upland areas of the Pennines and the Lake District.

As the late 15th-century work has affinities with the brick and terracotta of East Anglia, so too the remodelling of the house in the 16th century as a palace for the Council of the North also shows affinities with contemporary work in Essex and the south-east. A particular connection with Essex may be attributed to Thomas Radcliffe, 3rd Earl of Sussex, President of the Council from 1568 to 1572, who was connected with the FitzWalters of Essex and built himself a house at Boreham in Essex in 1573. The new wing begun in 1560 by the Earl of Rutland was mainly built of reused stone with stone-mullioned windows but the new windows inserted in the old building were made with brick jambs and mullions plastered to simulate stonework. This device was not uncommon in Essex and S.E. England but in the York area the only other example recorded is at Heslington Hall, just outside York, which was built by Thomas Eynnis, Secretary to the Council of the North in 1568. A plaster frieze at Heslington has been identified as coming from the same mould as one at Albyns in Essex (Jourdain, Fig. 10).

The use of plaster to simulate stone was a more reasonable economy in Essex where stone is not available locally. In York it appears to be less justifiable but the imitation of more expensive materials in cheaper substitutes is a regular feature of the royal buildings of Elizabeth I (E. Mercer, 'The Decoration of Royal Palaces 1553–1625' in *Arch. J.*, CX (1954), 150).

The new range completed by Radcliffe was built without attics whereas Sheffield's range built against it some forty years later, c. 1610, had the roof space used for rooms from the start. Sheffield's range is contemporary with college buildings at Oxford and Cambridge which, for the first time, were being built with attic storeys, although attics had been contrived in the roof space of Corpus Christi College,

Self Portrait. c. 1770.

Justice in a Triumphal Car beneath the Arms of the City. 1753.

(13) CITY ART GALLERY. Stained glass panels by William Peckitt.

Cambridge, for instance, as early as Henry VIII's reign, and had been added to existing buildings at Oxford after *c.* 1570 (Exeter College, St. John's College and New College. VCH, *County of Oxford*, III (1954), 116–17, 261).

The decoration of the Huntingdon Room, carried out for Lord Huntingdon *c.* 1580, is among the most interesting features of the Manor, and the windows of this period are notable for the early use of the ovolo (quarter-round) moulding on jambs and mullions instead of a hollow chamfer. The earliest known uses of the ovolo form occur during the previous decade at Kirby Hall, Northants., and Leicester's buildings at Kenilworth Castle.

Among the early 17th-century features a number of surviving stone doorways are remarkable; the great semi-circular bay windows erected by Lord Sheffield are known only from drawings and a few fragments of stonework reused in various buildings. Later in the century the contemporary use of brickwork is illustrated on the pilastered upper storeys of the central N. block and in the S.E. gable added to the mediaeval building with oval bull's-eye windows and tumbled brickwork for the coping.

The development of the abbey site is completed by the former headmaster's house of 1899 and the City Art Gallery, monument (13), which contains a collection of topographical prints and drawings, important for the study of the buildings of York. It includes some three thousand items within the period *c.* 1680–*c.* 1900. The core of this collection was presented by Dr. W. A. Evelyn of York in 1934. Local artists who are represented include Francis Place, William Lodge, John Haynes, Joseph Halfpenny, Henry Cave and George Nicholson; visiting artists include Hollar, the Buck brothers, Rooker, Marlow, Girtin, Cox and John Varley. The gallery also owns works by several artists who worked in York *c.* 1727–78, such as Drake, Hauck, Doughty and William Etty. A self-portrait in the Art Gallery must be considered among the best work of William Peckitt, the York glass-painter, whose work is also to be seen in the church of St. Martin-cum-Gregory (*York* III, 24), in the Minster, at New College, Oxford, and at Stamford (Lincs.).

OTHER PUBLIC AND INSTITUTIONAL BUILDINGS

Ingram's Hospital (23) and Wandesford House (26) are two pleasing ranges of almshouses in simple brickwork of the 17th and 18th centuries respectively. The early mental hospitals of York are of considerable interest to the social historian but Bootham Park Hospital (21) is also of architectural importance, having been designed by the leading York architect, John Carr. In contrast to Carr's Palladianism, the Greek Revival of the early 19th century is represented by the Yorkshire Museum (12) with its Doric portico, designed by William Wilkins, R.A. This was followed ten years later by St. Peter's School (29) where John Harper dressed a building of classical symmetry in Tudor gothic with a multiplicity of turrets and pinnacles.

DOMESTIC BUILDINGS

The houses recorded are all of fairly late date: one (monument 76) retains a fragment of timber framing perhaps of the 16th century but there is no complete domestic structure earlier than the second half of the 17th century. In the Bootham and Monkgate areas this no doubt reflects the damage done in the siege of York in 1644. A few houses survive from the late 17th century, brick-built and mostly of two storeys; the arrangement of these houses on plan falls into two types: those with a main range parallel to the street giving two front rooms with subsidiary wings at the back, and those that are built end-on to the street with one front room and one back room and a massive chimney-stack between the two rooms. In No. 21 Bootham (37) there is an interesting survival of late 17th-century painted wall decoration.

There are a few 18th-century houses which are only one room deep; the great majority of house plans of this period are based on a through passage with a staircase, and front and back rooms either on one side only or on both sides, corresponding to the Class U plan identified in RCHM *West Cambridgeshire* (xlvii) and RCHM *South-East Dorset* (lxiii). The same plan types continued in use through the first half of the 19th century and the later houses in the Inventory, not described in detail, generally conform to them (*see also York* III, xciv–xcvi). The larger houses in Bootham show a considerable variety of layout developed from the basic four-square Class U plan, while a small 19th-century house in Clarence Street (monument 60) shows how an ingenious designer can introduce interest into the same basic plan in a very confined space. As in *York* III a number of houses have a staircase placed transversely between front and back rooms either as the only staircase in the house or as a secondary staircase for servants.

Kitchens seem generally to have been on the ground floor. Basements in Bootham were suitable only for storage; the basement kitchen at No. 56 (57) was replaced by a kitchen above ground very soon after the erection of the house. The same house has an important reception room on the first floor (in 1972 the Council Chamber of Flaxton Rural District Council) showing that the first-floor saloon had not been abandoned even after 1840. Some of the larger houses in Bootham were built with small projecting wings for closets but how these were fitted remains conjectural. The development of the indoor water closet is illustrated by Fishergate House (105) of 1837, where the plan is designed to accommodate original water closets, and the small houses in Penley's Grove Street where one has an original internal water closet for which the plan is not adapted, but others still rely on privies at the far end of a low back wing, beyond scullery, coalhouse, etc. The original names of rooms in these houses have been reproduced in Fig. 81 from the architects' plans.

The external appearance of the 18th-century houses depends on good red brick with stone bands, and timber cornices carrying concealed gutters at the eaves. Parapets hardly exist in this area. Stone is used in the form of plain string-courses or bands dividing the storeys and sometimes also joining the window sills. The combination of wide storey bands and narrower sill bands was a Palladian feature used frequently by John Carr and continued by other architects. In few houses do the corners receive emphasis: at Nos. 39 and 45 Bootham (41) appear the unusual rusticated quoins in which alternate stones project with long faces on both sides of the corner (*see York* III, fig. 14, p. lxxx); a more usual type of quoin appears on No. 49 Bootham (43). Windows are simply treated, without architraves; the window arches are usually of rubbed brick, only occasionally interrupted by stone keys. No. 47 (42) and Nos. 53, 55 (45) Bootham have stone cornices above the brick arches without the support of brackets, frieze or architrave, an idiosyncratic treatment used also at No. 56 Skeldergate (*York* III, (117) 103, pl. 189) which can probably be attributed to John Carr.

House design of the later 18th century owes much to the influence of John Carr and his partner Peter Atkinson. Among the builders of the time whose work is represented in this area, Robert Clough stands out. His houses in Gillygate (Nos. 26 and 28, monument (128)) are rich in plasterwork, which was presumably executed by his son Robert who was a plasterer. Only one house in this area shows the use of

Fig. 8 (opp.). Staircases.

a. (40) No. 35 Bootham, late 17th-century.
b. (249) St. Olave's House, No. 48 Marygate, late 17th-century.
c. (41) No. 39 Bootham, 1748.
d. (39) No. 33 Bootham, 1754.
e. (38) No. 25 Bootham, 1766.
f. (128) No. 28 Gillygate, 1769.
g. (127) Nos. 16–20 Gillygate, early 19th-century.
h. (280) No. 42 Monkgate, early 19th-century.

Fig. 8.

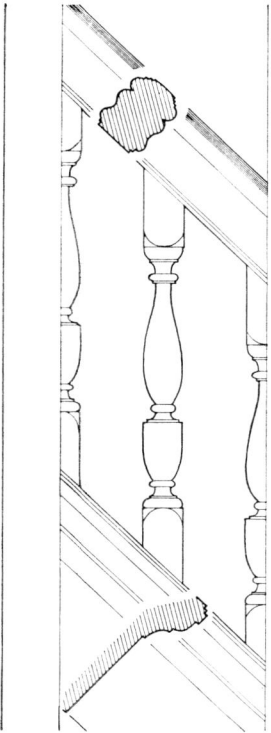

(a)

(b)

(c)

(d)

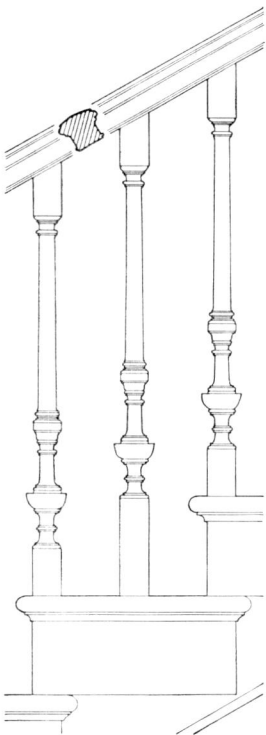

(e)

(f)

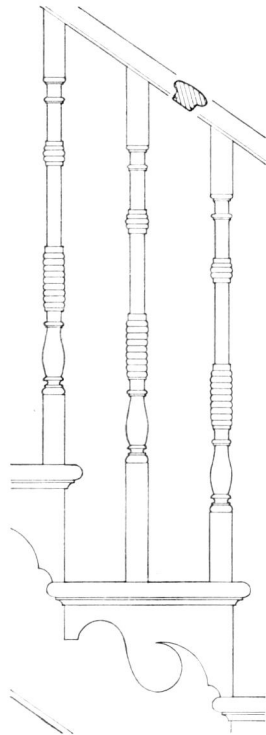

(g)

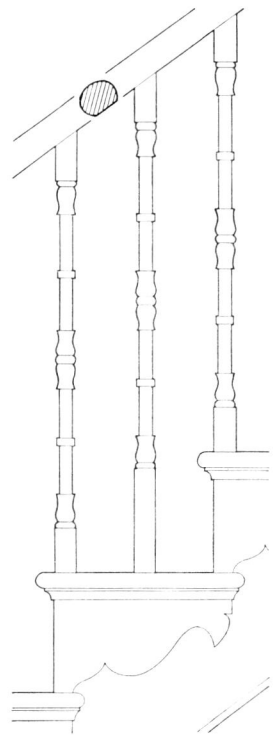

(h)

6   0   6   12   18 *Inches*

10   0   10   20   30   40   50 *Centimetres*

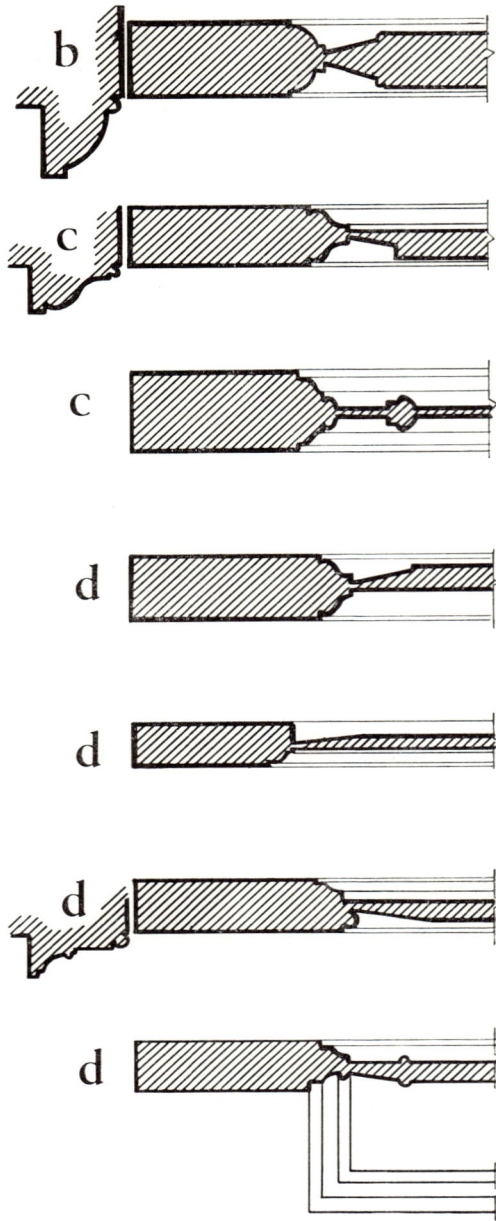

Fig. 9. Timber Mouldings, 18th-century.
a. (39) No. 33 Bootham, door architrave, 1754.
b. (128) Nos. 26, 28 Gillygate, door architrave and door, 1769.
c. (241) No. 29 Marygate, door architraves and door, c. 1780.
d. (117) Nos. 3, 5 Gillygate, door architraves and doors, 1797.

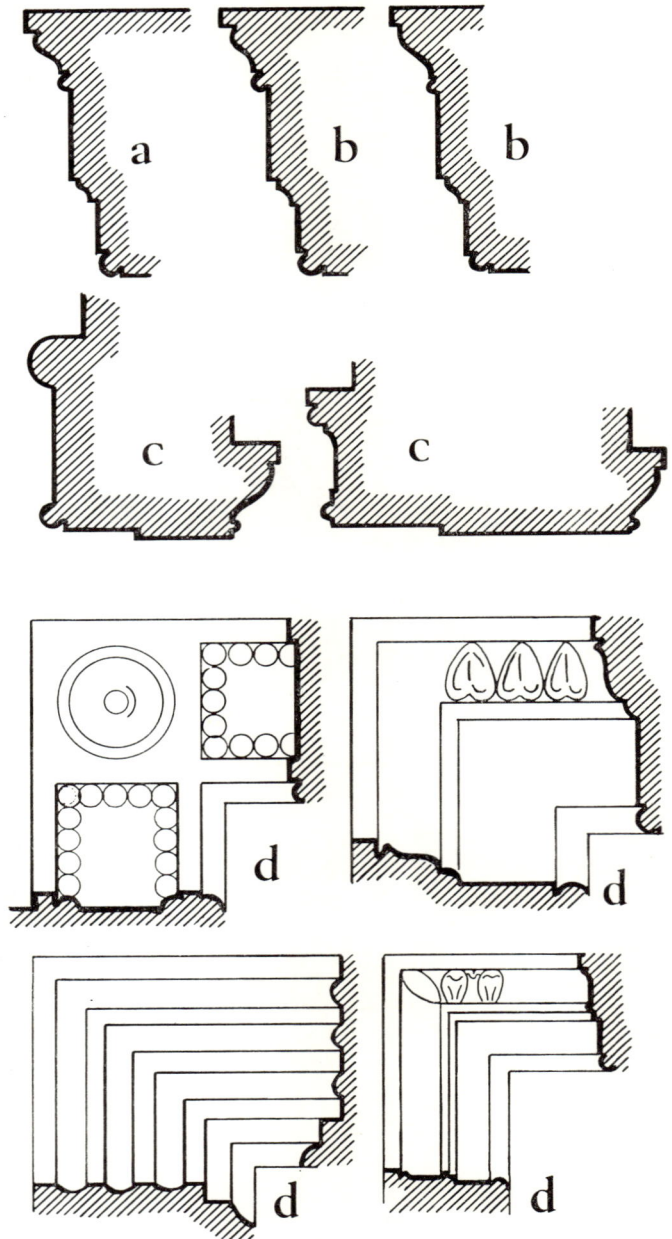

Fig. 10 (opp.). Timber Mouldings, early 19th-century.
a. (40) No. 35 Bootham, window architrave.
b. (263) No. 49 Monkgate, door architrave.
c. (263) No. 49 Monkgate, door.
d. and e. (109) Fulford Grange, door and window architraves.
f. (258) No. 15 Monkgate, door architrave and door.
g. (254) Almery Garth, Marygate Lane, door architrave.
h. (258) No. 15 Monkgate, door architrave and door.
i. (263) No. 49 Monkgate, fireplace surround.
j. (280) No. 42 Monkgate, door architrave and door.

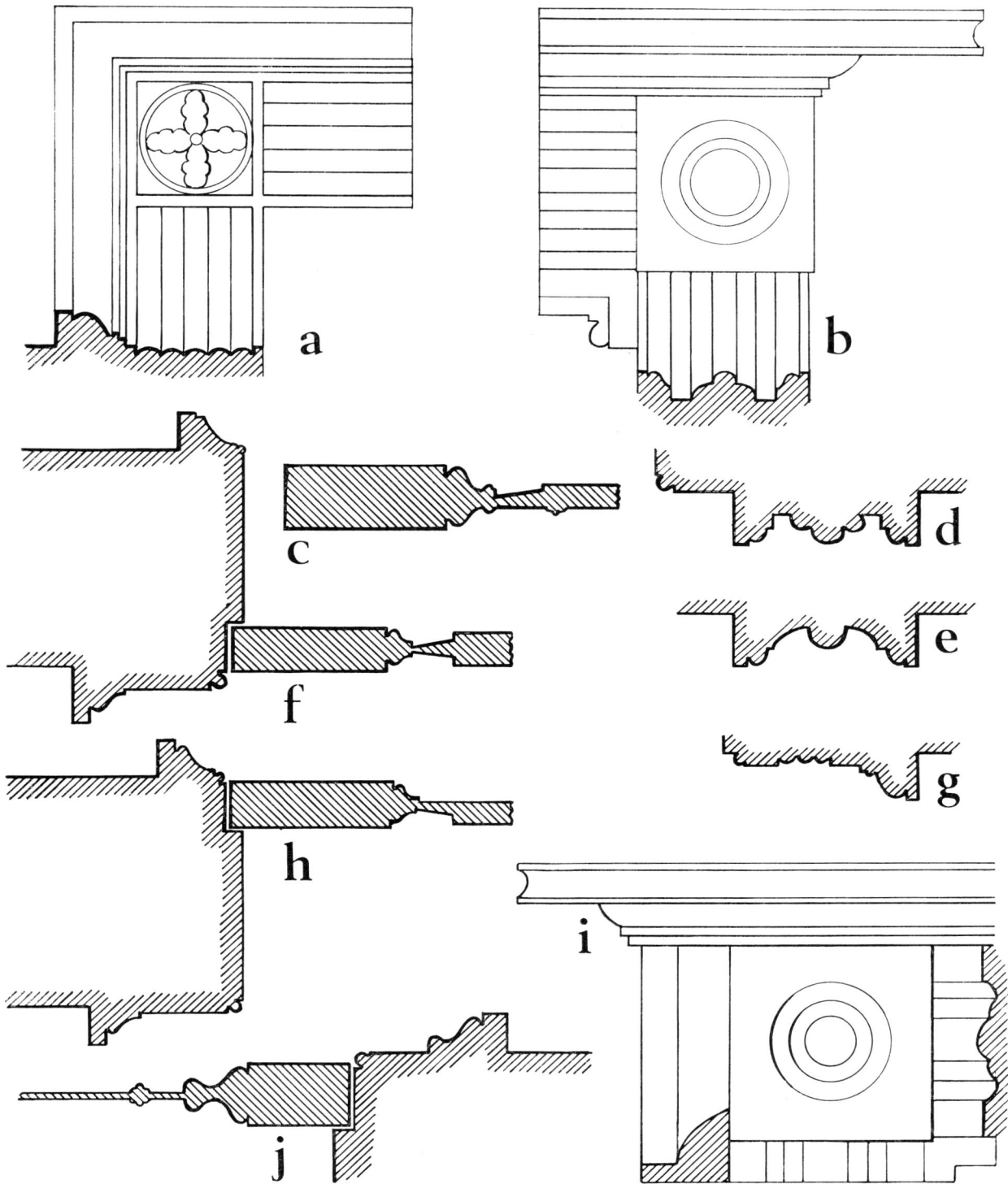

liii

Fig. 10.

decorative plasterwork around the staircase window; this is at No. 33 Bootham (39) also built by Robert Clough, but the plasterwork shows none of the profusion of decorative detail seen in the best houses in Micklegate described in *York* III. None of the Bootham houses have the richness of interior appointments found in some Micklegate houses; the most important features are generally the staircases which follow the same development discussed in *York* III, lxxxvii–xcii.

In 19th-century exteriors moulded architraves to windows make an occasional appearance (Nos. 51 Bootham (44) *c.* 1804 and 56 Bootham (57) *c.* 1840 and No. 44 Heworth Green (167)) but plain brickwork without dressings is more usual. The early years of the century produced little new building, but in the second quarter of the century there was extensive development of new streets of terrace houses, and the same period also saw the erection of a number of villa residences for the successful tradesmen and the professional class. The sale of the de Grey estate in Clifton in 1836 provided land for some of the larger residences but the biggest house to be erected at this time was Fishergate House (105), on the southern side of the City, built for a Mr. Laycock about whom no information has come to light apart from the bare record of his death. Fishergate House, designed by Messrs. J. B. & W. Atkinson, the grandsons of John Carr's partner Peter Atkinson, is a severe building of grey gault brick with brick pilasters at the angles. It is one of a number akin to the rather earlier grey gault brick houses of Cambridge, which have been noted for their originality, uncompromising severity, and bold articulation (RCHM, *City of Cambridge*, I, xcv). The pilastered treatment is also to be seen at No. 54 Bootham (56), No. 37 Monkgate (261) and Heworth Croft (164), and is repeated at Bootham Grange (80) where a certain grossness of scale makes it less acceptable. Internally Fishergate House owes much to the influence of Sir John Soane: the plan appears to derive from Tyringham and the elaborate three-dimensional design of the central arcaded light-well, with its interplay of arches without imposts, leaves little doubt as to the Atkinsons' source of inspiration. The arrangement of the staircase at Fishergate House is echoed on a smaller scale at No. 54 Bootham (56) where there is a similar lobby leading off the half landing to give access to a closet.

These severe houses carried out in gault brick show no detail which can be described as Greek, but the style may be derived from the ideas of simple architectural massing practised by such architects as Soane, Henry Holland and William Wilkins. Belle Vue House (145) seems to owe its gothic decoration to the trade of its owner as sculptor and monumental mason rather than to any general trend in architectural fashion. The only romantic house of any size, Glen Heworth (149), has been pulled down. Smaller buildings designed to be picturesque are illustrated in Plate 101.

In their interior fittings these 19th-century villas display an imaginative use of rather coarse modelling in ceiling cornices (Plate 122), while a number of doorcases and fireplaces are decorated with modelling from the moulds of Thomas Wolstenholme. Included in Wolstenholme's repertoire were a number of rectangular panels of mythological figure subjects and other decorative features which appear repeated in different houses (Plates 110–15). Throughout all the early 19th-century houses the most common form of architrave moulding is reeding butted against rectangular blocks at the angles, a motif used by Soane as early as 1790 (Stroud, 19) and incorporated into many of Wolstenholme's designs. Thomas Wolstenholme had achieved sufficient success by 1790 to purchase property at the junction of Gillygate and Bootham, on which he built houses. He died in 1812 and the business was carried on by his brother Francis who died in 1833 and his nephew John who died in 1865. They continued to use Thomas's moulds, with the same figure panels and the same idiosyncratic use of segmental shapes, after Thomas's death.

A striking feature of the larger 19th-century interiors is the number of staircases with decorative cast-iron balustrades. Those at No. 37 Monkgate (261) were certainly supplied by the firm of John Walker of Walmgate (Design Bk. 1, YCM 365/41) and those at No. 61 Bootham (48) and Bootham Grange (80) are so close to one of Walker's designs as to leave little doubt that they too came from the same works. The firm specialised in railings, many of which are still to be seen in York, and their work went to many country estates, to Sandringham, Kew Palace and the British Museum, as well as overseas (VCH, *York*, 273). Cast iron was also used as ceiling decoration for the centrepieces from which gas chandeliers were suspended; one at Burton Cottage (82) has ventilation ducts behind the foliage (Plate 121).

Much of the 19th-century development comprises terraces of small houses. New Walk Terrace (290) is of a quality above the average both in the size of the houses and in the standard of finish. At Grove Terrace (202) the designer has used a central pediment and end pavilions to compose a row of houses into one architectural unit, now obscured by trees in the front gardens. Elsewhere the terrace houses show considerable uniformity in design, and the planning of a block of the humbler class of dwelling is shown in Fig. 71 illustrating Redeness Street and Bilton Street, both now demolished.

INDUSTRIAL BUILDING

The only surviving industrial building in the area erected before 1850 is the former flax mill in Lawrence Street (209). Advertised as of fireproof construction, it appears to have vaulted brick floors similar to those in Bootham Park Hospital (21). For the early history of fireproof construction *see* H. J. Johnson and S. W. Skempton, 'William Strutt's Cotton Mills 1793–1812' in *Transactions of the Newcomen Society*, xxx, for 1955–7.

ARCHITECTS, BUILDERS and CRAFTSMEN mentioned in the Inventory.

John Carr, architect, born 1723, son of a mason and quarry owner, practised in York from 1754, served as Lord Mayor in 1770 and 1785, and died in retirement at Askham Richard, near York, in 1807 (Colvin, 122).

Peter Atkinson, senior, born 1735, trained as a carpenter, became assistant to John Carr and carried on the latter's practice after his retirement. Died 1805. (Colvin, 45.)

Peter Atkinson, junior, born c. 1776, son and pupil of the above, became his father's partner in 1801. Died 1842. (Colvin, 45.)

John Bownas Atkinson (1797–1875) and William Atkinson (1811–86) succeeded their father Peter in the architectural practice founded by John Carr. This practice is now carried on in York by Messrs. Brierley, Leckenby and Keighley.

Thomas Atkinson (c. 1729–98), architect, of York, was not connected with the above family (Colvin, 46).

John Harper, architect (1809–42), was born in Lancashire, became a pupil of Benjamin and Philip Wyatt, and practised in York (Colvin, 266).

Robert Clough, master builder, 1708–91, was the son of Robert Clough, bricklayer, who died in 1712. He owned the houses he built in Bootham and Gillygate (monuments 39, 128), as well as other properties. At the time of his death he was living in Low Petergate.

Robert Clough, plasterer, 1736–1800, son of the above, free of York 1758.

William Abbey Plows, sculptor and stonemason, 1789–1865, was the son of Benjamin Plows of Acaster Malbis (Gunnis, 308), who set up his own business in 1811 after 27 years as a journeyman (*YC*, 25/3/1811).

William carried on his father's business at Foss Bridge after the latter's death in 1824, offering monuments, chimney-pieces and side tables in stone, marble, alabaster, Roman cement, etc., as well as copings,
ridges, troughs, flaggings, etc. (*YG*, 1/5/1824). Designs for funeral monuments are preserved in YCL
(Y718 Plows). He lived at Belle Vue House (145), decorated with gothic detail presumably from his own
workshop, from *c.* 1834 till 1852. (*See also York* III, lvii, lviii.)

John Walker, ironfounder, apprenticed to the firm of Gibson of Walmgate 1815, became partner 1829 and
sole proprietor 1838. Succeeded *c.* 1847 by his son William Thomlinson-Walker, who later owned
Clifton Grove (monument 84).

Thomas Wolstenholme, joiner carver and maker of composition ornaments, born *c.* 1759, acquired
property at the corner of Bootham and Gillygate 1790, died 1812 leaving his 'business in Composition
ornaments . . . with all the stock in hand, moulds . . .' etc. to his brother Francis. Francis was succeeded
in 1833 by his son John (1794–1865) who worked as a carver in the Minster. His signature appears on some
of the bosses in the nave.

Fig. 11. (11) The King's Manor.
Carving from 18th-century staircase.

PLATE 17

From W.

Interior from N.W.

(4) ST. MARY'S ABBEY. Water tower. *c.* 1324.

PLATE 18

Gatehouse and Lodge from S.E. Drawing by F. Place, c. 1700.

Lodge from S.E. c. 1470.

(4) ST. MARY'S ABBEY.

PLATE 19

Passage from S.E. Late 12th-century.

S.W. wall of passage. Late 12th-century and later.

(4) ST. MARY'S ABBEY. Gatehouse.

PLATE 20

Abbey wall, Marygate, with Tower C and St. Mary's Tower, from S.W. 1266 and c. 1320.

(4) ST. MARY'S ABBEY.

PLATE 21

From S.E. *c.* 1324.

From N.E. *c.* 1324 and *c.* 1650.

(4) ST. MARY'S ABBEY. St. Mary's Tower.

PLATE 22

Abbey wall and Tower D, Bootham, from N.E. 1266 and *c.* 1320.

Abbey wall and Tower E, Bootham, from S.E. *c.* 1320.

Tower E, Bootham. Interior from S.W. *c.* 1320.

(4) ST. MARY'S ABBEY.

PLATE 23

From E.

From S.

(4) ST. MARY'S ABBEY. Postern and Tower, Bootham. 1497.

PLATE 24 'THE GARDENS OF THE YORKSHIRE PHILOSOPHICAL SOCIETY, YORK.' By J. Storey, c. 1860. York City Art Gallery)

MAP 2. Monuments in the Clifton area.

MAP 3. Monuments in the Bootham area.

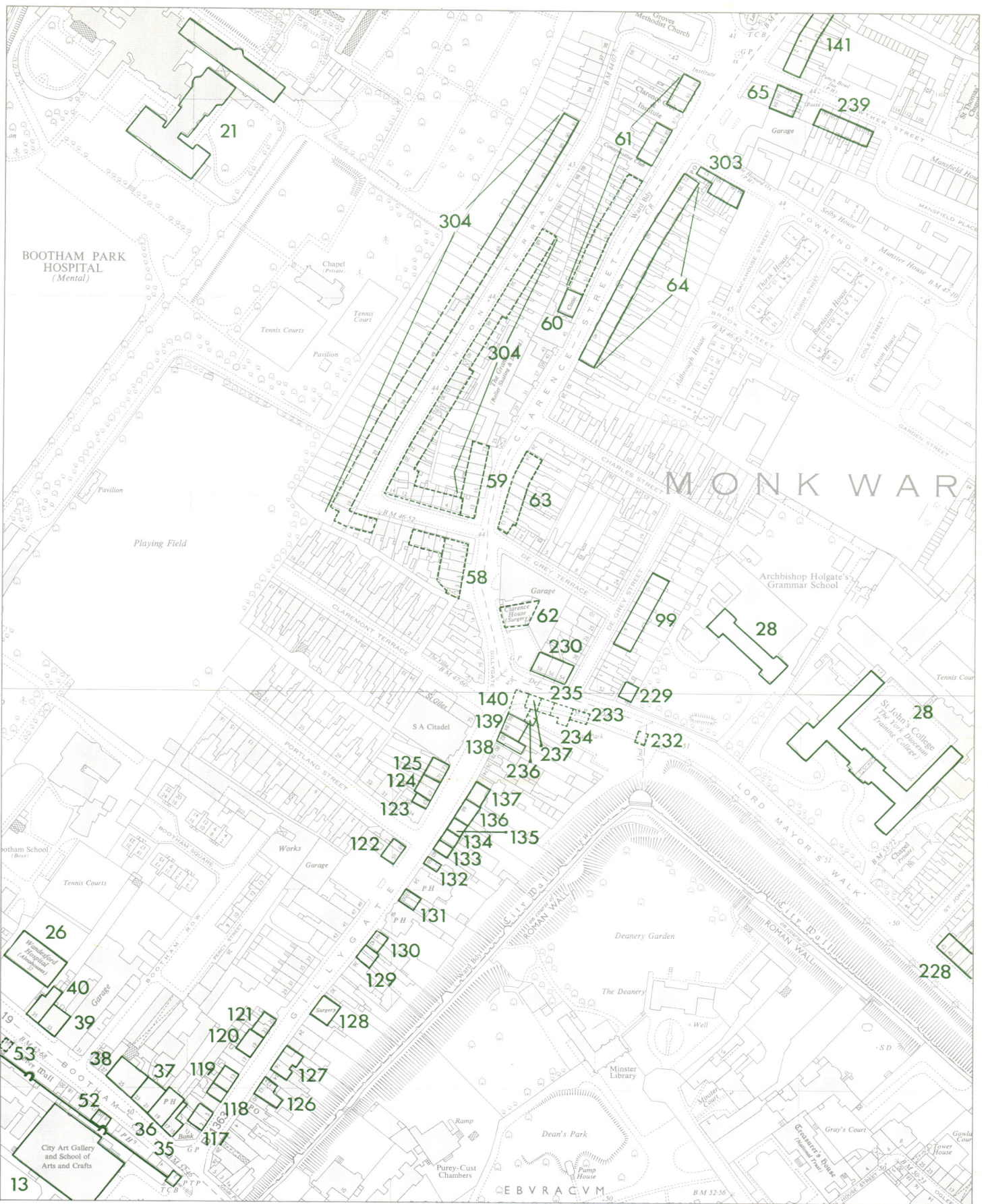

MAP 4. Monuments in the Gillygate area.

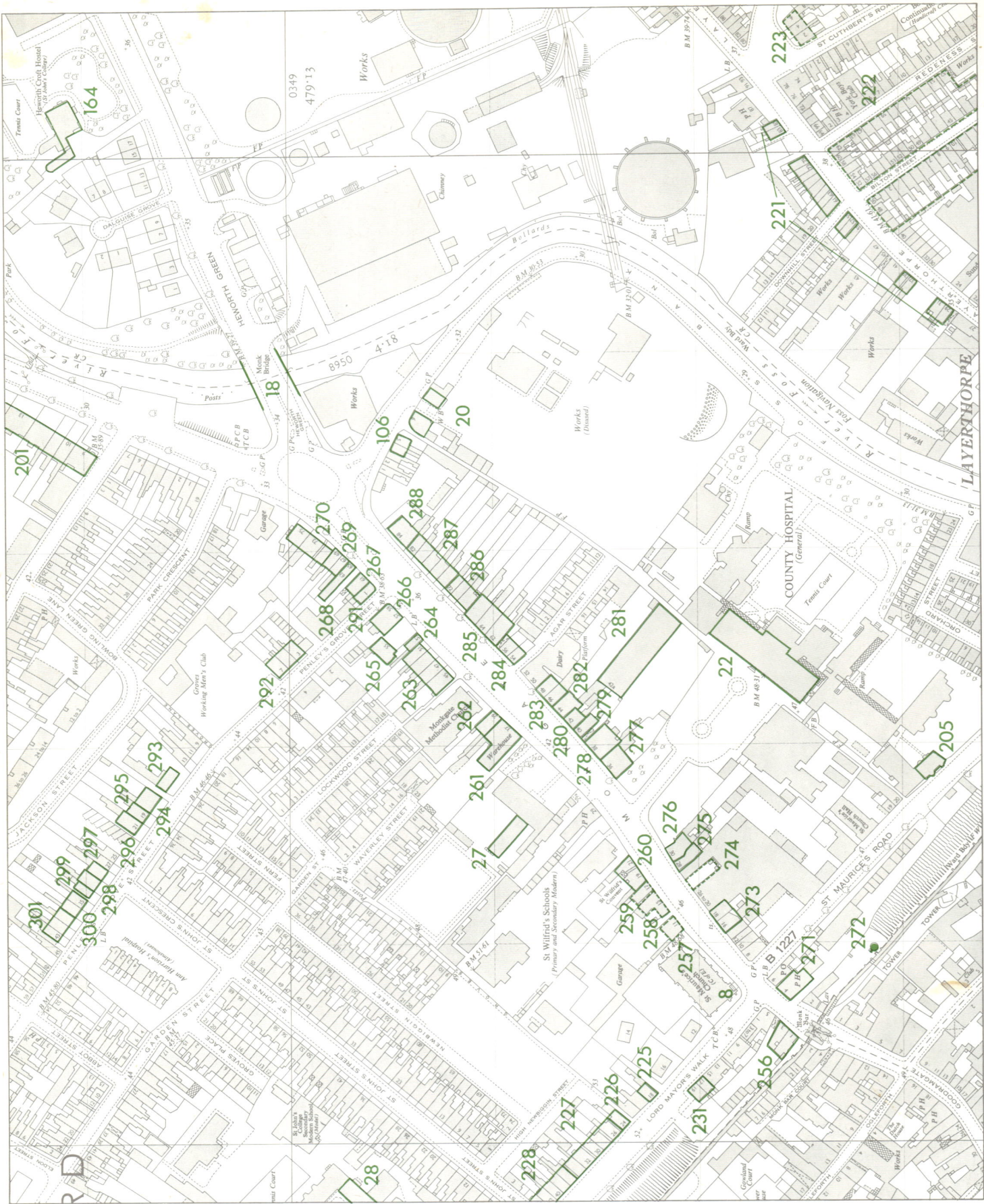

LAYERTHORPE

County Hospital (General)

Heworth Croft Hostel (St John's College)

HEWORTH GREEN

Monk Bridge

Works

Groves Working Men's Club

Monkgate Methodist Club

St Wilfrid's Schools (Primary and Secondary Modern)

St John's College (Secondary Modern School) (for Men)

Ann Allerton's Hospital (for Almswomen)

PARK CRESCENT

BOWLING GREEN LANE

JACKSON STREET

FERN STREET

ABBOT STREET

GARDEN STREET

PENLEY'S GROVE STREET

LOCKWOOD STREET

WAVERLEY STREET

NEWBIGGIN STREET

ST JOHN'S STREET

HIGH NEWBIGGIN STREET

GARDEN STREET

AGAR STREET

ST MAURICE'S ROAD

LORD MAYOR'S WALK

GOODRAMGATE

ST CUTHBERT'S ROAD

REDENESS

FOSS NAVIGATION

RIVER FOSS

TOWER STREET

AN INVENTORY OF THE HISTORICAL MONUMENTS IN THE CITY OF YORK OUTSIDE THE CITY WALLS AND EAST OF THE RIVER OUSE

(The dimensions given in the Inventory are internal unless otherwise stated and read first from E. to W. The National Grid References are in 100-kilometre square SE. The dates given in the description of memorials are of the death of the person commemorated.)

EARTHWORKS AND ALLIED STRUCTURES
AND
CULTIVATION SYSTEMS

EARTHWORKS

(1) LAMEL HILL (NG 61455095) occupies a commanding position overlooking the City, in the grounds of The Retreat (24), near the Heslington Road. It stands some 14 ft. to 22 ft. above the surrounding ground and measures 110 ft. to 125 ft. across. A summerhouse now stands on the top and the sides have been cut into for a wide path. The mound, excavated by Thurnam in 1849 (*Arch. J.*, VI (1849), 27–39), incorporates in its lowest 3 feet, which are undisturbed, part of an Anglo-Saxon inhumation cemetery. The whole of the upper part was thrown up to form a platform for a gun battery for Lord Fairfax at the siege of York in 1644 and contains disturbed remains of burials, showing that both the cemetery and the original, probably natural, mound were of much greater extent than at present. Both before and after the Civil War the site was used for a windmill.

The burials were in wooden coffins regularly laid on a W.–E. alignment, without grave goods, and were probably Christian. The coffin fittings are similar to those from an Anglo-Saxon cemetery at Garton in the E. Riding (J. R. Mortimer, *Forty Years Researches ...* (1905), 254–7, burials 31–60) which also had W.–E. burials and like Lamel Hill is at the boundary of the settlement it served.

(2) MOUND near Monk Bridge (61095277) standing 6 ft. high and 80 ft. in diameter has been excavated twice (Yorks. Geological and Polytechnic Soc. *Procs.*, VII (1881), 425; YAYAS *Procs.*, II, iv (1936), 44) and has been shown to be not older than the 18th century. It was probably erected as a garden feature to carry a summerhouse, of which the foundations can be seen on it.

CULTIVATION SYSTEMS

(3) Traces of former cultivation survive or are visible on air photographs; they consist of scattered parcels of mediaeval broad plough ridges, usually about 30 ft. wide, and fields of later narrow plough ridges, usually 12 ft. to 15 ft. wide. Areas of pasture, apparently never ploughed in the history of the city, also form an important part of the city's pattern of husbandry.

Citizens of York held strips in open fields which by 1546 had been enclosed, and forty-one closes were available for leasing within the city. Citizens also had rights of stray (i.e. of pasturing cattle) on nearby moors and commons, often involving arrangements with neighbouring townships. This pasture was supplemented by fodder from 'ings' or water-meadows by the rivers. The expansion of the city and townships led to a gradual reduction in the area of arable land and common pasture. The city gradually acquired plots of land over which it had sole rights of stray, in return for surrendering its rights of intercommoning in other townships. From these it created the present strays, organised within the four wards, three of which are on the N.E. side of the river.

Remains occur within the mediaeval city limits, and

I

in areas developed in adjacent townships but now taken into the modern city.

(a) York

In *Walmgate Ward* in 1772 (J. Lund, Map of Walmgate Ward Stray) there were North Field, Haver Garths, and Mill Field N. of Lawrence Street, and Chapel Flat to the S., all now obliterated by housing and refuse tips. Part of North Field, around 61455185, which existed until *c.* 1960, with broad ridges 30 ft. wide, was in selions in 1297 (ERAS, *Trans.*, XIX, for 1912, 276). All the fields were subject to half-year commoning (i.e. common grazing after harvest) until 1824. On Heslington Road, the grounds of The Retreat were Siwards How Field in 1484.

South of Heslington Hill, the city's rights of stray formerly extended S. onto the Tilmire and rights were shared with both Heslington and Fulford. Rights within Fulford township were delimited in 1484 (YCA, E30, 74), and in 1759 Low Moor was given to the city in lieu of its rights. An additional area was added to this, bought from compensation obtained for loss of rights in the four fields off Lawrence Street, to give Walmgate Stray its present shape and size of 77 acres. Much of the stray has never been ploughed, but two enclosures were made in Napoleonic times (around 61665005 and 61305046), and the temporary cultivation has left narrow plough ridges 15–16 ft. wide. Traces of narrow ridges on one of the approaches to the stray (61425094) suggest that this was once one of a group of small fields off Heslington Road.

Monk Ward in 1736 (J. Lund (1772), Map of Monk Ward Stray, unaltered map based on George Smith, 1736) had half-year grazing rights in the closes formerly called Grange Field and Hall Fields (Layerthorpe). Tang Hall Field is excluded on Lund's map; it was formerly the township of Tang, held by the prebendary of Fridaythorpe who constantly disputed the city's rights to common until the late 18th century. The surviving remains are in Grange Field. Here broad ridges, 30 ft. across, survived until 1965 between the railway and the River Foss (61075286), and they are recorded on the site of the railway in 1879 (J. Raine, Plan of Anglian Cemetery, Heworth, in YM). Narrow ridge-and-furrow survives on the lawns of Yearsley Bridge Hospital, and to the S. of it (61085359), all formerly one close.

The city's rights of stray formerly extended 6 miles N.E. of the city to Sandburn Cross (669586), a 17th-century boundary stone. These rights were exchanged at the time of the enclosures for 118 acres in Heworth, and 132 acres in Stockton, all of which remained outside the city boundary until its extension in 1884 and 1934 took in part of Monk Stray. This part is situated between two groups of Heworth's ancient enclosures along the Malton road. The wet clay land E. of the road (environs of 61675305) has always been pasture, but W. of the road the slightly better drained clays, now Heworth golf course, were ploughed in Napoleonic times and two large areas of narrow ridges, 14 ft. broad, and straight, survive to the N. (around 61655360) and S. (around 61555322) of Muncaster House

In *Bootham Ward* in 1772 (J. Lund, Map of Bootham Ward Stray) no rights of half-year commoning over the 'old

enclosures' existed within the ward. No remains of cultivations survive but the layout of fields on the 1853 O.S. maps suggests consolidated strips. Bootham Stray lies beyond and N. of York and Clifton, and was formerly pasture intercommoned with Clifton, Huntington, Rawcliffe, and Wigginton, which began with gifts of pasture in the Forest of Galtres by William Rufus and Henry I to York and Clifton. In 1633, 60 acres, now called The Intake (environs of 60305495), was given by Huntington to York to end York's rights in the township. In 1769 Clifton Enclosure Award gave 91 acres, called The New Intake in 1772, adjacent to The Intake, in lieu of the city's rights in Clifton, together with $21\frac{1}{2}$ acres for an outgang through the new closes from the Horsefair, preserved in part in fence lines and the shape of Clarence Gardens (60405292). Today the stray, based on these awards, has 180 acres, extensively covered with narrow ridges aligned on the 1769 boundaries and overlaid by the 1845 York–Scarborough railway line, again suggesting temporary ploughing in Napoleonic times.

(b) Parts of Incorporated Townships

Part of *Fulford* Field is now covered by the Broadway estate, in part developed within the limits of former hedge lines, and broad ridges survive in a block of four fields, formerly a parcel of strips aligned N.E.–S.W. and abutting onto Walmgate Stray (616498). The ridges are 27–30 ft. across, 800–1400 ft. long, and up to 2 ft. high with a slight aratral curve. The playing field off Cornwall Drive (612496) with reduced broad ridges 30 ft. across is part of the same field. A headland at the N. end of this field is referred to in 1484 implying a N.–S. direction to the furlongs at that end of the field.

Osbaldwick. Part of the open field called Slack Field is now within the city. Whernside Avenue and Penyghent Avenue (environs of 62455210) now cover parts of five fields which had groups of broad ridges in 1951. A playing field at Grange School (62855155) has broad ridges, 32 ft. wide and 2 ft. high, and was also part of the same field.

Parts of the fields of *Huntington* are incorporated near Yearsley Bridge, where fragments of narrow ridges survive on an overgrown plot (61055368), and on part of Rowntree's sports field (60925393) there were narrow ridges until 1966.

Formerly in *Heworth* township, fragments of broad ridges survived at Westlands Grove (62035335) but are now built on, and on a nearby playing field (61855325). These were part of old enclosures shown on the 1819 Heworth Enclosure Award map. A bungalow estate on Whitby Avenue (environs of 62355310) has recently obliterated a small parcel of broad ridges, each 30 ft. across, part of a parcel of ridges still extant in Heworth Without.

Formerly in *Clifton*, the largest surviving block of broad ridges within the city is on playing fields N. of Asylum Lane (around 60205300) measuring at least 300 by 200 yds. This clay area, at one time Laithe Close, has slightly sinuous ridges, 30 ft. wide and 1 ft. high. One parcel of ridges is aligned E.N.E. and another N.W. A less well preserved parcel of ridges survives off Kingsway (59655320). W. of Burdike and along the River Ouse are playing fields which may have been meadow. Between Burdike and Almery Garth there was a farm as late as 1796.

ECCLESIASTICAL

(4) ST. MARY'S ABBEY (Plates 2–23) lies between Bootham and the river Ouse immediately outside the City wall. The existing remains include the foundations of a church begun in 1089, ruins of the nave and N. transept of a church begun in 1270 (Plate 4), some fragments of the E. arm of the church and of the entrance to the Chapter House, part of the enclosing wall with its towers and gatehouse, and a building known as the Hospitium.

The later history of the abbey is largely recorded in three documentary sources. *The Chronicle of St. Mary's Abbey* (Surtees Society, vol. CXLVIII (1934)) is a detailed record running from 1258 to 1325 compiled probably by one of the monks in the dependent cell of St. Bees. The latter part is contemporary. There is a lacuna covering the years 1284–92 inclusive together with a part of 1293. The Chronicle contains details of the great rebuilding under Abbot Simon de Warwick and his successors. The *Anonimalle Chronicle*, running from 1333 to 1381 (ed. V. H. Galbraith, Manchester, 1927), was compiled at St. Mary's Abbey. It includes four entries concerning the abbey, only one of which—the fire of 1377—is pertinent to the building history. It is a legitimate deduction that no major new works were undertaken during the period. The *Ordinale of St. Mary's Abbey* (Henry Bradshaw Society, vols. LXXIII (1936), LXXV (1937) and LXXXIV (1951)) records the usage for the celebration of Divine Office throughout the year. It was drawn up by a commission appointed in 1390 (*Ord.* 1) and contains no material later than 1398. The extant MS. is a copy belonging to the Abbot's Chapel. The Ordinale contains many incidental references to the church and other buildings of the abbey, from which it has been possible to deduce something of the liturgical arrangement of the church, outlined below after the architectural description, and the uses of the various claustral buildings.[1]

St. Mary's Abbey was founded in 1088 by William II. Earl Siward of Northumbria, who died in 1055, was buried in the Minster of St. Olaf which he had founded at Galmanho outside the City of York on the site later occupied by part of St. Mary's Abbey. The minster with its possessions was given by Alan, Earl of Richmond (*EYC*, I, 265), to monks who had fled first from Whitby and then from Lastingham because of a quarrel with William de Percy who had refounded Whitby

Abbey, and with his brother Serlo who had become Prior of Whitby. King William found Alan's provision for the monks inadequate and granted them additional land, and in 1089 laid a foundation stone for a new abbey church of St. Mary. Of this church the plan (p. xl) has been recovered by excavations of 1827, 1912 and later; it was a cruciform building with an aisled nave, central tower and a short apsidal presbytery flanked by aisles apsidal internally but square-ended externally; two apsidal chapels set in echelon projected from each transept. That this church was nearing completion in the period 1120–35 is suggested by a gift to the abbey for the roofing of the church (*EYC*, XI, 100–1). It is stated that in 1137 extensive damage was done to the City and the abbey by a great fire (Trinity College, Dublin, MS. E 6 4 (503), f. 130; J. Stowe, *Annales* (ed. 1631), 144; quoted by J. H. Harvey in *YAJ*, XLI (1966), 365) and decorative details of mid and late 12th-century date found on the site show that building work was going on during the second half of the 12th century; there are late 12th-century fragments built into the foundations of the W. wall of the church (*YAYAS, Report 1953–4*, 13) but only in the vestibule of the Chapter House is any work of this period preserved *in situ*. Artistically the most important remains from this period are the great series of sculptures of *c.* 1200 which were connected with the Chapter House.

In 1270 the rebuilding of the church was begun under Abbot Simon de Warwick, a new E. arm being the first work to be undertaken; the Chronicle records that the foundations reached in places to a depth of 26 ft. The first stone of the walling was laid in that year and of the columns of the choir in 1277 (*Chron.* 15) and the eastern arm must have been structurally complete by 1283 when Archbishop William Wickwane consecrated six altars in the new choir on 2 February and on the following day the altar of St. Catherine in the vestry (*Chron.* 22). Demolition of the central tower, which had been endangered by the carelessness of the masons, began in 1278 (*Chron.* 19) and the whole church took twenty-four years to build (*Chron.* 65). There is however no record of its completion or dedication in the Chronicle, which starts again in September 1294 after a gap of ten years. The monk Hugh de Compton, whose death is recorded in 1314, apparently acted as master of the works (*Chron.* 65). Except for the wall of the S. aisle against the cloisters, building seems to have proceeded according to one consistent, uniform design until the W. end was reached. Breaks in the coursing of the masonry of the nave suggest that the nave was built in two stages, and there are some small variations in detail

---

[1] Mr. A. B. Whittingham has also made a study of these documents and the Commissioners are grateful to him for having made available to them the results of his researches, prior to publication in *The Archaeological Journal*, CXXVIII, where some alternative interpretations are presented.

between the E. and W. parts. Nevertheless the nave arcades and the N. aisle wall were evidently completed before the rebuilding of the W. wall was undertaken and this then followed, to a stylistically more advanced design. The buttresses to the S. aisle of the nave, with canted sides, do not match the main part of the church but appear to have been similar in style to the W. front, and the greater thickness of this S. wall against the cloister suggests that the Norman wall may have been retained and refaced. A new bell was installed in the tower in 1306 (*Chron.* 41).

In 1377 the abbey church was struck by lightning and the central tower and the transept were damaged; fire spread to the S. choir aisle, the nave and the cloister, but the choir, the nave of the church and the monastic buildings were saved (*Anonimalle Chronicle*, 95).

The cloister lay to the S. of the church and fragments preserved in the Yorkshire Museum show that the cloister arcade was built in stone as early as the 12th century. In the E. range an early chapter house was converted at the end of the 12th century to form the vestibule to a new chapter house built wholly to the E. of the range. This was probably done after the election to the abbacy of Robert Longchamp, brother of the Chancellor, in 1198 (Radulfus de Diceto, II, 151, RS, LXVIII). It is certainly unlikely that the work would have been started under Abbot Clement (1161–84) who was described as 'lupus rapax' wasting everything that others had accumulated. In 1297 Prior William of Derby had built, partly at his own expense, a hall for 'Wlays'[1] (strangers), in which Abbot John died in 1313 (*Chron.* 29, 58). This probably occupied the upper storey of the W. range. In the following year he began work on the N. end of the Dormitory in the E. range but complete rebuilding of the Dormitory followed under Abbot John de Gilling, 1303–13 (*Chron.* 36). The provision of a 'long room for the recreation of the brothers' is recorded in 1314 (*Chron.* 66) but its position is not indicated. The illustrations to Wellbeloved's account of the excavations in 1827–8, combined with the few surviving remains, make it clear that later rebuilding took place, probably after the fire of 1377, affecting at least the Inner Parlour and the Chapter House Vestibule in the E. range, and the E. part of the S. range. In 1455 an altar was dedicated in a newly constructed chapel in the Infirmary (*York Fabric Rolls*, SS xxxv (1859), 239–40).

Of the building known as the Hospitium (Plate 15), between the abbey church and the river, the lower storey was built in the 14th century and may have formed the *sartrina* or tailors' shop of the abbey (*Ord.* 87; *Chron.* 67). The upper storey was built in the 15th

---

[1] Anglo–French Waleis, a variant of the Germanic Walh; *cf.* Welsh.

century, when it probably served some other purpose, and the whole was restored and partly reconstructed in modern times.

The Abbey Gatehouse (Plates 18, 19) was built in the late 12th century and remained fairly complete until the early 18th century. It stands directly over the line of a Roman road (*York*, I, 2, no. 5). Doorways in the side walls indicate that there were flanking buildings of the 12th century but these have entirely disappeared and the buildings now flanking the gatehouse are of c. 1470–80. A chapel at the gate was built by Hugh de Compton, who died in 1314, presumably before his appointment as Prior of St. Bees in 1296 (*Chron.* 65). Restoration of the chapel was planned or in progress in 1376 when indulgences were offered to those contributing to the work (Papal Letters, IV, 511; Raine, 266–7). The chapel is described as 'supra portam' (*Chron.* 65) and 'juxta portam' (*Ord.* 319); it probably occupied the upper floor of the building between the gatehouse itself and St. Olave's Church (Plate 14).

The earliest mention of the building of a stone wall around the precinct was during the abbacy of Simon de Warwick, in October 1260, when an enquiry concluded that 'to build a stone wall below the abbey of St. Mary as far as the infirmary of St. Leonard would strengthen and improve rather than damage the city of York' (*CIM, 1219–1307*, 20, no. 255). On 9 December of the same year the king granted permission to the abbot and convent to 'construct the said stone wall within their abbey up to the aforesaid infirmary according as shall seem most advantageous to them' (*Close Rolls 1259–61*, 315). The phrase 'up to the . . . infirmary' of St. Leonard's Hospital must mean up to a point outside the city defences opposite the infirmary, which lay within the city wall but was probably visible above it. Work did not, however, start immediately, and in the meantime a dispute between the citizens and the abbey erupted into violence in August 1262, resulting in the killing and plundering of some of the abbey's tenants and the burning of houses in Bootham (*Chron.* 6).

The stone wall was started in 1266: 'Pridie Kal. Junii eiusdem anni inceptus est (murus) petrinus circuiens Abbatiam Sancte Marie Eboracensis, incipiens ab ecclesia Sancti Olaui et tendens versus portam civitatis eiusdem loci que vocatur GALMANLITH' (*Chron.* 8). The city gate called Galmanlith has been identified as Bootham Bar. The wall probably started from the gatehouse near St. Olave's church, since at that time the church, before the widening of the N. aisle, lay entirely within the precinct. The new wall faced towards Marygate and Bootham. The end of the length of wall parallel to Bootham has been demolished, but maps of the King's Manor estate made in 1770 and 1798 (PRO,

MPE, 344, 575) show that it was about 36 ft. S.E. of the Postern Tower.

The wall begun in 1266 was simply a boundary to the precinct and served no defensive function in a military sense but on 12 July 1318 a licence was granted to the abbot and convent to crenellate the abbey 'which is without the city of York, but is contiguous thereto, provided that the wall to be constructed between the wall enclosing their abbey towards the great street of Bouthum and when the walls of the abbey need repair they shall have easement in the High Street there by the dykes and walls which extend from Seintemariegate to Bouthumbarre, to repair these at their will, and easement also in the place extending from Bouthumbarre to the Ouse, between the abbey walls and the city dyke, for such repairs'. The city would not build upon the

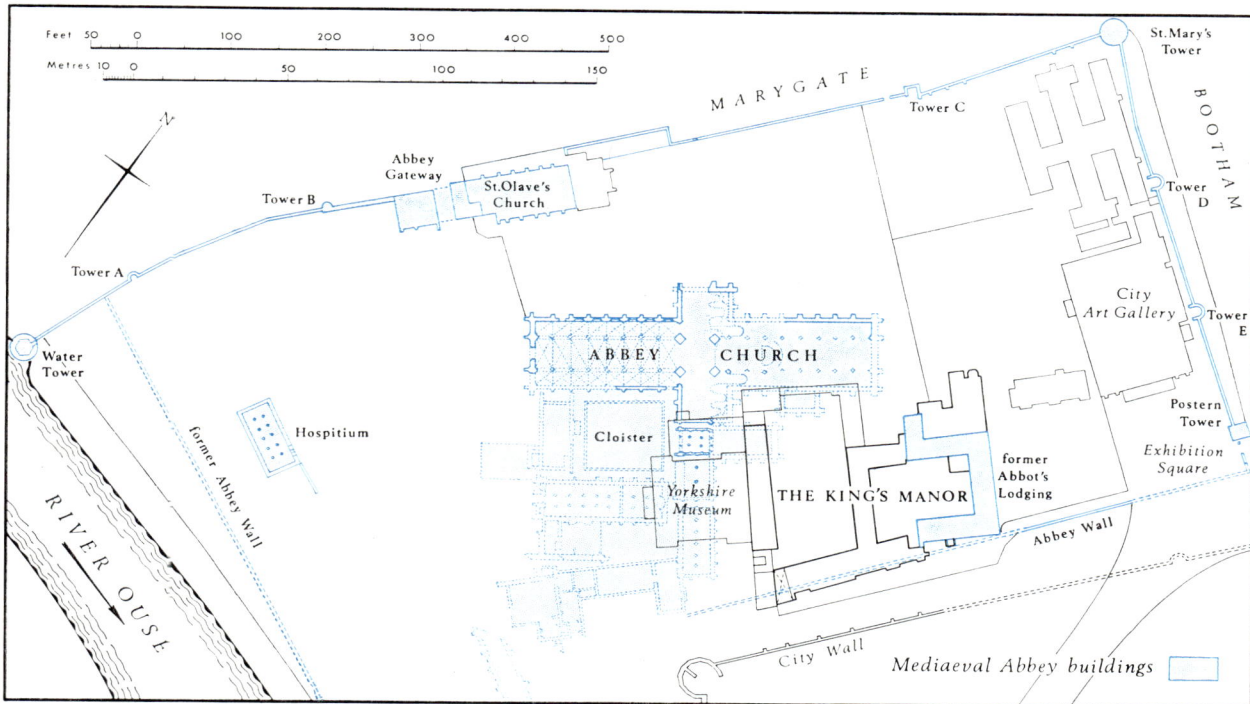

Fig. 12.   ST. MARY'S ABBEY AND THE KING'S MANOR

abbey and the wall of the city shall not exceed 16 feet in height and shall not be crenellated' (*CPR, 1317–21*, 190). Crenellation involved the raising of the wall facing Marygate and Bootham by an additional 5 ft. to 6 ft.

On 24 June 1354 an agreement was concluded between the abbey and the city which was intended to settle their perennial dispute over Bootham. This provided 'that it shall be lawful for the abbot and convent to scour a dyke extending from the said Ronde Tour at the end of Seintemariegate towards the gate of the city called 'Boothumbarre', which dyke is within the said suburb whenever they please, for the safety of their abbey ditch along Bootham, and if the abbey did so 'with houses opening on the said street of Bouthum', the built-up area was to pass into the city's jurisdiction. Another provision was 'that it shall be lawful for the abbot and convent to make their wall on the said water (of Ouse) in the manner in which it has been commenced' (*CPR, 1354–8*, 84–6).

This river wall, already started in 1354, appears on maps by Speed (1610), Archer (*c.* 1682), and Horsley (1694), but not on that by Cussins (1722), or on subsequent maps. Speed shows it as crenellated, so does Place in an engraving of *c.* 1700 (Drake, 331) in which it is shown as pierced by an archway with a path or ditch leading to the water's edge. No remains of this wall survive, nor is its course marked by any features on the ground. According to Drake 'the foundations of the wall which faced and ran parallel to the river were of late years dug up, which I myself saw run very deep in the ground, and all of Ashlar stone' (*ibid.*, 577). As shown on the old maps it was close to, but clear of, the Hospitium, and joined the wall in Marygate about 18 ft. S.W. of Tower A. Wellbeloved claimed that there were two such walls 'built by Abbot Thomas de Malton in 1534 [*sic.*, 1354 is intended]; the one proceeding from the tower at the end of the Abbey-wall in Marygate, along the margin of the river till the Abbey-wall from near Bootham Bar, and the other parallel to it, near the Water-gate' (*YMH* (1854), 20). None of the maps show a wall immediately by the river bank, although traces on the Water Tower suggest that a wall may have abutted against it. A 17th-century drawing shows a short stump of wall here and a mid 19th-century photograph reveals that rubble core then exposed has since been replaced by a patch of facing stone; S.E. of the Water Tower the drawing shows a gently sloping bank with no trace of a wall in line with the tower (BM 1850.2.23.832 and NMR, CC61/12; *cf.* Drake 331, after Place). The abbey used a landing place at the end of Marygate, N.W. of the Water Tower; there is no record of quays to the S.E.

In 1497 a postern gate (Plate 23) was made in the precinct wall near Bootham Bar. This is commonly called Queen Margaret's Arch, due to an erroneous belief that it was made for the convenience of Margaret Tudor, the daughter of Henry VII, who visited York on 13–15 July 1503 on her way to be married to King James IV of Scotland. Details of her visit are preserved in the city records (*YCR*, II, 184–9). There is, however, no doubt of the real date of the gateway. William Sever, Bishop of Carlisle and Abbot of St. Mary's, in a letter of April 1500 to the Mayor of York during a renewed dispute over building beside the precinct wall in Bootham wrote: 'when we brake our walle thre yeres past to make our postrone . . . ther was founde at that tyme no contradiccion by any maner of evidence shewed . . . by the said Mayre . . . and then I, with thadvice of my bredern and our Councell proceded furthe in makieing of our postrone and the toure ther, for the ease of us and our monasterye and honour of the same and for the strenghe and defence of the Citie' (*YCR*, II, 149–50). The Mayor's reply reveals that the reason given by the Abbot at the time of building was 'that the Kyngs good grace then in his noble viage toward Scotland wuld rest within your monastery and for his pleasure and passage to the mynster ye wuld make ye said posterne' (YCA, B8, f. 84v). The date is

further confirmed by the fact that the work occurred during Thomas Gray's mayoralty (Feb. 1497–Feb. 1498). There may already have been a postern on this site, since 'Great Bootham with the curtilages, posterne and all appurtenances' is mentioned in 1350 (Widdrington, 123; *CPR*, *1348–50*, 550). The new work probably supplied a need for better access to the abbot's house, which became the nucleus of the King's Manor. The postern and adjoining tower still remain in a relatively unaltered state. The tower is interesting for the use of brick on the interior faces of the walls.

The Dissolution. The abbey was dissolved in 1539, pensions being found for the abbot and 49 monks (*L & P Henry VIII*, XIV(2), 603; Dugdale, *Mon.*, III, 569). The bells were taken down in 1541/2 (PRO, SC 6 Henry VIII, 4644). The chapter house and the S.E. part of the church were taken down or blown up (Harvey Brook, YMC, III, 164) to make way for improvements to the Abbot's Lodging which became the palace of the Council in the North (*see* The King's Manor (11)). Other abbey buildings were used on occasions as royal lodgings. An undated report of *c.* 1545 on the State of the Palace of York records the walls of the church with the steeple standing wanting roof in length 333 ft.; the King's Hall, alias the Frater, all uncovered; the Queen's Lodging, alias the Dorter, with an upright roof all uncovered. The Gatehouse was in good state. Some apartments were occupied. Some barns and outbuildings were in a fair state. A stable building had been repaired by order of the Lord President (PRO, E 101/501/17).

After the dissolution of the abbey the precinct walls probably remained unaltered for some time. The earliest known plan of the city, made in *c.* 1545 (PRO, MPB, 49, 51; *cf.* *YCR*, IV, 63; RCHM, *City of York*, III, fig. 1, p. xxviii), describes the former abbey as 'The Kinges Maner of Seynte Marys with oute the Cittie Walle, Inclosed with his owne Walls' and marks the main gateway, the postern, and the two round towers at either end of Marygate. Speed's plan of York of 1610 shows the complete circuit of the wall still standing. From about 1540 until the siege of 1644 St. Mary's Tower, then usually described as 'the round tower at St. Marygate end', was used to hold a large collection of records of Yorkshire monasteries made by officers of the Court of Augmentations and with an official keeper, appointed by the Crown (*YAJ*, XLII (1968), 198–235; (1969), 358–86; (1970), 465–518).

During the siege this side of the city was invested by the forces of the Earl of Manchester, who 'raised a battery against the mannor wall that lyed to the orchard; he begins to play with his cannon and throws down [a] peice of the Wall. We fall to work and make it up with earth and sods; this happned in the morning' (Slingsby, *Diary*, 109). The morning was that of Trinity Sunday, 16 June 1644, and the length of wall so damaged was probably near St. Mary's Tower. At noon on the same day the mine under the tower, on which work had started at least ten days before, was exploded with considerable effect. Many civilians were killed and the records were buried or destroyed (S. Ash and W. Goode, *A Continuation of True Intelligence . . . from the 16th of June, to Wednesday the 10th of July 1644* (1644). For other references to this episode see *YAJ*, XLII (1968), 198–9, and L. P. Wenham, *The Great and Close Siege of York 1644* (1970), 57–74). The tower was subsequently

parts of the abbey grounds were granted to the Yorkshire Philosophical Society who erected their museum on the site of the buildings of the E. cloister range and carried out the first of several excavations to recover the plan of the church and other buildings (Plate 3); further excavations were carried out in 1901, 1912 and 1952. In 1840 St. Mary's Lodge, the building W. of the abbey gate, was thoroughly renovated for occupation by John Phillips, keeper of the newly erected Yorkshire Museum. The precinct walls were restored in 1950–7 and in 1971 the site of the E. part of the abbey church was levelled, the remains of the church consolidated and the plan marked out on the ground with new stonework.

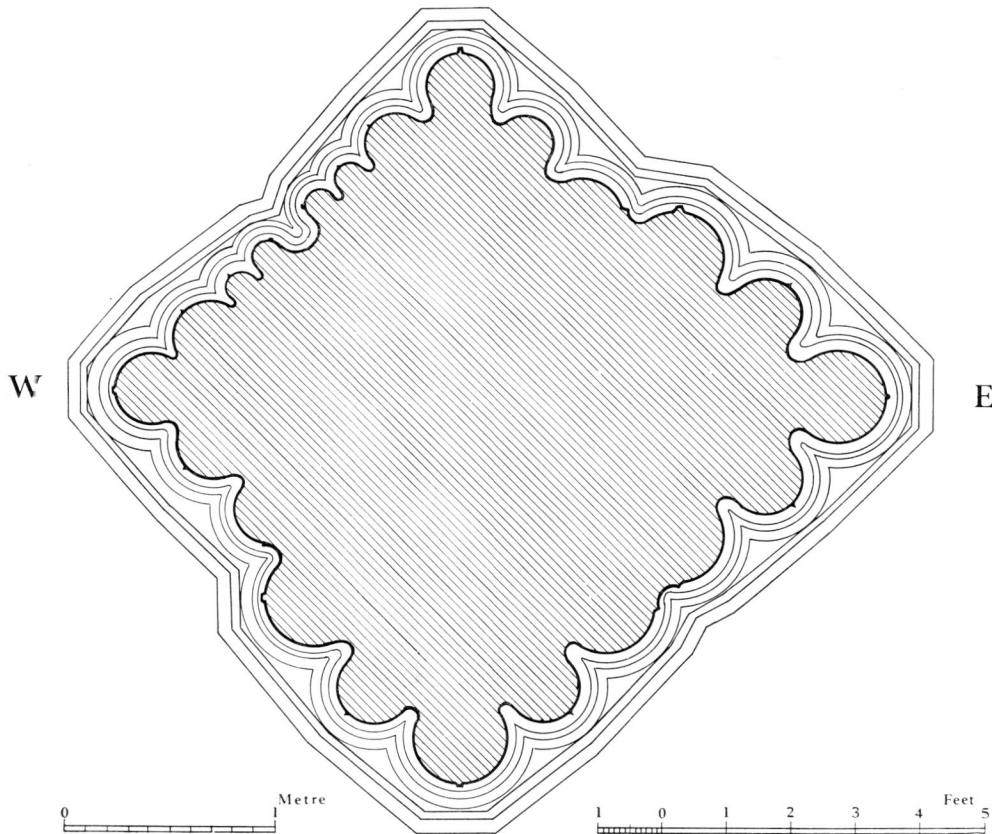

Fig. 13. (4) St. Mary's Abbey. Main crossing pier. Plan.

rebuilt on approximately its former lines, using old materials, and with a conical tiled roof.

In the first quarter of the 18th century the abbey was used as a source of stone for the County Gaol, for the Ouse Bridge, for the reconstruction of St. Olave's church, and for repairs to Beverley Minster. In 1822

Architectural Description. The buildings are all of white magnesian limestone except for the earliest work, of the late 11th century, which is mostly of brown gritstone. This gritstone is probably all reused material of Roman origin, some of the blocks showing the marks of Roman tooling and of Roman cramps.

THE ABBEY CHURCH. The plan of the *Church of 1089* has largely been recovered by excavation. The short choir terminated in a semicircular apse projecting beyond the choir aisles which had small apses finished square externally; transepts were the same length as those of the later 13th-century church and had two apsidal chapels of varying projection on the E. side of each (Fig. 7, p. xl). Some of this arrangement is displayed on the ground by stonework built up on the original foundations. At the N.E. corner of the N. transept a few courses of the original wall remain, partly embedded within later stonework (Plate 9). The original work at this point is of the dark brown gritstone, contrasting with the white limestone of the 13th-century work. Towards the W. end of the nave excavation has shown that the S. wall of the 13th-century church stood on the foundations of the original church, the arcades were built overlapping the earlier foundations but further N., and the N. wall stands still further to the N. of the earlier wall.

Of the *Church of 1270* the *Eastern Arm* was of nine bays and is now represented by part of the E. respond of the N. arcade, the base of part of the E. and N. walls of the N. aisle and the bases, rebuilt in 1912 and in 1971, of the four western piers of the S. arcade. The positions of the other piers are outlined in modern stone built-up on the sleeper walls which survive below.

The E. respond of the N. arcade consists of five filleted or keeled roll mouldings separated by hollows (cf. Fig. 16, p. 11). These mouldings are typical of the main structural members remaining throughout the church. The N. wall is divided into equal bays externally by buttresses and internally by triple engaged shafts rising from a wall-bench. The re-erected lower courses of the S. arcade piers show an octofoil plan with roll mouldings alternately keeled and filleted, except on the aisle side where the S. member consisted of a triple roll matching the engaged shafts on the wall of the N. aisle opposite.

Of the *Crossing*, the N.E., S.E. and S.W. piers have been rebuilt to a maximum height of five courses. The N.W. pier stands to the springing of the crossing arches (Plate 4). The piers are of irregular plan with a complex outline of keeled and filleted rolls (Fig. 13, p. 7); the main members have the points of the keels or the flat fillets at a position of maximum projection but in the minor members these are moved round to the side producing, in the case of adjacent fillets, a composite moulding presaging a form common in the 14th century. The bases to the piers are of simple triple-roll form.

The *North Transept* is in three bays with an aisle on the E. side. Base courses only remain of the E. wall and to the N. a fragment of the respond to the arcade. Of the arcade itself the base of the S. pier remains. In line with this arcade, in the surviving fragment of the N. wall and partly covered by 13th-century masonry, stonework of the transept of 1089 remains to a height of some 4 ft. Enough of the 13th-century work remains to indicate that the N. wall was arcaded to match the W. wall of the transept and the N. wall of the nave aisle to be described below. At the N.W. corner of the transept are the remains of a vice. The W. wall is in three bays, the S. bay containing the archway opening to the N. nave aisle. The two N. bays survive in the lower stage only and were arcaded, each in two bays with labels over the arches, and under each arch

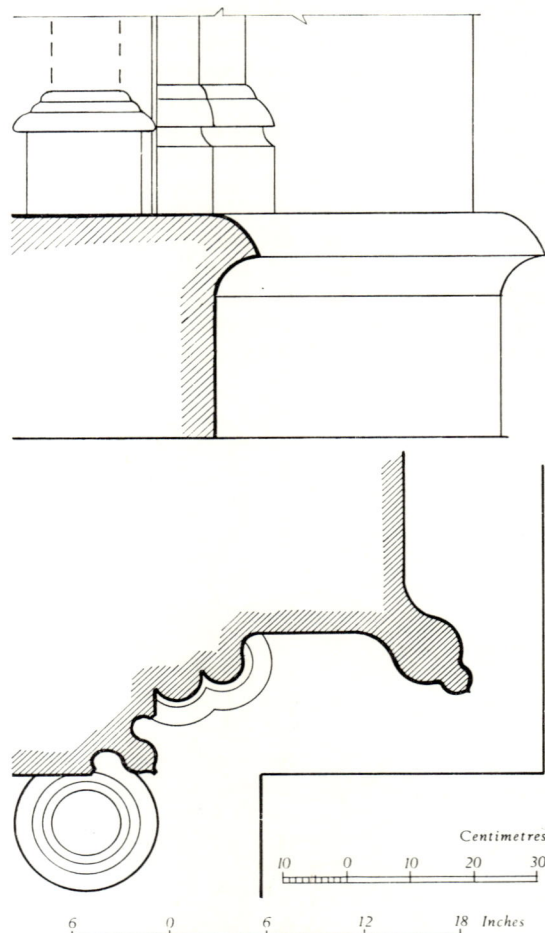

Fig. 14. (4) St. Mary's Abbey. S. Transept. Buttress and arcading.

a roundel above two subsidiary arches, moulded and without cusping, and springing from the moulded, corbelled capitals of shafts which are now missing but which stood on a stone wall-bench. In the middle bay the S. jamb of the window in the upper stage remains, against the N. wall of the nave aisle, with the springing of the window arch. Externally the arch moulds die into the aisle wall; internally the window jamb has engaged shafts with decayed moulded bases and capitals. The arch to the N. aisle is of three moulded orders, very decayed but comprising filleted rolls and hollows. Over the haunches of the arch decayed cone-shaped corbels carry small double and triple shafts which presumably carried vaulting ribs. Over the arch is a bay of blind triforium arcading, of four trefoiled lights with tracery consisting of three quatrefoiled circles, not now complete. The mullions are treated as shafted piers with foliated capitals, repeated in the jambs under the inner order of the window arch; the outer order was carried on free-standing shafts, now missing, in front of hollows with foliated edges.

9

N

50 Cms   0   1 Metre

1   0   1   2   3   4 Feet

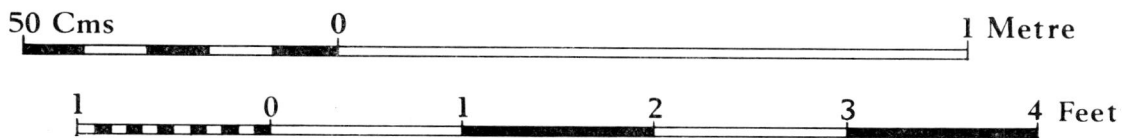

Fig. 15. (4) St. Mary's Abbey. Nave pier. Plan.

The *South Transept*. Of the E. wall, with its buttresses, only the base courses survive. The base of the S. wall, partly rebuilt, appears in the basement of the museum, forming the N. side of the passage to the cemetery or vestry and continuing E. to terminate in a mass of masonry representing the bottom of the S.E. buttress.

The bottom of the W. wall, partly rebuilt, has the E. face just showing above ground but more is exposed by the lower ground level to the W. The wall is divided into two bays by buttresses with triple shafts on the external angles and twin shafts in the angles with the wall (Plate 9). In each bay is a recess with moulded jambs and originally arcaded; bases for the corner shafts remain and in the S. bay rough projections represent three out of five original intermediate shafts.

The *Nave* is of eight bays. Of the archway to the crossing the N. pier stands to the springing of the arch and the bottom of the S. pier was reconstructed in 1827. Of the S. arcade only the base of one pier, some rough foundations for a second, and the W. respond remain; the latter stands to a height of some 18 ft.

The N. arcade has all been removed except for the E. and W. responds. Over the E. respond the haunch of the first arch of the arcade remains; it is of three moulded orders with a hood-mould towards the nave. In the angle E. of the hood-mould is a foliated corbel of conical form carrying a double engaged shaft which presumably carried the vaulting ribs. Over the main arcade and below the triforium the spandrel walling is of plain ashlar. The E. respond of the E. triforium arch remains with the springing of its arch. A drawing by W. Lodge of *c.* 1677 (York Art Gallery) (Plate 2) shows a triforium of semi-circular arches each enclosing four arched openings with tracery of three quatrefoils, the mullions between the openings being treated as piers, with capitals. The existing respond has an inner order corresponding to those mullions. The outer order carrying the main arch had a free-standing shaft, now missing, with moulded base and foliated capital in front of a hollow with foliated edges. The spandrel walling above the triforium is of plain ashlar.

At the W. end nothing remains of the triforium stage and at neither end is there anything left of a clerestory. The W. respond to the N. arcade has foliated caps and part of the springing of the arch remains (Plate 5).

The *North Aisle* opens into the N. transept by a moulded archway; the plinths of the responds are cut to take a screen. Above the archway plain walling rises to a raking stone weathering, giving the outline of the lean-to roof which covered the aisle. The N. wall is fairly complete except for the window tracery, and is divided into two stages by string-courses inside and out. Externally (Plate 5) there is a bold plinth and the bays are divided by two-stage buttresses. The lower stage of the walling is plain, interrupted only in the seventh bay by a doorway with moulded two-centred head and a hood-mould with stops, springing from jambs with alternate attached and detached shafts with decayed capitals. To each side of the doorway the walling is recessed under a narrow pointed arch forming, with the outer order of the doorway, an arcade of three arches. In the upper stage the wall-face is set back under a triplet of arches in each bay, the centre arch being wider and open to form a window, and the flanking arches blind; in the E. bay the E. blind arch is omitted.

The flanking arches are built tight up against the buttresses so that the labels cannot be carried down to springing level but stop over the haunches of the arches. The arches were carried on free-standing shafts, all now missing, and die out into the buttresses. The capitals to the shafts remain, some moulded and some foliated; of those that are foliated most are derived from the 'stiff leaf' of the earlier part of the century but two are naturalistically treated.

The windows, all of equal size, are alternately of three and two lights with tracery, now largely missing, of one or three circles foliated with pointed soffit cusping. Of the spandrels round the circles, some are pierced, some are blind. The mullions, where they remain, are moulded as multiple piers with capitals, now badly decayed.

Internally the wall is divided into bays by triple shafts with moulded bases and foliated capitals (Plate 6). The bases interrupt a wall-bench below which is a quantity of reused gritstone; above the bench the lower stage is arcaded, each bay having three arches each enclosing two subsidiary arches under a roundel, all without cusping. Shafts to carry the arcading are missing. In the seventh bay is a stilted segmental-pointed rear-arch to the doorway flanked by narrow pointed arches which range with the arcading in the other bays. At the W. end is a plain blocked doorway to a vice.

In the upper stage the design of the exterior is repeated, with the wall recessed on each side of each window under a narrow blind arch but where on the outside the arches die into the buttresses, on the inside they spring from detached shafts (now missing) with capitals ranging with those of the vaulting shafts that divide the bays (Plate 9). Shafts are also missing from the window jambs. The springers of the aisle vaults remain above the capitals of the wall-shafts. The capitals provide a seating for the wall-ribs and the transverse ribs but the diagonal ribs emerge from behind the other ribs with which they mitre and in some cases interpenetrate (Plate 7).

The *South Aisle* is represented by the lower part of the wall of the five eastern bays of which three were reconstructed in 1913 by W. Harvey Brook, 'using new stone only where absolutely necessary'. To the E. the lowest course visible externally is largely of dark gritstone. The bays are divided by buttresses with splayed sides which had engaged trefoil shafts on the angles. In the E. bay is a doorway of which none of the facing stone is original. Further W. are fragments indicating arcading between the buttresses, and internally the wall was also arcaded between the vaulting shafts that mark the bay divisions.

The *West Front*. The stonework of the W. wall of the nave and N. aisle is separated by straight joints from the N. wall and the responds of the arcades. The wall is divided internally into two stages, of which the lower continues the arcading of the N. wall across the aisles and part way across the nave, but flanking the W. doorway are blind arches similar to those on the outside of the W. front. Reused in the base of the wall are parts of a Norman gritstone cornice. Externally the W. front has buttresses with splayed re-entrant angles presumably designed to support octagonal turrets (Plate 6). The whole of the W. front is treated with blind arcading, the arches being trefoiled under crocketed gablets; the arches rise from engaged trefoil shafts with moulded or foliated capitals and the gablets

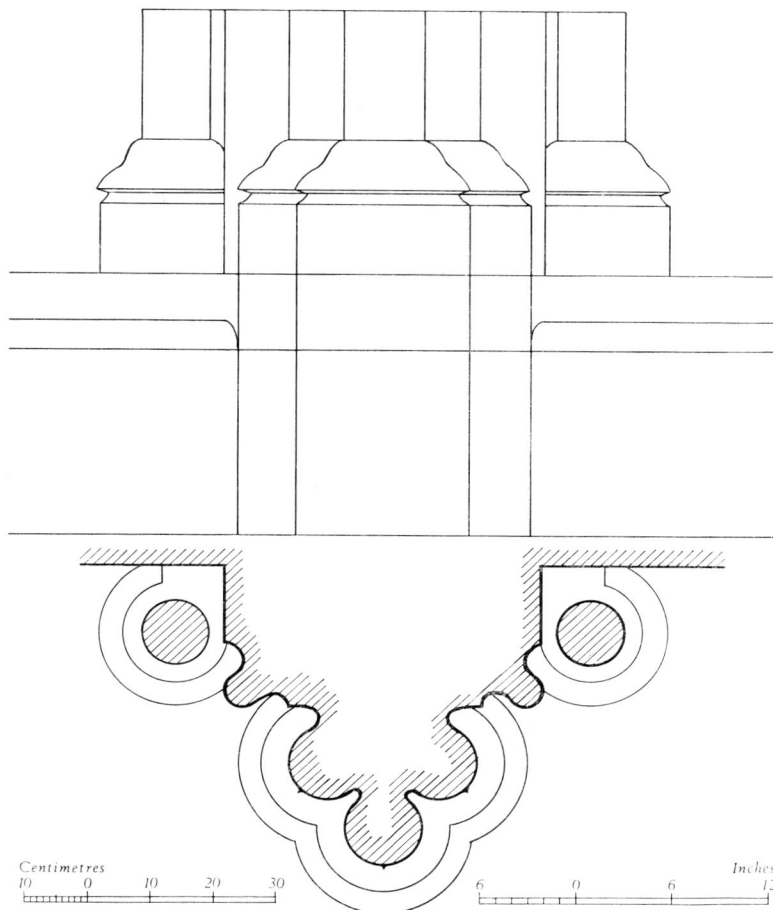

Fig. 16. (4) St. Mary's Abbey. N. aisle. Internal wall-shafts and
arcade shafts. Plan and elevation.

are carried on small corbels. In the middle of the nave the arcading is interrupted by the W. doorway; the jambs had four free-standing shafts, now missing, and one engaged multiple shaft, with hollows between them carved with un- dulating vine-trails (Plate 8). The haunches only of the arch to the doorway still remain, for a two-centred arch of five orders, three deeply moulded, one carved, and one now missing. Above the doorway the N. jamb of the W. window survives, pierced by a passageway in the thickness of the wall. The W. window to the N. aisle remains complete except for jamb-shafts and tracery. The window was of three lights, and the jambs, pierced by the wall-passage, are richly moulded with engaged shafts on moulded bases and with capitals carved with naturalistic foliage which also forms a continuous band across the thickness of the wall (Plates 7, 8); the whole design of the window gives a much richer composition than the windows in the N. wall.

Liturgical Arrangement. Some details of the arrangement of the church in the late 14th and the 15th centuries can be deduced from the instructions for the conduct of the daily ritual of the abbey in the Ordinale, and from the Chronicle, interpreted in the light of existing remains. At the E. end of the nave three modern steps represent an original rise of three steps running the full width of the church. E. of these steps and under the W. crossing arch was the pulpitum. The crossing and six bays of the eastern arm were enclosed by screens to form the monks' choir. Within these screens the monks' stalls occupied the space under the tower and extended into the two bays next E. of the crossing, providing room for the 48 monks recorded in 1284 (*Chron.* 23–4) or the 49 to whom pensions were granted in 1539 (Dugdale, *Mon.*, III, 569). E. of the stalls was the first of three steps leading up to the high altar and beyond this step were buried Abbot Simon (*ob.* 1296) and Abbot John de Gilling (*ob* 1313), as builders of the church

(Ord. 71). Abbot Simon's tomb was marked by a slab level with or only slightly above the floor, since trestles were placed across the tomb to form a temporary platform for the relics of St. William which were brought to the abbey in solemn procession annually on Whit Monday. The second step must have been in the fourth bay from the W., beyond the choir altar named in honour of the Holy Trinity (Ord. 70) where the morning mass was usually celebrated. Between the two lower steps doorways in the screens led to the N. and S. aisles. The third step also lay in the fourth bay. The high altar stood against a screen between the fifth pair of piers, and above it was a great cross (Ord. 102). To each side of the high altar was a doorway leading into the chapel of the Holy Trinity behind the high altar, occupying the sixth bay (Ord. 68). There were thus two altars dedicated to the Holy Trinity.

The seventh bay formed the ambulatory used by processions around the choir, with the chapel of Our Lady to the E., occupying the eighth and ninth bays. Here was celebrated the Mass of Our Lady, the monks being accompanied with a choir of boys from the Almonry. Flanking the Lady Chapel in the eastern bays of the two aisles were the altars of St. Peter and St. Stephen; at the latter was said daily the mass for the dead, for the souls of the founders and all the faithful departed. In the sacristy which projected southwards was the altar of St. Catherine which served as an Easter Sepulchre, being dressed on Maundy Thursday like a tomb to receive the Lord's Body.

The altars of Holy Trinity behind the high altar, St. Catherine, St. Peter and St. Stephen were the four upper altars of the church; there were four lower altars in the chapels in the eastern aisles of the transepts, dedicated to St. Nicholas, St. Thomas of Canterbury, St. Mary Magdalene and, probably, St. Benedict. Relics of St. Thomas were exposed for veneration before his altar on 29 December. Abbot Benedict of Malton was buried in front of the altar of St. Benedict, a departure from the normal rule that abbots should be buried in the Chapter House.

In the nave the eastern bay formed the retrochoir in which weak or convalescent monks heard the services. On the W. it was closed by the Rood screen over which stood or hung the great Crucifix or Rood. Such a screen was usually pierced by two doors flanking an altar placed against its western face. The St. Mary's Ordinal does not specifically mention an altar in this position but the Chronicle speaks of the lighting of a lamp by the altar of St. Mary in the nave of the church. Westward a chapel seems to have extended two and a half bays down the nave to a step which ran the full width of the church, corresponding with the step down in the wall-bench and in the plinth of the N. wall. As well as the central W. door, there were doors in the nave aisles, in the seventh bay on the N. side and in the first and sixth bays on the S.

CLAUSTRAL BUILDINGS. Of the buildings around the cloister most are known only from excavation; some stood on the site now occupied by the Yorkshire Museum where fragments are preserved in the basements. The uses to which the various buildings were put can be deduced with some certainty from the Ordinal.

On the E. side of the cloister the Dormitory range ran S.

from the end of the transept, extending well beyond the line of the S. range. Next to the transept is a narrow room with doorways to E. and W., its N. wall formed by the surviving base of the wall of the transept, divided into three bays by attached shafts of c. 1280–90 (Plate 13). The burial of monks in the abbey cemetery is recorded in the Chronicle (Chron. 54) and the Ordinal makes it clear that the way to the cemetery must have been through this room (Ord. 386), the cemetery lying E. of the cloister. It may also have served as a vestry and be the room where refreshment was prepared on Whit Monday for the bearers of the relics of St. William (Ord. 334).

Next southward is the Vestibule to the Chapter House, which was known as the Galilee (Ord. 22). Before the end of the 12th century it had probably formed the Chapter House itself, and it was probably after the election of Robert Longchamp as abbot in 1198 that a new Chapter House was built, entirely E. of the range. The foundations of the new Chapter House were exposed in the excavations of 1827; Wellbeloved records that only the lowest courses of these foundations were still in position. The Chapter House measured 61 ft. by 26 ft. and the buttresses indicate that it was designed in five bays. The vault to it would thus have twelve springing points which would accommodate twelve figures of the apostles, with Old Testament figures and St. John the Baptist in the vestibule (see pp. xlii–xliv), forming a sequence leading to a great Christ in Majesty at the E. end (Ord. 75), comparable perhaps to the Majesty at Worcester in the Refectory, now the School Hall. Abbots were buried within the Chapter House (Ord. 275), the floor of which was at two levels separated by a step (Ord. 22–3). The vestibule was separated from the cloister walk to the W. by an arcade of three arches, for which the lower parts of two piers and two responds of c. 1200 remain; they have water-holding bases and free-standing shafts (Plate 16, Fig. 5, p. xxxiii). At the E. end the entrance to the Chapter House remains in part, with some reconstruction (Plate 13); it comprised a central arched doorway between flanking arches above dwarf walls. The piers between the doorway and the flanking arches are cruciform and enriched with chevron mouldings and formal leaf ornament behind free-standing shafts with water-holding bases with angle spurs, all of white limestone. The responds to the flanking arches, less complete, match the piers. A composite capital, wrongly reset but belonging to this composition, shows leaf decoration of a heavy fleshy type and also of swirling 'stiff-leaf' type. None of the arches are in situ but one has been reconstructed (incorrectly) on the ground. The arches were visually of three orders and very heavily enriched. A number of voussoirs from vaulting shafts preserved in the museum are thought to have belonged to the original vault of c. 1200 covering the vestibule; they are of trefoil section with rolls crossed by a trellis of sunk bands. The dormitory on the first floor above was rebuilt during the abbacy of John de Gilling, 1303–1313 (Chron. 36), but the vestibule was remodelled later in the 14th century when it was vaulted in nine compartments, the lower parts of the E. and W. vaulting shafts and of the four free-standing piers remaining (Plate 16). The mouldings of this work may be compared with work being executed in the second quarter of the 14th century at Exeter and Lichfield where the work is advanced for its date; they are not so advanced as work in York Minster of the last quarter of the century. These comparisons could point to a mid 14th-century date but had there been any major construction work at that time it might be expected to be recorded in the Anonimalle Chronicle which was written at St. Mary's and covers the years 1333 to 1381. It seems probable that the vaulting of the vestibule was part of a restoration programme carried out after the fire of 1377.

S. of the vestibule lay the *Inner Parlour* where speech was allowed (*Ord.* 74), a vaulted room of three bays, of which the base for a N. vaulting shaft of *c.* 1310 remains *in situ* in the museum (Plate 16). Similar bases to the piers are shown by Wellbeloved (Plate 3) but the E. wall can be seen from the plinth to have been refaced, probably after 1377. The W. wall of the Parlour was of sufficient thickness to include within it the day stairs to the Dormitory above; the narrowness of the stair is brought out by the rule that monks descending must give way to those coming up (*Ord.* 31). S. of the Inner Parlour a passage led from the cloisters towards the Abbot's Lodging, and further S. lay the *School* (*Chron.* 85–6) which retained the traditional name *Scola Infantum* although after the early 13th century oblate children were no longer taken; it was of six bays, and projected beyond the S. range. The plinth to this part, seen in 1813 (Plate 3), was conformable to a date of *c.* 1310 and was presumably part of Abbot Gilling's rebuilding.

The S. range, as rebuilt in the late 14th century, contained at the E. end a passage leading from the cloisters to the court-yard beyond. Against the wall of this passage was built the fireplace of the *Warming House* which occupied the first three bays of the range. Here a fire burnt and conversation was permitted (*Ord.* 158). The fireplace is preserved in the museum basement (Plate 13). Flanking the fireplace are vaulting shafts, now carried up well beyond their original height, with late 14th-century mouldings matching those of the Vestibule. The lower part of the N.E. buttress to the Warming House and part of the adjacent wall are preserved in the Hospitium (Plate 13). They are finely moulded, of late 14th-century date.

A doorway in the W. wall of the Warming House, not shown on Wellbeloved's plan but clearly indicated by Ridsdale Tate, led to a large room of six bays vaulted in three aisles; at the end of the room the W. door was recessed between flanking projections, the N. one forming a lobby from which a circular stair led to the upper floor. This large room may have been the Common Hall in which the abbot entertained strangers (*Ord.* 151), perhaps occupying the site of part of the 'long room for the recreation of the monks' built in 1313 (*Chron.* 66). A hall in this position would be convenient for the regulation that those eating not in the refectory but in the hall or the prior's chamber should await the abbot or the prior in the Warming House.

On the upper floor of the S. range was the *Refectory*. It was reached by a flight of stairs of which the foundations were uncovered, rising from the S.W. corner of the cloister across the width of the W. range. At the foot of the stairs and set against the wall of the S. range opposite the W. walk of the cloister was the foundation of the lavatory where the monks washed their hands and combed their hair (*Ord.* 142). The *Kitchen* and *Servery* were built against the four W. bays of the S. range, with fireplaces to N. and E.

The W. range comprised a vaulted undercroft forming the *Cellarer's Store*, entered from the cloister by a doorway in the S.E. corner, and the guest house on the first floor, which included, perhaps, the Hall of the Wlays[1] built in 1297 by Prior William de Derby (*Chron.* 28–9).

The rectangular projection to W. may have contained the latrine of the guest house or alternatively the Cellarer's Chequer. The projection into the cloister walk, apparently for a stair, is unprecedented in monastic planning and must be a post-Reformation accretion. The N. bay of the lower storey, against the church, was the *Outer Parlour* and formed the entrance through which all visitors to the abbey passed. Here was the Gate of Tobias, named in allusion to the stranger who revealed himself as the angel Raphael (Tobit, xii, 15) and here three poor men were fed each day and other alms and hospitality dispensed (*Ord.* 136–7).

S. of the Refectory range was an irregular court bounded on the E. by the Dormitory range and on the W. by the Kitchen. To the S. it was closed by a long room running E.–W. and the end of a vaulted room at right angles, with a fireplace in the E. wall. This was the *Prior's Room* to which he might invite senior monks for a drink and for warmth in winter (*Ord.* 151); the longer room to the W. would then, by analogy with other abbeys such as Ely, be the *Prior's Hall*, but it is not specifically mentioned in the York texts.

The *Infirmary*, in which a chapel was referred to as 'newly built' in 1455 (SS, xxxv, 239–40), lay S. of the monastic kitchen; part at least was in two storeys with a vaulted under-croft. The chapel was probably on the first floor at the S. end where there is a small projection to E.

The *Hospitium* (Plate 15), some 90 yds. W. of the cloister, is a two-storey building of six bays; a sketch of 1840 shows most of the two S. bays missing but the lower part of these had been rebuilt before 1930 when the upper storey was completed and the roof of the whole building reconstructed to a steeper pitch than the original. The lower storey is of the 14th century and has walls of ashlar; the upper part is of the 15th century and timber-framed. In the E. wall, in the fifth bay from the N., is a rebuilt stone arched doorway. Further N. are windows of two square-headed lights with chamfered jambs and mullions. The floor of the upper storey is carried on two rows of octagonal stone columns with moulded bases and shaped corbels projecting N. and S. under the main beams.

The timber-framed upper storey has a modern N. wall, and the two S. bays are of 1930–1. The original timber-work is exposed; the main wall-posts are strutted off sill-plates and from them curved braces rise to the eaves plates and to the tie-beams. A vertical stud in the middle of each bay is narrowed in the middle to form the mullion of a two-light window, but the windows have all been restored. In the E. wall, in the fourth bay, is a doorway with timber two-centred head. The main timbers between the third and fourth bays from the N. show mortices for the rails of a partition.

Adjoining the Hospitium to the S. is a length of walling containing a *Gateway*, a smaller doorway, and windows, all of *c.* 1500 (Plate 15). The wall stands to a height of one and a half storeys and is of ashlar with brick backing to the W. above

---

[1] See p. 4.

Fig. 17. (4) St. Mary's Abbey. Hospitium.
Timberwork of upper floor.

first-floor level. The gateway and the doorway are both four-centred, of two chamfered orders with a label; the windows are of one and two arched and trefoiled lights in square heads.

Chapel of St. Mary at the Gate. Between the gatehouse and the W. end of St. Olave's church is a space, now roofless, which must have contained the chapel on the first floor (Plate 14). On the E. side a lofty recess in the W. wall of the tower of St. Olave's may have been at the back of the chapel altar. On the S. side is a thick wall continuous with the S. wall of the tower and containing a straight staircase in its thickness, with doorways at the foot of the stairs opening to the undercroft and to the churchyard. Further W. is a ground-floor window of two cinquefoiled lights. The W. end is enclosed by the gatehouse, and the N. side by the precinct wall built up at the E. end to the height of the aisle wall of St. Olave's church and stepped down westwards, where only the lowest courses of the wall are mediaeval.

PRECINCT WALL AND TOWERS. In the following account the wall is described in a clockwise direction starting at the W. angle of the precinct beside the river Ouse at Marygate Landing. Unless otherwise stated, all the masonry is of magnesian limestone.

Water Tower (Plate 17, Fig. 18, p. 15) was built after the licence to crenellate of 1318, while Stephen de Austewyk was sacrist. It is circular outside but hexagonal inside and built of ashlar stone in courses about 1 ft. 4 ins. high. The upper part of the wall is set back slightly, the change being marked by a small chamfered weathering. A battered base towards the river has been concealed by the modern embankment. On the S.E. in the lower half of the wall are four stones in alternate courses cut so that they project as if to provide bonding for a wall running along the bank to the S.E. A patch of renewed facing stone with very narrow mortar joints has replaced a former area of rubble core. This wall which abutted on the tower was about 7 ft. thick, probably with a narrow parapet, and was possibly the revetment for a quay.

There are six openings in the tower wall, one corresponding to each side of the interior. Four of these are cruciform arrow slits with a round oillet to each arm, all in a damaged state; this form of loop occurs frequently in the early 14th-century work on the precinct wall. Facing S., directly over the patch of renewed facing where the river wall abutted the tower, is a rectangular opening, set a little higher than the arrow slits; this has a small chamfer all round and square sockets for iron bars in the jambs and lintel. To the N.E. where the precinct wall begins is a shoulder-headed doorway with a plain chamfer all round; this must originally have led to a wall-walk, now destroyed.

Drawings made in the 17th century show that the tower was then crenellated, but the parapet is now broken down to below the level of the embrasures and is of irregular height. On the N.E., directly over the doorway just described, are the remains of a second doorway with chamfered jambs and also formerly shoulder-headed (Cave, pl. 28); this can only have been accessible by a staircase from the wall-walk. Two stone spouts draining the tower roof remain on the S. and E., and there is a hole for the same purpose on the W.

The inside (Plate 17) had a floor supported on an offset. In the sides are deep recesses, some not centrally placed, leading to the six openings. The one leading to the doorway has splayed sides, but all the others have parallel walls sharply splayed at the ends. All these recesses have flat lintels carried on quadrant corbels. The tower was roofed behind the narrow parapet, leaving space for a walk on top of the wall.

The precinct wall from the Water Tower to St. Mary's Lodge is about 420 ft. long with three small changes in alignment. In origin this stretch is wholly early 14th-century work but has been much restored and partly reconstructed. Part of the wall immediately adjoining the Water Tower was removed in the early 19th century and replaced with a stone archway to provide access from Marygate to the riverside walk. This opening has a four-centred arch with a double splay to each side.

The wall as far as Tower A and for about 50 ft. beyond has been reduced in height and is now only about 7 ft., without crenellation. The lower part, of ashlar in large courses,

WATER TOWER

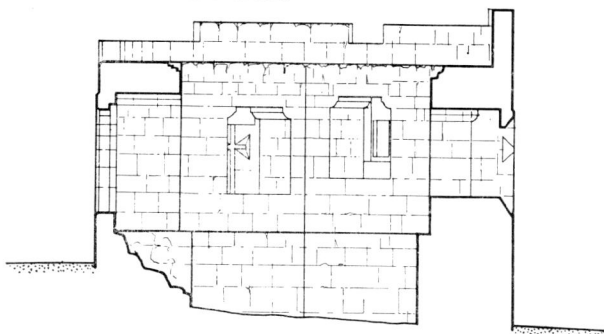

Section

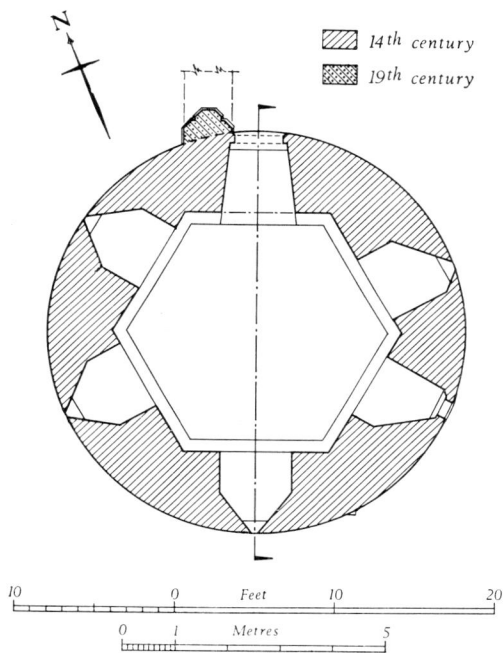

///// *14th century*

▨▨ *19th century*

10   0   Feet   10   20

0   1   Metres   5

Fig. 18.

including a few gritstone blocks, is original, but the upper two or three courses, of much smaller stones, are probably a replacement. On the inside the ground level within the Museum Gardens has been raised and only the upper 3 ft. of the wall is visible, all rebuilt in coarse rubble. There is an original postern doorway, now blocked, 4 ft. S.W. of Tower A. This is 2 ft. 8 ins. wide and 5 ft. 10 ins. high with a shouldered head and a small chamfer all round.

Tower A (NG 59775208), 120 ft. from the Water Tower, is semicircular, 10 ft. in diameter and projecting 5 ft. Outside the masonry is largely original. The inside is mostly filled with earth, but where the wall is visible it has been stripped of facing stone and repaired with rubble.

Tower B (NG 59815214), 215 ft. from Tower A and similar to it in size and plan, is entirely of new stone of the 19th century. It appears on Archer's map of *c.* 1682, but not on subsequent maps, and there is little doubt that it was wholly or partly demolished in about 1700 and rebuilt in the late 19th century, after the houses built up against the outside of the wall here had been removed.

The rest of the wall up to St. Mary's Lodge stands to the original height of about 13 ft. It is 1 ft. 8 ins. thick, but 121 ft. N.E. of Tower A a wall-walk 2 ft. 10 ins. wide begins, supported upon a thickening of the wall beneath. The walk gradually narrows to 1 ft. 7 ins. at a point 30 ft. S.W. of St. Mary's Lodge where it ends. Thence the wall has been rebuilt. The parapet has embrasures, which are mostly restored except for a few immediately S.W. of Tower B; these have L-shaped slots on the reveals, intended for housing wooden shutters, a feature which occurs elsewhere on the wall where the original embrasures have survived unrestored. (Fig. 19, below.)

The length of wall immediately S.W. of St. Mary's Lodge was rebuilt when the Lodge was erected, and the moulded plinth of that building continues for 20 ft. along the outer face of the wall. There is an inserted 19th-century doorway in this length.

The Gatehouse (NG 59835216. Plates 18, 19; Figs. 20, 28, pp. 16, 26). The group of buildings forming the gatehouse range consists of parts of the side walls of the late 12th-century gate-hall, joined at the Marygate end by a large contemporary

St. Mary's Abbey: sections through wall to Marygate

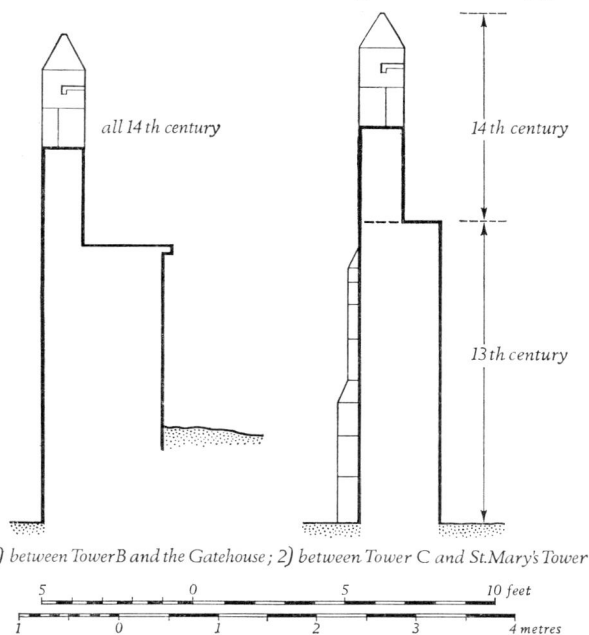

all 14th century   *14th century*

13th century

1) between Tower B and the Gatehouse; 2) between Tower C and St. Mary's Tower

5   0   5   10 feet

1   0   1   2   3   4 metres

Fig. 19.

ST. MARY'S ABBEY GATEHOUSE

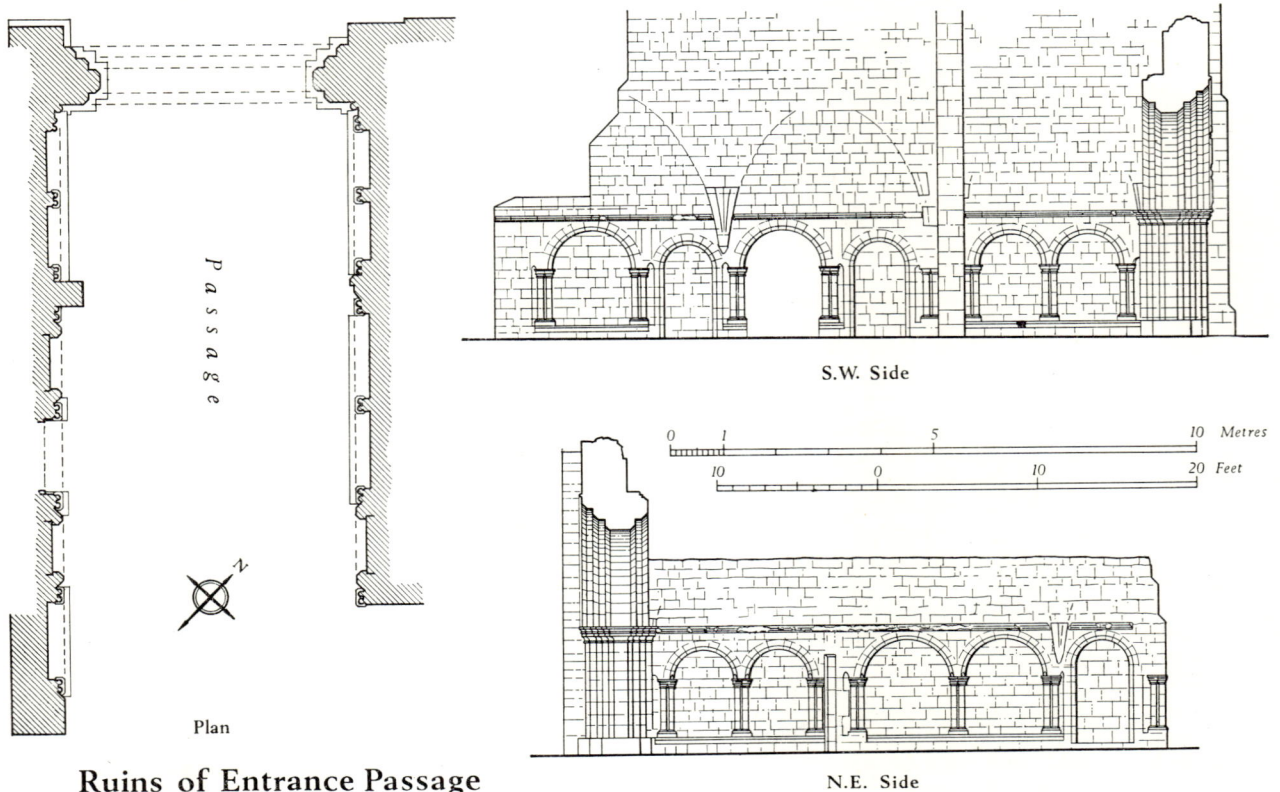

S.W. Side

N.E. Side

Plan

Ruins of Entrance Passage

Fig. 20.

archway, and of additions on each side of those built in *c.* 1470, at the time of the rebuilding of the N. aisle of St. Olave's church. The addition to the S.W., known as St. Mary's Lodge, is complete, but that on the N.E., between the original gatehouse and the church, is ruined, only the outer walls of the ground floor surviving.

The gate hall is $18\frac{1}{3}$ ft. wide and was probably a little over 40 ft. long; the wall on the S.W. side, which is the more complete, survives to this length, though rebuilt at the S.E. end. At the N.W. end of the passage is a round-headed archway, of three chamfered orders on the front and two on the rear; the mouldings of the impost caps continue as a string course along each side of the gate hall. Drawings made before demolition show that there was a similar archway at the S.E. end. There was also an intermediate archway, of which part of the N.E. jamb survives, which could be closed by doors.

The N.W. end of the gate hall beyond the intermediate archway was vaulted in one bay; springers of the vault survive in three angles of the bay, from which it appears to have had diagonal ribs only, and the two in the N.W. angles rose from vaulting shafts. The rest of the gate hall was vaulted in two bays, with heavy chamfered springers, probably of the 14th century, cutting across the moulded strings. Above the string course on the S.W. side enough of the original wall-facing survives to show that the transverse vaults were pointed. This same wall, which continues upwards to form one side of St. Mary's Lodge, was largely refaced in the 19th century, but the core is probably that of the original side wall of the late 12th-century gatehouse. At each end of the wall the thickness is represented by pilaster buttresses at the corners of the Lodge. The N.E. wall of the gate-hall survives only to a height of 13 ft. and above the string is entirely refaced.

Both side walls are decorated with original blind arcading, of two round arches in each bay, standing on paired shafts with moulded caps and bases of attic form. In the inner part of the gatehouse are two round-arched doorways to the S.W. and one to the N.E., each of these taking the place of one arched recess. All three doorways are now blocked with ashlar masonry, but one recess of the S.W. arcade has been opened up to make a doorway into the Lodge. An offset in the wall over the archway at the N.W. end of the gate-hall probably marks the level of the original upper floor.

St. Mary's Lodge (Plate 18; Fig. 28, p. 26) is a two-storey building with basement. The walls, except on the N.E. side, are of ashlar with narrow joints, and the low-pitched roof is covered with lead and slates. The principal elevations are to the N.W. and S.E. and have moulded plinths and moulded strings at the upper floor level. The plinths have a moulding similar to that on the N. aisle of St. Olave's church. The N.E. elevation, originally built up against the 12th-century gatehouse, was refaced in the 19th century. The other three elevations are each divided into two bays by narrow buttresses of deep projection. At the W. angle is a large square buttress, only the top part of which conforms to the proportions of the others. The windows, one or two in each bay, are generally stone-mullioned and of two pointed cinquefoiled lights contained within a splayed rectangular reveal. They are considerably restored and some on the N.W. side have had the sills lowered in the 19th century. On the S.E. there are also three narrow round-headed openings at different levels which light a staircase in the thickness of the wall. All the basement windows and all the windows on the S.W. elevation are of 1840. There has been some restoration of the external masonry, and an inserted chimney, later removed, accounts for a narrow strip of brickwork up most of the N.W wall. A low stone parapet around the whole building was added in 1840. In the early 19th century there was a hipped tiled roof (Halfpenny, pl. 30).

The interior was modernised in 1840, and all the internal walls on the ground and first floors are probably of that date. The only original features visible are a chamfered ceiling beam in a basement room and the staircase to the first floor contained in the thickness of the S.E. wall: the latter has stone treads and an arched stone roof rising with the stair. A short flight of stairs from the entrance lobby to the inner hall has bulbous balusters of the late 17th century. The fittings of 1840 are in the Tudor style. There is a variety of fireplaces, but the doors are more uniform, and have tall, narrow panels. The ground-floor rooms have ceilings divided into square panels by moulded wooden ribs with carved bosses.

The building to the N.E. of the original gatehouse stands between it and St. Olave's church and is joined to both. It was built at the same time as St. Mary's Lodge, *c.* 1470, and was also of two storeys. It still stood intact in the early 18th century but is now ruined, lacking the walls of the upper storey, and the interior is partly occupied by modern structures.

The outer walls to the N.W. and S.E. have moulded plinths like those of St. Mary's Lodge, and on the S.E. side a short length of matching moulded string course survives. In the N.W. wall is a tall doorway 4 ft. wide and 9 ft. high with a two-centred arched head and chamfered jambs. The S.E. elevation, like that of St. Mary's Lodge, was divided into two bays by a narrow buttress, now mostly gone. There is one badly preserved two-light window and a small arched doorway with wave-moulded jambs, perhaps reset 14th-century work. From just inside the doorway a staircase ascends within the thickness of the wall; this stair has an arched stone roof like that in St. Mary's Lodge. No indications remain of the internal arrangements, but two stone corbels on the S.E. side mark the original first-floor level.

The N. aisle wall of St. Olave's church probably incorporates

6

masonry of the precinct wall of 1266. This is visible internally below the window sills, but the exterior was refaced in *c.* 1470. The piece of wall so incorporated is about 30 ft. long.

To the N.E. of the church the circuit of the wall was interrupted by a building 74 ft. long and about 21 ft. wide which lay entirely on the N.W. side of the general line of the precinct wall. This building, which has been identified as the Almonry, probably dates from 1318, and the ground-floor walls of ashlar masonry survive on the S.W., N.W., and N.E., partly built over by a late 18th-century house (No. 29 Marygate). The maximum height of the wall above the pavement of Marygate is 11¼ ft. at the W. corner. It has a high chamfered plinth, interrupted on the S.W. by a doorway 3 ft. wide and 6¾ ft. high with a corbelled head. This doorway was defended by a portcullis, for which the wide slot remains, with rebates for a door behind; it is now blocked. The N.E. wall has a similar door, not so well preserved, visible in a cupboard opening off a ground-floor room in No. 29 Marygate. The N.W. wall has four tall narrow openings, each 6 ins. wide and now blocked with brick. In the corner of the S.W. wall against the church is a blocked window of two lights with cinquefoiled heads; though 14th-century, it is a later insertion into the wall of 1318 but antedates the aisle wall of the church, which is splayed back to clear it. The interior of this building has been largely filled with earth to form a raised garden for No. 29 Marygate, but in the S.W. corner is a chamber where the springers for vaulting ribs are visible, indicating a vault of at least two bays.

The precinct wall of 1266 forms the core of the S.E. ground-floor wall of No. 29 Marygate. From this house it then runs N.E. for a distance of about 250 ft. to Tower C and, after a small change of alignment, continues for a further 200 ft. to St. Mary's Tower.

The part of the wall up to Tower C has been much restored and only the last 45 ft. remain largely unaltered, with the wall of 1266 standing to a height of 11½ ft. and a crenellated parapet of 1318 superimposed upon it. The rest was extensively restored with new facing stone on the side towards Marygate, especially at the S.W end, after the demolition of the houses built against it. Parts of the brick walls of these houses still stand on top of the mediaeval wall. The side towards St. Mary's Abbey is divided into three sections by straight joints, signifying rebuilding at various times. The N.E. half, basically the wall of 1266, has buttresses of two stages with weathered offsets and unusually placed here both inside and outside.

Tower C (NG 59945228. Plate 20; Fig. 21, p. 18) is rectangular and open at the back. It was originally higher than the adjacent precinct wall but has now been reduced to a height of 18¾ ft. It had a floor 11 ft. from the ground, supported on an offset, and a roof supported by a second offset 7 ft. 8 ins. higher. The walls are of ashlar masonry in courses generally over 1 ft. high. The four cruciform arrow slits, one each in the N.E. and S.W. walls and two in the N.W. wall, are of the type found elsewhere in the work of 1318; internally they have very widely splayed openings at first-floor level with joggled lintels. Survey drawings of 1952 show that before the restoration of that year the tower was a little higher and that the lowest parts of an upper tier of arrow slits then existed, one in each wall; these must have opened off the roof platform. In the S.W. wall

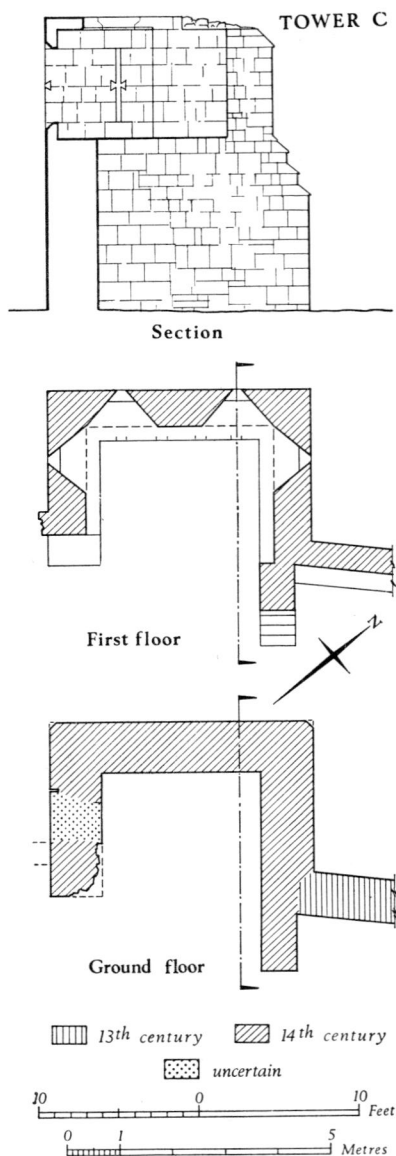

TOWER C

Section

First floor

Ground floor

| ▥ 13th century | ▨ 14th century |
| :::: uncertain | |

10   0   10
                  Feet
0   1   5
                Metres

Fig. 21.

13th-century wall up to a height of 11 ft. with the crenellated parapet added in 1318, although the latter has been destroyed towards the N.E. end. The outer face originally had nine buttresses, not quite regularly spaced but averaging 22 ft. apart; they are now mostly robbed, leaving scars on the wall. Tooling marks are especially well preserved on the masonry of this stretch of wall. On the inner side are five large buttresses, irregularly spaced and not bonded into the wall, which were probably added for stability in the later Middle Ages. One of the wooden shutters has been restored in the fifth embrasure N.N.E. of Tower C.

St. Mary's Tower (Plate 21; Fig. 22, p. 19), at the N. angle of the precinct, was built in *c.* 1324 as a tall circular tower about 34 ft. in diameter, with thick walls and an octagonal interior of two storeys; the original height was over 30 ft. It is no doubt the new tower which the sacrist Stephen de Austewyk caused to be built. After part destruction in 1644 it was rebuilt, preserving the octagonal interior, but with thinner walls on the outer side, with the result that the reconstructed portion follows an irregular curve. When sketched by Place in *c.* 1715, the tower had a conical tiled roof resembling the present roof, but with a central finial. Lean-to brick and timber buildings which concealed much of the base of the tower were removed in 1896 and 1920.

The original part consists of the whole S. quadrant facing into the precinct and extends some distance outside the precinct wall towards Bootham. This original quadrant has a chamfered plinth and a doorway on the ground floor 4 ft. 8 ins. wide with a two-centred pointed arch. On the upper floor are two doorways which originally provided access to the wall-walks along the Bootham and Marygate walls. The one opening on the Bootham side has a corbelled head and is placed a little above the level of the wall-walk whence it must have been reached by a short wooden staircase. The other doorway appears to have been at about the same level as the Marygate wall-walk; it has below it an area of rubble masonry which probably replaces a former corbelled support for the timber wall-walk. The latter doorway has been altered at the top, probably from a corbelled head similar to the others which occur in the 14th-century work on the wall, to a straight lintel. The part of the wall facing Bootham has a cruciform arrow slit on the first floor.

The walling of the rebuilt quadrant was clumsily joined to the original work; on the W. the joint is marked by a strip several feet wide of exposed rubble core, and on the E. there is a setback. The masonry, generally of large squared stone, is of varying quality, a difference being most marked on the N.E. side where the lower half of the wall is well built, comparable to the work of 1318, but above is poorly laid; this may perhaps indicate rebuilding in two stages. The new wall is built of reused masonry, some being from the original tower, and incorporates on the ground floor a 15th-century window of two lights with pointed trefoiled heads; the latter probably comes from elsewhere in the abbey. On the first floor are three windows facing N.; each one has a single mullion and transom with ovolo mouldings; these are reused dressings from a large bay window in the outer S.W. range of the King's Manor built in *c.* 1610–20 by the Lord President, Lord Sheffield, which was also ruined during the siege of 1644. A short length of

of the tower is a blocked doorway, now visible only from the inside where the opening was 2¾ ft. wide. The outside at this point has been rebuilt in rubble masonry, but there is a tall slit, one side of which is possibly a jamb of the doorway. The precinct wall immediately to the S.W. has been removed for a modern gateway, but at the S. angle of the tower are bonding stones for the crenellated parapet of the wall.

The wall between Tower C and St. Mary's Tower is one of the best preserved parts of the *enceinte*; in places it has been carefully restored in recent years. It consists of the original

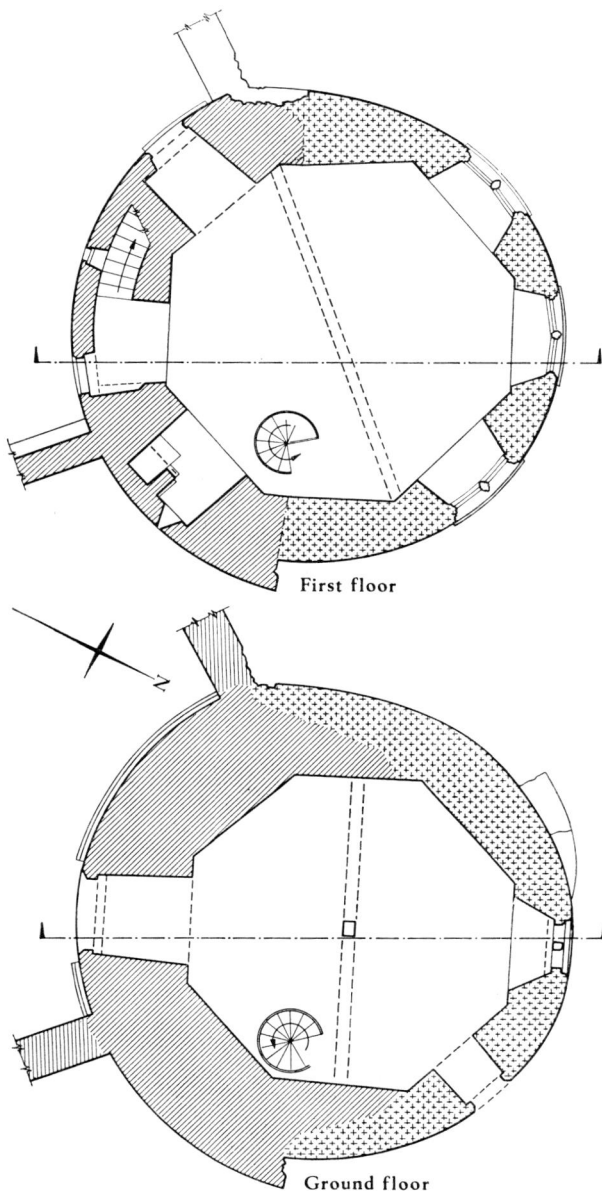

First floor

Ground floor

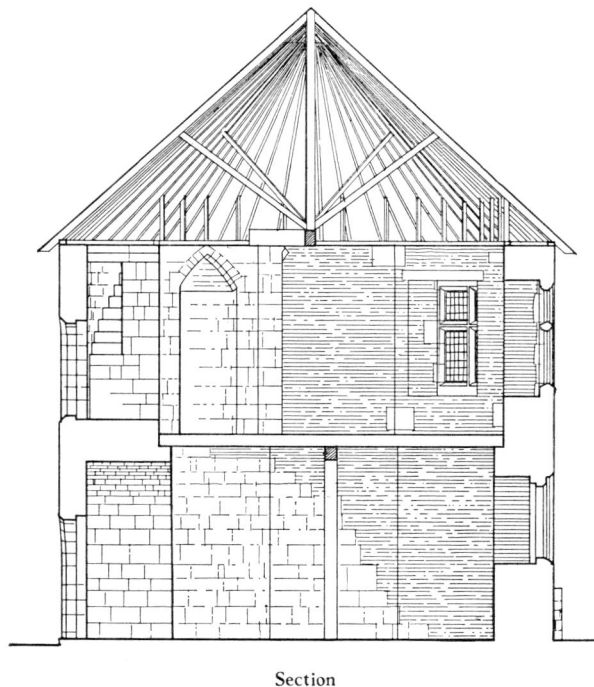

Section

ST. MARY'S TOWER

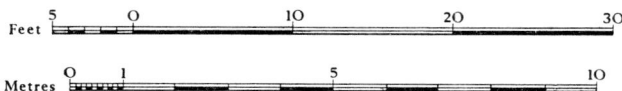

13th century

14th century

17th century

Feet 5 0 10 20 30

Metres 0 1 5 10

Fig. 22.

fluted frieze and some of the masonry too is from the same source, all, including the windows, being cut to a sharper curve than that of the wall of the tower. Facing N. is a partly restored 17th-century doorway with a four-centred arch with key-block also perhaps from the King's Manor.

Inside, the octagonal room on the ground floor has in the S. wall the original doorway with a two-centred pointed rear arch. The wall in which the doorway is set and the two

adjoining walls are of ashlar stone and original. Two other sides are partly of stone faced with 17th-century brickwork, and the rest, wholly rebuilt in the 17th century, are entirely faced with brick. Access to the first floor is by a 19th-century cast-iron newel stair.

The first floor, like the ground floor, has three original walls of stone on the S. and three 17th-century walls faced internally with brickwork on the N., the other sides being partly of each period. In each of the original walls is a wide recess, and the junctions with the rebuilt work show in straight joints which are the reveals of two other recesses. Each of the eight sides in the original tower probably had such a recess. Of the three which survive, two lead to the doorways giving access to the wall-walks: the one to the Marygate side has a depressed pointed arch; the other to Bootham was higher, but the arch is now destroyed and only the springing survives. Leading from the second recess is a stone staircase rising within the wall thickness, originally to the parapet walk of the tower. The third recess has an arrow slit opening from it and contains a garderobe in one corner; the chute downwards is blocked, but there is an upward continuation in the wall thickness which must have been for another garderobe on the parapet walk. This third recess also lacks its original arch. The three 17th-century windows in the N. side of the tower have stone frames, but the reveals and sills are of brick. The roof construction is of the 19th century.

The precinct wall continues parallel to Bootham from St. Mary's Tower to the Postern Tower, a distance of about 435 ft., with a slight change of alignment about 98 ft. S.E. of St. Mary's Tower and with intermediate Towers D and E at distances of 147 ft. and 296 ft. respectively. The whole of this length consists of the wall of 1266 heightened in 1318; Towers D and E are entirely of the later date. Much of the outer side is obscured by 18th- and 19th-century houses and shops, facing Bootham, which have been built up against the wall. One length of 100 ft. including Tower D was exposed in 1914, but otherwise only small parts are visible adjoining St. Mary's Tower and Tower E, uncovered in 1896; the latter tower is still half obscured. In the parts which are exposed the facing is poorly preserved and most of the buttresses have been robbed. The inner face of the wall is visible along the whole of this length. At a point about 120 ft. S.E. of St. Mary's Tower the courses in the 1266 work break bond, indicating perhaps a pause in the building of the wall.

The parapet between St. Mary's Tower and Tower E is unrestored, though damaged; some merlons have gone completely. In four of the merlons there are cruciform arrow slits with widely splayed reveals internally. Several embrasures immediately to the S.E. of Tower E are completely restored; nearer to the Postern Tower they are original but filled in by the rear walls of buildings facing Bootham.

Towers D and E were equal in size and identical on plan, being half-round to the front facing Bootham, semi-octagonal inside and open at the rear and with two short projecting stub walls. They were of two storeys, roofed, with an open crenellated parapet walk. Tower D (NG 60025230. Plate 22) is the less well preserved, although the curved front has been cleared of accretions. There are three cruciform arrow slits at first-floor level, all much damaged, and the parapet and side walls are greatly broken down. Tower E (NG 60065227. Plate 22; Fig. 23, below) stands to its full original height, but the front is partly obscured and only one arrow slit is visible. The interior walls on the ground floor are plain, although each of the stub walls is splayed on the inner angle to a height of 5¼ ft. from the base. At first-floor level is an offset to provide seating for joists. There are three rectangular openings each for an arrow slit with splayed reveals and set in a shallow recess

TOWER E

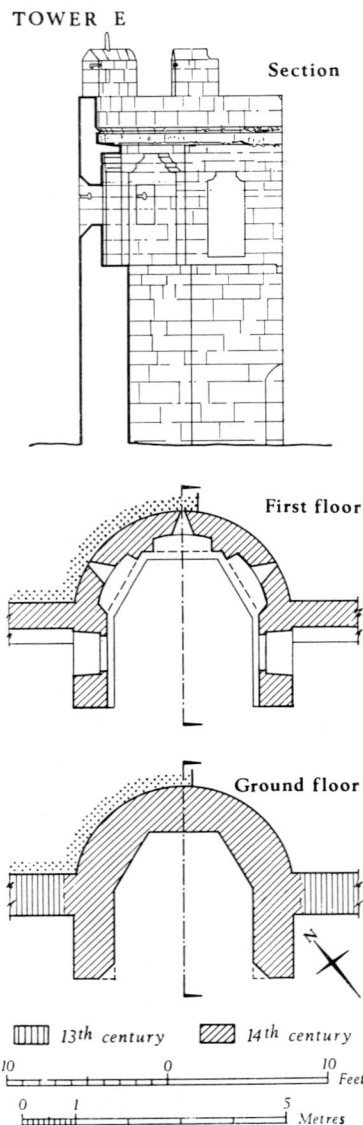

Fig. 23.

with a double-corbelled head. In the other two sides are doorways; these clearly indicate the former presence of timber wall-walks; they have corbelled heads inside and joggled lintels externally, and the doors opened outwards from the tower. On the outside, just below the doorways, are small patches of brick infilling, probably representing the original sockets for the timbers of the wall-walk. A second offset supported the roof. The parapet wall is much thinner than the walls below and is semicircular on plan internally as well as externally; in it are three plain embrasures, and on one merlon a very eroded pinnacle survives. This last is an unusual feature also occurring at Conway Castle, built in 1283-9.

Postern Tower (Plate 23; Fig. 24, below), built in 1497 together with the adjoining archway, is rectangular, projecting outside the line of the precinct wall. The walls are of brick faced with ashlar. Originally two storeys high, the tower was converted to three storeys, probably in the 17th century, by the insertion of a floor in the upper part. It stands 26¾ ft. high, excluding the hipped tiled roof which is probably of 17th-century origin. The N.W. wall is partly masked by a later building.

The N.E. wall has a moulded plinth, mostly modern restoration, which is continued on the S.E. wall, and a moulded eaves cornice, which is carried all round the tower. On the ground floor is a modern N.E. window of three lights, which replaces an 18th-century bay window shown in views by Price and Cattle of 1805 and by C. Dillon of *c.* 1840. Above, on the first floor, there was formerly a rectangular window of which no trace remains outside. In the S.E. wall is a doorway, partly restored since at one time it was partly blocked to form a window; it has boldly moulded jambs and a four-centred arched head with sunk spandrels under a moulded label and a four-centred brick rear arch. In the same wall on the first floor is a window with a four-centred head, sunk spandrels, and label, and just below the eaves is a small square window inserted in the 17th century. In the S.W. wall a doorway

similar to that on the S.E. but with simpler mouldings has been partly filled in to form a window. Above it are two small rectangular windows, probably original.

The inside has one room on each floor. Access between them is by a narrow newel stair in the W. angle, brick-built, vaulted in brick, and reached through doorways with four-centred brick arches. In the N.W. wall and now only visible from inside the tower are blocked cruciform slits with large oillets at the ends of the arms (Fig. 2, p. xii); they are different in proportion from the loops in the 14th-century wall. They have widely splayed reveals and four-centred rear arches. There is a rear arch, probably for a similar opening, in the N.E. wall at first-floor level. The inserted second floor is reached by a modern timber staircase, and here in the S. corner are signs of an inserted chimney, subsequently removed. Padstones at the head of the walls no doubt supported beams of the original roof.

The Postern of 1497, popularly known as Queen Margaret's Arch, consists of a stone archway 10 ft. 7 ins. high with a segmental head; it is rebated for doors which opened inwards and is flanked internally by buttresses. Between it and the tower is a pedestrian way 7 ft. high with a corbelled head, which was cut through the wall in 1836. The wall above these openings has a parapet with plain embrasures; one merlon is pierced by a

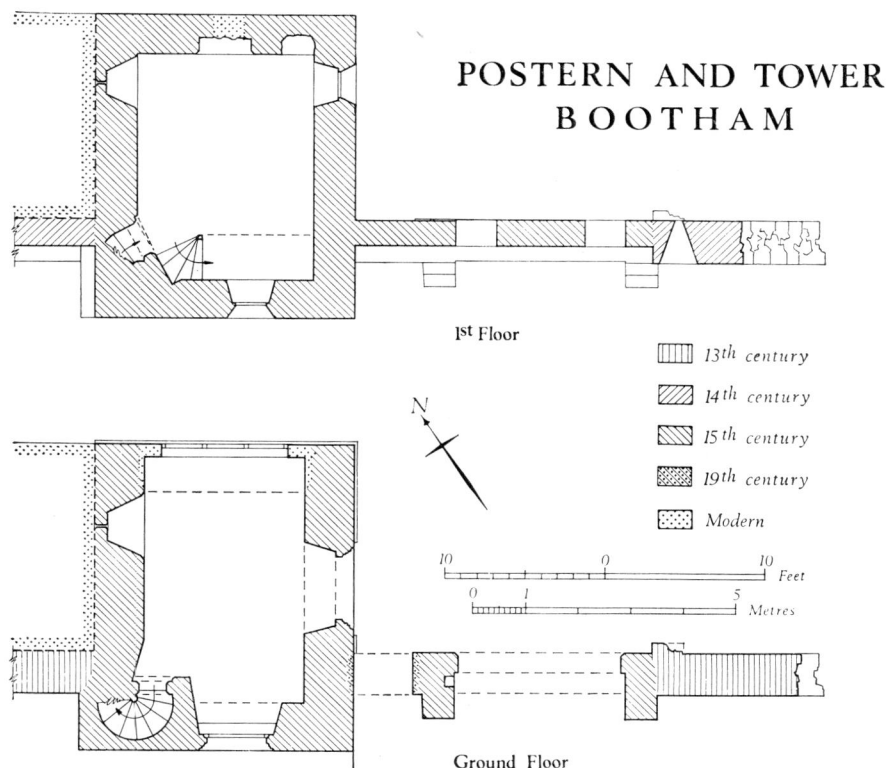

POSTERN AND TOWER
BOOTHAM

1st Floor

N

	13th century
	14th century
	15th century
	19th century
	Modern

10   0   10 Feet
0   1   5 Metres

Ground Floor

Fig. 24.

slit. To the S.E. of the postern is a short length of the wall of 1266 with later heightening and including the remains of an external buttress. A bronze plaque set up on this wall in 1899 by the Yorkshire Philosophical Society perpetuates the misconception that the archway was made in 1503 for the use of Margaret Tudor.

A length of about 60 ft. of the precinct wall, aligned towards the S.W., now forming the boundary to a car park beside No. 6 St. Leonard's Place, begins about 200 ft. S.W. of the Postern. This wall, originally facing the city defences, retains a triangular coping 1¾ ft. high and three characteristic 13th-century buttresses on the outer side. It stands to a height of 7¾ ft. The continuation, deflected S., though also in magnesian limestone, is of 19th-century date. The original line crossed the lane leading to the Museum Gardens beside the King's Manor, and two short fragments of the wall, only one course high, adjoin and are partly overbuilt by the S.E. wall of a wing of the King's Manor built by Lord Sheffield in c. 1610.

The only other fragment of the abbey wall on this side of the precinct is in the Museum Gardens, N. of Lendal Tower. It is about 15 ft. long, 2 ft. 8 ins. thick, and 6 ft. high, although largely obscured by the raised ground surface around it. There are no visible remains of the wall on the river side of the abbey grounds.

Masons' marks from St. Mary's Abbey wall

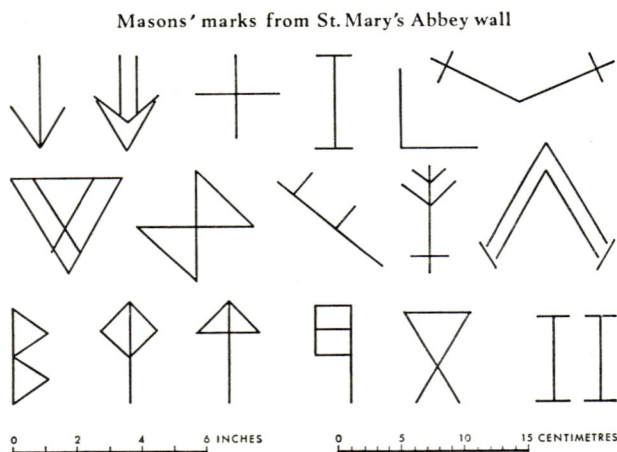

0 2 4 6 INCHES 0 5 10 15 CENTIMETRES

Fig. 25.

ARCHITECTURAL FRAGMENTS AND FITTINGS IN THE YORKSHIRE MUSEUM. There is a large collection of carved and moulded stonework much of which is the product of the excavations in the Abbey but much of the material has been brought from elsewhere. A MS. catalogue prepared in 1921–1933 does not give the provenance of all the exhibits and many of the numbers by which they were then identified have been lost. Fragments of the claustral buildings preserved *in situ* have already been described above. In the descriptions that follow only those stones which are believed to be connected with the

Abbey have been included; other sculptures are listed in the Sectional Preface, p. xliv.

Fragments of the 11th and 12th centuries:

(1) String-course and bay division shafts enriched with lozenges probably from the abbey church, probably late 11th-century.

(2) String-course with bold hollow mould containing spherical pellets, probably late 11th-century. Further pieces of this string are built into the bottom of the N. wall of the abbey church.

(3) Voussoirs of an arch having, within a hollow, a roll-mould interrupted by circular bands of roll and nailhead ornament, 12th-century.

(4) Capitals for shafts and responds including (Plates 27a, c, d, 29): two primitive capitals with corner volute treated as faces, 11th-century; scalloped capitals, early 12th-century; water-leaf and sophisticated acanthus capitals, late 12th-century; capital elaborately carved with struggles between men and monsters including an eagle and a fish (Gardner, 93, fig. 158).

(5) Paired capitals and bases for small shafts, from cloister arcade, late 12th-century (Plate 29d, e).

Fragments from the abbey church of 1271–95 and others later:

(1) Capitals with foliage of developed 'stiff-leaf' form, and with later, more naturalistic foliage (Plate 10). The provenance of some of the capitals is not certain.

(2) Fragments of window jambs including carved foliated capitals, most of a developed 'stiff-leaf' character.

(3) Vaulting springers and ribs including ribs matching the bosses described below.

(4) Vaulting bosses (Plates 11, 12), of c. 1300 and early 14th-century, mostly found in the Warming House, probably collected there in the 16th century, include;

5 large bosses 2½ ft. in diameter for the intersection of eight ribs, carved with foliage, one with *Agnus Dei*, and one with the bust of a human figure framed in foliage.

8 bosses for the intersection of four ribs, carved with monsters, with a monk playing a viol, with birds and figures embowered in foliage, and with foliage only.

3 bosses for the intersection of three ribs, carved with foliage, one including a running deer.

Many of the bosses are carved with rather large-scale undulating formalised foliage but some are more delicately carved with finely modelled natural forms.

(5) Wall arcading of two-centred arches enclosing trefoil tracery under cusped gablets enriched with crockets and finials, and two enriched with naturalistic flowers and foliage (Plate 12). These last are said to have come from the E. cloister walk and may be dated to the last years of the 13th century. Those of more formalised design are probably slightly later.

Coffin Lids: (1) coped, broken, inscribed Alanus d.., 13th-century; (2) coped, broken, inscribed Helis persona, 13th-century; (3) coped, broken, inscribed Thomas, 13th-century; (4) flat, broken, with foliated cross inscribed Ema de Ben... perhaps for the widow of Adam de Benfield of Morton in Cleveland, a benefactress of the abbey, 13th-century; (5) flat, tapered, inscribed ..dulfus filius Joh.., 13th or 14th-century; (6) flat, shaped, with moulded under-edge, and indents for marginal inscriptions, crocketed canopy and figure with helmet and shield, found in the choir of the abbey over a brick tomb containing the skeleton of a boy, 15th-century.

Floor-slabs: (1) fragments inscribed H..iac..de..., 13th-century; (2) probably of Thomas Spofford, abbot 1405–22, bishop of Hereford 1422–48, died at St. Mary's Abbey 1456; broken and incomplete, retaining parts of marginal inscription '... [Her]eford sacre p(a)gini p(ro)fessor et quonda(m) Abba[s] hui[us] ... cuiu(s) a(n)i(m)e p(ro)pic[ietur] ...' with corner medallion of Lion of St. Mark; within the border representation of bishop in mass vestments, holding crozier and book, the head flanked by two doctor's caps; partly incised, partly in very low relief (Fig. 26. *Cp.* YPSR for 1902, 75; *Antiquaries Journal* XVIII (1938), 290); (3) large slab to two brothers, William Hewick, magister, and dominus John Hewick, capellanus, with incised cross on octagonal base drawn in perspective; traces of a kneeling figure each side of cross; 15th-century; (4) fragment with black-letter inscription to Frater Thomas, 15th-century; (5–7) three fragments of slabs with incised crosses, mediaeval.

Mortar (Plate 44), of bronze, with two handles of twisted form; body decorated with pattern of quatrefoils containing beasts, between bands at top and bottom; inscribed Mortaria sci Johis Evangell de Infirmaria be Marie Ebor, Fr Wills de Towthorp me fecit AD MCCCVIII.

Piscina (Plate 42d), with hexagonal bowl supported by a half-length figure flanked by a smaller figure, 14th-century.

Sculpture

(1) A series of 13 stone statues (Plates 1, 30–37), about life size, of *c.* 1200, of which seven were dug up in the S. aisle of the abbey church from under a layer of broken 13th-century window tracery, two were recovered from St. Lawrence's church, one from Clifton Bridge, and two from Cawood (*see* p. xlii). The post-mediaeval history of the 13th figure is not known. The figures from the S. aisle, when first uncovered, showed considerable remains of colour. Exposure to damp and flood water in the museum has obliterated almost all traces of it. These figures are discussed in the Sectional Preface, p. xlii. Figures:

i. Moses (Plates 30a, 34). Complete figure with feet damaged, with forked beard and horned; holding tablets of the Law and serpent, with bird-like head, coiled around staff.

ii. John the Baptist (Plate 35b). Figure with feet missing and very weather-worn, bearded, holding roundel with *Agnus Dei*.

iii. Probably St. John the Evangelist (Plates 1, 30b). Figure of young man, feet missing. Book in left hand, knot of drapery over right arm.

iv. Apostle (Plates 31b, 33). Complete figure, with forked beard. Left hand holding drapery and book, open right hand.

v. Apostle (Plates 31a, 32). Complete figure with feet damaged, bearded, holding book between the two hands.

vi. Apostle (Plate 36b). Figure without head, feet damaged. Holding drapery and book in left hand, right hand gone; from Cawood.

vii. Figure (Plate 36c), headless, very damaged and decayed, with rounded back.

viii. Figure (Plate 35a), complete except for hands but very decayed; holding some object, now shapeless but possibly a book, in left hand.

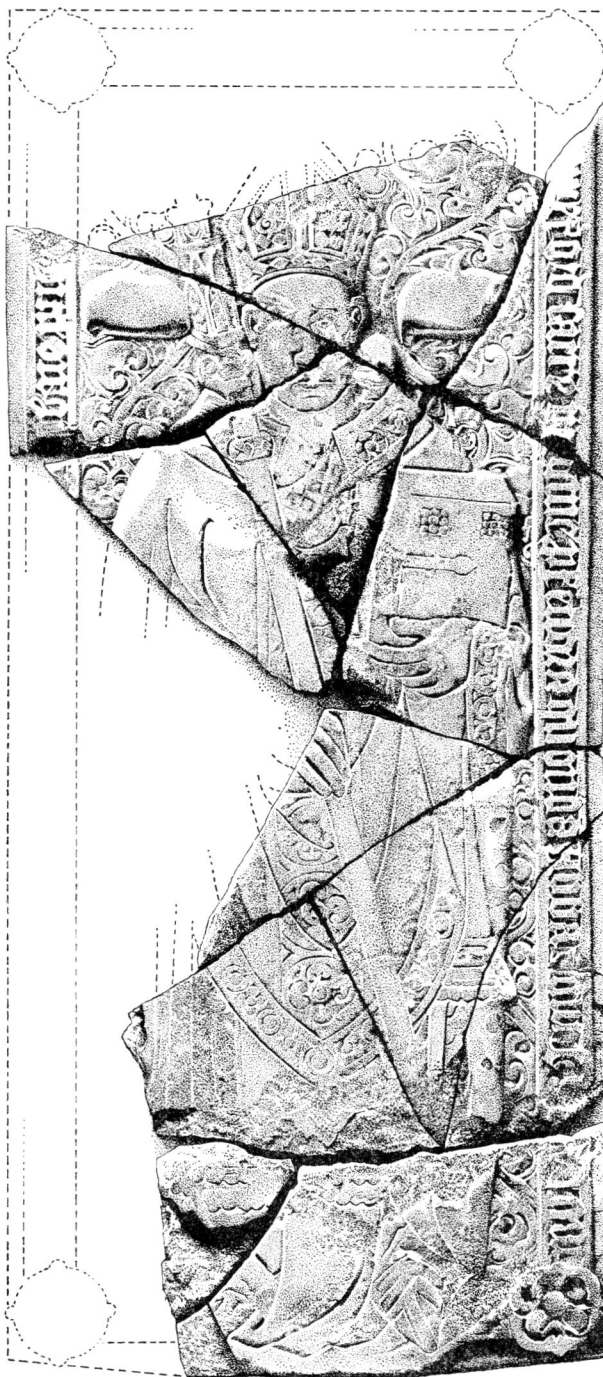

Fig. 26. (4) St. Mary's Abbey.
Floor-slab of Thomas Spofford (?).

ix. Figure (Plate 36a), headless, right side of body and right arm defaced; feet damaged; object held in left hand defaced, with angled back.

x. Apostle (Plate 37c). Figure complete except for head; holding book in left hand, right hand with forefinger extended to book, with angled back.

xi. Apostle (Plate 37b). Figure lacking head and lower part of legs. Holding book in left hand, knot of drapery in right, from Cawood.

xii. Apostle (Plate 37a). Figure without head or feet, holding book in left hand and knot of drapery in right.

xiii. Fragment of torso, similar in style and in scale to the other figures; origin unknown.

xiv. Head, fragment only (Plate 43c).

At the back of each head, where complete, there is a 7 in. shaft to which the head is attached. The backs of nine figures are flat, two are angled as for setting in a re-entrant corner, and one is rounded.

(2) Virgin and Child (Plate 41d). Draped figure of the Virgin, seated, holding the infant Jesus on her lap. Both figures headless; present height 3 ft. 4 in. The back of the group is rough and slightly rounded, presumably to be set into a wall. Similar in style to the thirteen life-size figures described above; recovered from Cawood with two of the figures; c. 1200.

(3) Scenes from the New Testament (Plates 38, 39). Seven voussoirs from an arch of several orders carved with figures representing scenes from the New Testament. The figures, which average 13 in. in height, are carved in a deep hollow moulding; c. 1200.

i. The Visitation. Two figures, one nimbed but headless, the other fragmentary.

ii. The Nativity. The infant Jesus, nimbed, in a cradle or manger and in the background a figure now headless behind a barrier with drapery.

iii. The Wise Men before Herod. Three standing figures, one crowned, one headless, and a seated figure also now headless.

iv. The Epiphany. Seated figure of the Virgin, now headless, holding the Child, and the three Wise Men, one kneeling.

v. Herod ordering the Massacre of the Innocents. Under two arches rising from a central column and with crenellated parapet above, a King crowned and two armed men.

vi. The Marriage at Cana (?). Apostle and two other figures, a pitcher in the background.

vii. The Raising of Lazarus. Three standing figures, one headless but with cruciform nimbus, behind a coffin from which a fourth figure is rising.

(4) Coronation of the Virgin. Fragments of figures, rather more than life size (Plate 40); late 13th-century.

i. Figure of Christ, broken into three pieces. The middle part of the figure was found c. 1952 built into the base of one of the 13th-century buttresses to the S. nave aisle where it had been used for repairs perhaps after the fire of 1377. The figure has wavy hair and curly beard, and wears a robe held by a belt. The head is inclined forward, facing the front; the right arm is missing. The left hand holds a book.

ii. Head of the Virgin. In Museum basement, with wavy hair similar to that of the Christ, and wearing a coronet

which is held by a hand which must belong to the missing right arm of Christ.

iii. Fragment of torso and right arm, probably of the Virgin.

iv. Fragment of legs and feet, probably of the Virgin.

(5) Corbel head (Plate 42a), face disfigured, with wavy hair restrained by a band, late 13th-century.

(6) Coping stone, carved with large figure of a lizard, late mediaeval.

(7) Coping stone surmounted by figure of man wearing surcoat over chainmail, with sword and shield, possibly from the abbey defences, late mediaeval.

Tiles include: (1) decorated with bell between sword and key; (2) with monkeys; (3) with ram and inscription Sol in Ariete (Figs. 1, p. iv, 27).

Fig. 27. (4) St. Mary's Abbey. Floor-tiles.

(5) PRIORY OF ST. ANDREW, Fishergate, is now represented only by some fragments of stone precinct walling on the N. side of Blue Bridge Lane. The church of St. Andrew is mentioned in Domesday Book; in 1202 it was given by Hugh Murdac, Archdeacon of Cleveland, to the Gilbertine Canons who built a house there (Raine, 299).

(6) CHURCH OF ST. HILDA, Tang Hall, is modern but contains the following fittings brought from elsewhere: *Chairs*, 17th-century, two, from church of St. Mary Bishophill Senior, *York* III (9). *Font*, c. 1850, and fontcover, 1638, from church of St. John the Evangelist, Micklegate, *York* III (6). *Plate*, cup, stand-paten and almsdish, all 1839, from Archbishop's Palace, Bishopthorpe; cup, cover-paten and salver, made by John Langwith in 1703, from 'the chappel in the Castle in the County of York' and dated 1706. (*York* II, 81, Pl. 15)

(7) PARISH CHURCH OF ST. LAWRENCE. The greater part of the mediaeval church (Plate 48) has been demolished, leaving only the W. tower standing. It consisted of a 12th-century aisleless nave with a chancel probably of later date and lit by 14th-century windows. The W. tower (Plate 45) was added in the late 12th

PLATE 25

a. Cresset lamp.

b. Anglian cross–head.

c. Grave–slab from Parliament Street.

d. Fragments of late Anglo-Danish grave-slab, probably from St. Olave's Church.

e. Head of Anglo-Danish grave-slab from Clifford Street.

THE YORKSHIRE MUSEUM. Pre-Conquest carving.

f. Grave–slab from Parliament Street.

PLATE 26

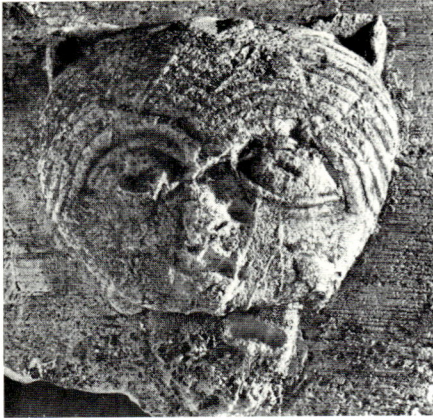

a. Corbel with head of lion.

b. Gritsone carving of wolf (?), from Micklegate.

c. Corbel heads.

d. Corbel head.

e. Corbel with head of muzzled bear, from Piccadilly.

f. Corbel with head of an animal.

THE YORKSHIRE MUSEUM. Carved stones. 12th-century.

PLATE 27

a. Capital with face, from St. Mary's Abbey. 11th-century.

b. Capital with head and dragons. 12th-century.

c. Volute capital with face. Late 11th-century.

d. Water-leaf capital. 12th-century.

e. Capitals with heads. 11th-century.

THE YORKSHIRE MUSEUM.

f. Label stops. 12th-century.

PLATE 28

a. Voussoirs.

b. Voussoir.

c. Capital.

d. Capital.

e. Nook shaft.

f. Voussoirs.

a–e. Probably from All Saints, Pavement. f. From a doorway found in York.

THE YORKSHIRE MUSEUM. Fragments from archways. 12th-century.

PLATE 29

a. Tympanum, probably from York Minster.

b, c. Respond capital, probably from St. Mary's Abbey. c.

d, e. Double capitals from cloister arcade, St. Mary's Abbey. e.

f, g. Carved capitals, from St. Mary's Abbey. g.

THE YORKSHIRE MUSEUM. Tympanum and capitals. 12th-century.

PLATE 30

a. Moses.

b. Probably St. John.

(4) ST. MARY'S ABBEY. Statues. *c.* 1200.

PLATE 31

a.

b.

(4) ST. MARY'S ABBEY. Statues of Apostles. *c.* 1200.

PLATE 32

(4) ST. MARY'S ABBEY. Statue of Apostle. *c.* 1200.

PLATE 33

(4) ST. MARY'S ABBEY. Statue of Apostle. *c.* 1200.

PLATE 34

(4) ST. MARY'S ABBEY. Statue of Moses. c. 1200.

PLATE 35

a.

b. John the Baptist.

(4) ST. MARY'S ABBEY. Statues. *c.* 1200.

PLATE 36

c.

b. Apostle. From Cawood.

a.

(4) ST. MARY'S ABBEY. Statues. c. 1200.

PLATE 37

c.

b. From Cawood.

a.

(4) ST. MARY'S ABBEY. Statues of Apostles. c. 1200.

PLATE 38

b. The Epiphany.

d. The Marriage at Cana (?).

a. The Nativity.

c. The Wise Men before Herod.

(4) ST. MARY'S ABBEY. Voussoirs with Scenes from the New Testament. *c.* 1200

PLATE 39

Herod ordering the Massacre of the Innocents; The Raising of Lazarus; The Visitation.

(4) ST. MARY'S ABBEY. Voussoirs with Scenes from the New Testament. c. 1200.

PLATE 40

(4) ST. MARY'S ABBEY. Fragments from the Coronation of the Virgin. 13th-century.

PLATE 41

a. Early 14th-century.

b. 12th-century.

c. Alabaster. Late 14th-century.

d. From Cawood. c. 1200.

THE YORKSHIRE MUSEUM. Statues of the Virgin and Child.

PLATE 42

a. Corbel head, probably from St. Mary's Abbey. 13th-century.

b. Label stop. Head of a bishop. 14th-century.

c. Figure of St. Margaret. Early 16th-century.

d. Piscina bracket, from St. Mary's Abbey. 14th-century.

THE YORKSHIRE MUSEUM.

e. Limestone head from Davygate. 15th-century.

PLATE 43

a. Statue of bishop. 14th-century.

b. Limestone head. Late mediaeval.

c. Fragment of head, from St. Mary's Abbey. *c.* 1200.
THE YORKSHIRE MUSEUM.

d. Statue of lady. Late mediaeval.

PLATE 44

ST. MARGARET'S CHURCH, Walmgate. Doorway. By J. Carter, 1791. 12th-century.

(4) ST. MARY'S ABBEY. The Infirmary Mortar. 1308.

century, the two lower storeys being of this date, but the second storey was partly rebuilt and new windows were put in in the 13th century. The top storey was added in the 15th century. The rest of the church was pulled down after the building of a new church in 1881–3 and the old 12th-century N. doorway (Plate 46) was re-erected against the tower.

The lower stages are of coursed rubble, the lower part of the E. wall being part of the W. wall of the earlier nave. A 16th-century window has been inserted in the W. wall cutting across a small 13th-century light. The top storey is of good ashlar with two-light 15th-century windows and a parapet that rises at the corners with open panels surmounted by stone finials. The doorway has a semicircular arch of four orders: the inner order has plain voussoirs springing from moulded imposts over scalloped capitals; the second order has an interlace above two monsters springing from capitals also carved with monsters (Plate 47); the two outer orders are carved with formalised foliage. Behind, a semicircular rear arch springing from chamfered imposts is probably the original tower arch of the late 12th century.

Fittings. *Bell* in belfry of new church, not hung, inscribed Deo Gloria 1739. *Bell-frame in situ*, in the old tower, not accessible but partly visible from the ground, of uncertain date. *Font*, moulded octagonal bowl with brattishing on rim, small foliage and grotesque carvings between mouldings, on octagonal stem with moulded foot, *c.* 1500. *Monuments* in churchyard include: (1) to four sons and two daughters of John and Ann Rigg who died in a boating accident, 1830, inscribed tablet framed by pilasters and entablature set against a brick wall and overlooking stone-covered grave surrounded by iron railings, by William A. Plows; (2) to . . . Allen, upright stone cylinder, probably early 19th-century; (3) to Elizabeth White, 1783, headstone. *Plate:* in new church, cup and cover by William Busfield, York 1681; cup and cover by Thomas Mangey, York 1682, with inscription of 1684; paten by Robert Abercromby, London 1738, with inscription recording the gift of Ann Yarburgh 1740; brass almsdish, with temptation of Adam and Eve, German, 16th-century (Fallow and McCall, I, 13, 14).

(8) PARISH CHURCH OF ST. MAURICE stood at the junction of Monkgate and Lord Mayor's Walk. A mediaeval church dating at least from the late 12th century was taken down in 1875 and a larger structure erected which was in turn demolished in 1967. The following architectural fragments and fittings have been preserved.

Arch (Plate 49) reconstructed probably from S. doorway, now of two orders but incorporating voussoirs of three types, as well as jamb stones reused as voussoirs. Decoration in form of roll mouldings, rosettes and beak-head ornament. Arch springs from capitals carved with foliage and (?) interlace. Third quarter of 12th century. Built into fabric of church of St. James the Deacon, Acomb Moor.

Window (Plate 48), of two round-headed lights, roll-moulded

7

heads and parts of jambs. Lights divided by round shaft attached to rectangular pier. Head pierced by circular hole forming embryonic plate tracery (J. H. Parker, *Introduction to Gothic Architecture* (1861), 52). Late 12th-century. In Yorkshire Museum.

Fittings. *Bells:* (1) inscribed Gloria in Altissimis Deus SS 1665; (2) venite exultemus Domino SS Ebor. Now at church of St. Hilda, Grangetown, Teesside. *Coffin Lid*, carved with raised cross, the head foliated against a plain circular background, 13th-century. At church of St. James the Deacon. *Monuments*, in churchyard, headstones of 1781 and later. *Panelling*, reset in 19th-century door, four panels carved with four Evangelists and their symbols, 16th-century. At church of St. James the Deacon. *Plate:* cup of 1568, cover-paten by John Oliver, York 1684, now in the care of St. Michael-le-Belfrey, York; cup and two patens by Barber and North, York 1842, now at church of St. Thomas, Lowther Street, York (Fallow and McCall, I, 20, 21; cup of 1568 pl. VI). The present whereabouts of a flagon by Robert and David Hennel, London 1797, is not known.

(9) PARISH CHURCH OF ST. OLAVE, Marygate (Plates 45, 50), has walls of magnesian limestone ashlar, and roofs covered with tiles and lead. The present church comprises a N. aisle partly of the 15th century, W. tower also of the 15th century, nave and S. aisle of the 18th century, and chancel, vestries etc. added after 1850.

The Anglo Saxon Chronicle records that Siward Earl of Northumbria died in 1055 and was buried 'in the minster at Galmanho which he himself had built and consecrated in the name of God and St. Olaf' (ASC, Text D); St. Olave was Olaf, king of Norway, who was killed in 1030 (*see* Bruce Dickens, *Cult of St. Olaf in the British Isles*, Viking Soc. Saga Bk. XII, pt. 2 (1940), 53–80). The original church must therefore have been built between 1030 and 1055 but of this building nothing remains identifiable today. Symeon of Durham speaks of it as a 'little church' (*ecclesiola*) which became a noble monastery (SS, LI (1868), 94). After the Conquest Alan, Count of Brittany, gave the church to Benedictine monks who came from Whitby and Lastingham. Gifts from William I and William II (EYC, I, 265) enabled these monks to build themselves a new church which became the great St. Mary's Abbey, the church of St. Olave remaining within the abbey precinct, appropriated to the monastery and serving the people of Bootham and Marygate. The abbey, which had acquired a valuable rectory, maintained that the status of St. Olave was that of a chapel. In 1313 the sacrist, to whose office the revenues of the church were allotted, was ordered to supply fitting furniture for the church and to provide for its other needs (Raine, 264). Further disputes arose in the course of the later 14th century, leading to a decision by the archbishop and confirmed by the pope (*Papal Letters 1396–1404*, 8) that the

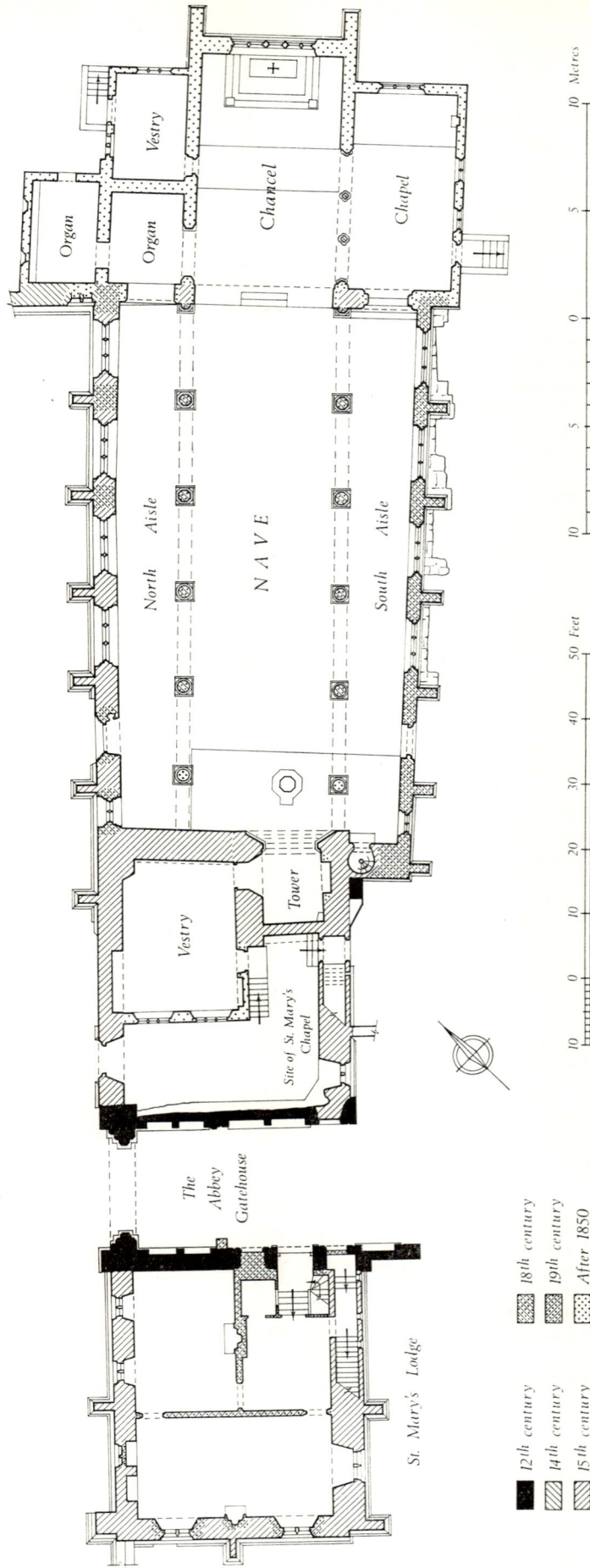

NAVE

North Aisle

South Aisle

Chancel

Chapel

Vestry

Organ

Organ

Tower

Vestry

Site of St. Mary's Chapel

The Abbey Gatehouse

St. Mary's Lodge

12th century
14th century
15th century
18th century
19th century
After 1850

10   0   10   20   30   40   50 Feet

10   0   5   10 Metres

Fig. 28. (4) St. Mary's Abbey. Gatehouse. (9) Church of St. Olave.

parishioners should repair the church. There is however no evidence that any work on the fabric was actually carried out at this time.

The original church was no doubt a rectangular building without aisles but by the end of the 12th century an aisle had been added to the full length of the church on the S. side, corresponding in width to the present S. aisle, and an aisle on the N. side of which the length is uncertain but it was probably also the full length of the church. The outer wall of this N. aisle stood in the position of the present N. arcade. Some of the stonework in the exposed footings of the E. part of the S. wall probably belongs to a late 12th-century wall with shallow pilaster buttresses over which later buttresses of greater projection were built.

In 1458 Roger Stanes left 6s. 8d. for glazing the window over the door (Raine, 266); this probably marks the completion of a new S. aisle. In 1463 Thomas Hornby, rector of Stokesley, left 5 marks towards the fabric of the nave provided that the parishioners would begin it within two years (*Testamenta Eboracensia*, SS, xxx (1855), cci, 257) but it appears that nothing was done, for in 1466 Archbishop George Neville gave orders for extensive rebuilding of the church (Reg. Geo. Neville, 86b quoted by Raine, 264); parochial status, which had been a subject for dispute with the abbey, was finally granted and the parishioners were ordered to rebuild and repair much of the church but the monks were to rebuild the N. side of the nave after the pattern of the outer wall on the S. side. This instruction reflects the fact that the church was to be widened northwards and the new N. wall was to be an adaptation of the existing precinct wall of the monastery. The N. arcade was moved to the line of the outer wall of the former N. aisle, leaving the tower eccentrically placed. Of the rebuilding that followed, the W. half of the present N. wall represents the precinct wall reconstructed with openings to the aisle; a change of direction in the wall probably indicates the position of a former interval tower in St. Mary's precinct wall. The lower courses of the E. part of the wall show original 13th-century masonry on the inside. The new N. windows were to copy those on the S. side, two of which are now reset to form the two eastern windows in the N. side. The footings of the mediaeval S. wall are now exposed below the present S. wall and extend the whole length of the present nave; the spacing of the former buttresses differs slightly from the present arrangement. The rebuilding of the body of the church including the clerestory was probably complete by 1471 when John Hartyng left 6s. 8d. towards making a Rood (Raine, 266). The rebuilding of the tower followed: in 1478 Robert Plumpton left 40s. towards the tower then being

built; further donations for the tower followed in 1483, 1485 and 1487 (Raine, 265). In 1498 Francis Foster left 6s. 8d. for bells, and in 1501 Christopher Johnson left 6s. 8d. for bells recently bought. The S. wall of the tower is of one build with the S. wall of the ruined building immediately W., of which the upper floor probably formed the chapel of St. Mary at the Gate and which must have been rebuilt at the same time.

During the Civil War the church suffered damage when the roof was used as a gun platform (Ward, II, 218; Hargrove, II, pt. ii, 598; Benson, III, 39) and the church was repaired in the reign of Charles II (Sheahan and Whellan, I, 514). Nevertheless by the opening of the 18th century it became necessary to rebuild most of the church (YML, Hornby MSS., II, 231) and stone for the work was granted from the ruins of the adjacent abbey (*Treasury Papers 1702–7*, pt. I, 134, 297, 358). A drawing of 1705 by James Poole (Bodleian, MS. Tanner 311, f. 170) shows that at this time the nave had a clerestory which was removed in the course of reconstruction. The main work was carried out 1721–2: the S. wall was completely rebuilt; both N. and S. arcades were rebuilt with some reuse of mediaeval stone; and the N. wall was partly reconstructed, with two 15th-century windows reused from the S. aisle in the two E. bays and other 15th-century material reused to give a uniform appearance to the N. side of the church; towards the W. a new doorway was made, and a new square-headed window further W. still. Differences between the two E. columns in each arcade and those to the W. possibly perpetuate a mediaeval feature, and a discontinuity in the roof suggests that the present nave represents a mediaeval nave and chancel, the latter occupying the two E. bays. A W. gallery was inserted in 1832.

The present chancel was added in 1887–9 to the designs of G. Fowler Jones of Stonegate, York. A vestry was added in 1898. In 1907 an organ motor chamber was built to the designs of G. F. Walker over the remains of a vaulted chamber in the almonry to N.E., and in 1908 the vestry was converted to a S. chapel, a new N. vestry was added and the chancel was enlarged, all to the designs of Francis Doyle of Liverpool. At the same time the W. gallery was removed and the W. pier of each arcade was rebuilt.

Architectural Description. The *Nave* (80 ft. by 21 ft.) is of 5½ bays. The chancel arch is modern. The N. and S. arcades have two-centred arches of two chamfered orders. The two E. piers in each arcade are octagonal with octagonal bases and capitals, and there are semi-octagonal responds to match; to the W. three columns in each arcade are circular with circular capitals under abaci alternately square and octagonal. The two W. piers have been rebuilt with modern stonework. Some of the other piers include reused mediaeval masonry, but the capitals and bases are all of the early 18th century with quasi-classical mouldings. The tower arch, in the S. part of the W. wall, is two-centred, with three chamfered orders to the E. and two to the W., springing from hollow-moulded imposts above simple splayed jambs. N. of the tower arch is a vertical straight joint marking the N. corner of the tower structure.

The *North Aisle* (8 ft.–9 ft. wide) has an E. wall which is modern above a 15th-century base and contains a modern opening to the organ chamber. The lower part of the N. wall is part of the 13th-century precinct wall of the abbey and is built in two lengths meeting at a slight angle between the second and third windows; this junction may perhaps represent the position of a mural tower, removed in the 15th century, since it comes at a distance from Tower B equal to the distance between Towers A and B on St. Mary's precinct wall. On the outside a moulded plinth of the 15th century has been added and between the bays are 15th-century buttresses rising to gargoyles and crocketed pinnacles. In the two E. bays are mid 15th-century windows each of three lights with hollow-chamfered jambs and vertical tracery in a four-centred head; in the third and fourth bays are windows of slightly later date which are generally similar to the first two but show slight variations in detail, having three-centred heads and larger foils in the tracery. In the fifth bay is an 18th-century doorway with moulded jambs and two-centred head. Further W. is a window of two lights with a square head. The sixth buttress is further W. than the W. end of the aisle, opposite the middle of the tower. The W. wall of the aisle is of plain masonry aligning with the W. wall of the nave but not of the same build.

The *South Aisle* (9½ ft.–7¾ ft. wide) was rebuilt in the 18th century with reuse of mediaeval masonry and a decorative fragment of early 17th-century work from the King's Manor. In the E. wall is a modern opening to the N.E. chapel. The S. wall stands above the footings of an earlier and thicker wall with buttresses laid out at a spacing different from the present arrangement. In the first three bays from the E. are 18th-century windows each of three lights with hollow-chamfered jambs and vertical tracery in a three-centred head. The tracery and mullions in the second and third windows are entirely modern. The fourth window has three cinquefoiled lights and a four-centred head with no tracery. The doorway has chamfered jambs and a two-centred head. The westernmost window has two cinquefoiled lights and a square head. The W. end includes the tower staircase.

The *West Tower* (10½ ft. by 8½ ft.) has a projection for a staircase on the S. side and another projection, further W., of irregular rubble masonry, which may represent a buttress of a 12th-century tower. The lower part of the tower, which is open to the nave, was originally divided into two storeys by a floor which has been removed. The upper room so formed must have served as a vestry for the chapel of St. Mary at the Gate which adjoined to W. It was reached by an external stair leading to a doorway, now blocked, in the S. wall and from it another doorway in the W. wall, also blocked, led into the chapel. In the W. wall of the tower, in the W. face, is a tall arched recess (Plate 14) which may have accommodated the altar of the chapel, with the doorway to the vestry beside it. On the inside of the S. wall of the tower is a recess, rebuilt in the lower part, which was probably a cupboard to contain the

banners and the dragon (representing the devil) which were carried by the monks in the Rogationtide processions; these processions started after a mass in the chapel (*Ord.* 319). In the N. wall, above the modern vestry, two projecting stone corbels suggest that there may once have been a higher roof built up against the tower on this side. In the W. wall, above the arched recess, is a window of three cinquefoiled lights with vertical tracery in a two-centred head, and above again a small rectangular light. At a corresponding level in the E. wall is a small doorway which formerly gave access to the nave roof, before the removal of the clerestorey. The top stage of the tower, above a weathered string-course, was rebuilt in the 18th century with an original 15th-century window reused in each face; these windows have four-centred heads and labels but no mullions or tracery. The *West Vestry* is a modern structure; its E. wall is formed by the W. wall of the nave and N. aisle and has a 15th-century string-course. The N. wall retains 13th-century masonry of the precinct wall in the lower part and some 15th-century walling with a moulded string-course above, but much of the upper part of the wall is modern.

Fittings—Bells: six, all bearing the maker's name 'Dalton maker York' and dated 1789, inscribed (1) With cheerful voice O Lord I'll sing to thee, (2) Have faith in Christ and live eternally, (3) We call, Come ye watch and pray, (4) In praise to God loudly we unite Halleluiah, (5) In concert I'll Jehovah's name resound, (6) To Father Son and Holy G'st eternal glory raise. William Dade, vicar. William Bayldon Christopher Bearpark William Cuthbert Richard Wood, churchwardens.

Benefactors' Tables: in N. aisle (1) framed by clustered columns at the sides and moulded arched head, recording benefactions from 1766 to 1871, early 19th-century, repainted; in S. aisle (2) in moulded frame with arched head recording benefactions from 1607 to 1740, 18th-century. *Coffin Lid,* in Yorkshire Museum, two fragments with cross bar of cross and top of panels below carved with interlace finishing in beasts' heads; found in excavations in front of the Museum and almost certainly from St. Olave's church; *c.* 1060 (Plate 25d; *see* Preface, p. xlvi). *Communion Table,* with turned legs, rails enriched with arabesques, superimposed modern top, 17th-century. *Doors:* in N. doorway (1) with arched head, plain panels externally, with original furniture, 18th-century; in S. doorway (2) with arched head, panelled both sides, with original furniture, 18th-century. *Font,* disused in churchyard, moulded octagonal bowl on octagonal stem with narrow waist, 15th-century very weather-worn. *Glass* in E. window, main lights contain fragments assembled to form five figures with heads representing a king (St. Olaf?) and saints. In tracery, in centre lights, the Annunciation, incomplete, and above, two kneeling figures, one with scroll inscribed Ave Maria, 15th-century. *Images:* in S. aisle at E. end, (1) Crucifixion in low relief; Christ crucified between two small figures holding scrolls, stone, badly worn, 15th-century; (2) Madonna and Child, marble, 18th-century Spanish.

Monuments and *Floor-slabs.* Monuments: In N. aisle (1) to David Russell, 1840; (2) to Sarah Eyre, 1825, wall monument with sarcophagus and urn, shield-of-arms below; (3) to Rev. Thomas Cripps, 1794, with sarcophagus, cross, and shield-of-arms; (4) to William Cattell, 1830, Sarah Cattell his wife, 1842;

(5) to David Poole, 1830, and two daughters, with sarcophagus, by M. Taylor, York. In S. aisle (6) to Charles Christopher Richard Hacket, 1849; (7) to George Hutchinson of Reeth, 1775, and Elizabeth his mother, 1774, erected by Walter Gray, with Adamesque urn (Plate 51), probably from the same workshop as (8) below; (8) to Alathea Jordan, 1741, Col. John Jordan her husband, 1756, and Anne Maria Alathea, their daughter, widow of James Maude, 1778, by whose will the monument was erected, with Adamesque urn, by Fishers (Plate 51); (9) to Frances Worsley, 1837; (10) to Michael Loftus, 1762, with shield-of-arms; (11) to John Roper, 1826, and Sarah his widow, 1835, with sarcophagus; (12) to George Stephenson, 1800, with draped urn; (13) to William Thornton, joyner and architect, 1721, and Robert his son, 1724 (Plate 51). In churchyard, S. of church, headstones include (14) to Sarah wife of John Wolstenholme, carver, 1834; (15) to Thomas Wolstenholme, sculptor, 1812; (16) to Francis Wolstenholme, carver, 1833; (17) to George Wolstenholme, 1822; by N. door of abbey church (18) table tomb to William Etty, R.A., 1849.

Floor-slabs: In nave (1) to Mrs. Anna Burgess, 1792, and John her husband, 1795; (2) to Robert son of William Thornton (architect), 1724; (3) to the Rev. Thomas Mosley, early 19th-century; (4) to Elizabeth Mosley, 1787; (5) to Capt. Isaac Moorsom, 1779; (6) to Rebecca widow of James Legard, 1783; (7) to Anne Mosley, 1782; (8) to Anthony Thorpe, 1830; (9) to Frances wife of the Rev. Lamplugh Hird, prebendary of York cathedral. In N. aisle (10) to Lt. David Naylor (1831). In Yorkshire Museum (11) two fragments found reused in St. Mary's Abbey but almost certainly from Earl Siward's church of St. Olaf and perhaps from Siward's own grave, carved with a monster with barbed tongue and spiral wing, mid 11th-century.

Niche, externally over N. doorway, with vaulted head, pinnacles and crockets, 15th-century reset, probably from Abbey gateway and now containing modern statue. Plate includes cup with 17th-century bowl mounted on earlier stem and flat sexfoil foot, 15th- or early 16th-century, cup of 1633 by Thomas Harrington, flagon of 1703 and stand-paten of 1715, both by Seth Lofthouse, paten of 1767 by Butty and Dumée, and a brass almsdish embossed with St. George and the Dragon and, on the rim, remains of inscription recording gift to church in 1707, almost completely obliterated. Royal Arms, in N. aisle, of Henry Prince of Wales, died 1612, stone shield held by a lion, repainted (Plate 51). Stoup, in N. jamb of N. doorway, round bowl reached from doorway and from aisle, mediaeval reset.

Miscellanea: under chancel arch, part of base of stone screen (Plate 52) having, above a base mould and between miniature buttresses, quatrefoil panels carved in relief with figures of angels playing musical instruments; 14th-century, found in excavating for the organ chamber.

(10) CEMETERY, with chapel 280 yds. S. of Walmgate Bar, and lodge to W., was opened in 1837 by the York Public Cemetery Company. The chapel (Plate 83) is built of Roche Abbey stone and was designed by J. P. Pritchett; the lodge was presumably by the same architect (New Guide (1838), 149; York Courant, 4 Jan. 1838). The cemetery originally comprised some 8 acres but was subsequently increased to 30 acres. The chapel is a pleasant example of the Greek Revival style, but in 1972 was standing derelict.

The Chapel is aligned on an E.–W. axis with a tetrastyle Ionic portico on the N. side. The entablature is continued all round the building. The E. and W. ends are each divided into three bays by Ionic half columns between pilasters under a pediment. On the S. is a small entrance porch over a flight of steps down to the basement. Windows on E., S. and W. sides have battered sides, and are enclosed by eared architraves.

The inside of the chapel is divided into seven bays in length and three in width by wooden pilasters, painted to simulate marble, above a boarded dado. Above an entablature with cornice, the ceiling is divided into 21 square compartments; two cast-iron ceiling roses served as ventilators and points of suspension for the former gas lights. The chapel was at one time divided into two parts, one for the Church of England and one for Nonconformists, but no trace of this division survives. Below the chapel is a brick-vaulted basement with compartments for burials.

The Lodge is a simple two-storeyed structure, faced with ashlar, and T-shaped on plan. The two ends of the main range are each divided into three bays by simple pilasters, and the entablature above is continued all round the building; the gable ends are treated as pediments. A single-storey addition on the N. side replaces an original tetrastyle portico (New Guide, 151) and the S. wing has been lengthened.

Cross, 50 yds. W. of chapel, of cast iron, stands about 8 ft. high above an octagonal stone pedestal. It was made by John Walker of Walmgate, probably in 1837.

Gates and Railings. The entrance, on the W. from Cemetery Road, has iron gates hung to stone piers with honeysuckle and Greek fret ornament; it is flanked by lengths of railing terminated by large stone piers, that at the N. end surmounted by a stone sarcophagus (Plate 81), that at the S. end by a sphinx.

Fig. 29. (9) St. Olave's. Wall monument.

SECULAR

(11) THE KING'S MANOR (Frontispiece, Plates 53–76) stands E. and S.E. of the abbey church of St. Mary; it represents the mediaeval abbot's house adapted and enlarged as the headquarters of the Council of the North, and now forms a part of the University of York.

The historical part of the text which follows is largely based on the account of the King's Manor by H. M. Colvin which will appear in volume IV of the *History of the King's Works*. The parts of the building are identified by letters A to K which are shown on the plan in the pocket at the end of the book. For the sake of clarity the buildings are described as though they lay on an E.–W. axis corresponding to the liturgical E.–W. of the abbey.

The Abbot's House. The first house on this site was almost certainly built *c.* 1270 for Simon de Warwick, Abbot of St. Mary's in 1258–96. Late 13th-century plinth mouldings, similar to those of the abbey church, still exist in enough places to suggest that the house was U-shaped and of the same extent as the later mediaeval rebuilding. The surviving 13th-century work is of white magnesian limestone ashlar and is confined to the lower storey; it does not appear on the E. elevation. It is likely that this represents a stone-built lower storey which had timber-framing above, as at St. Anthony's Hall, York, a building of the 15th century where the timber-framing was all replaced by brick in the 18th century, and in the Hospitium of the abbey. The quantity of reused heavy timbers to be found in parts of the flooring may be further evidence of such construction.

The house as it now exists is mostly a rebuilding of the late 15th century. The work, usually assigned to Abbot William Sever (1485–1502), was begun by his predecessor Abbot Thomas Boothe to whom a Crown licence was given on 30 September 1483 to retain Richard Che(r)yholme, bricklayer, and his four servants as long as it should please him (BM Harleian MS. 433, f. 119v). Cheryholme had been admitted to the freedom of the City of York as a 'tyler' in 1481/2 (SS, XCVI, 204); Cherryholme is a place near Drax, by Selby, and the centre of a brick-making district. The building work was continued by Abbot Sever until his translation to the bishopric of Durham in 1502 when he promised to pay workmen who were still owed money for work done during his abbacy (YCA, Memorandum Book B/Y, f. 173b). The walling of the new work was all of brick above a stone plinth. The windows were framed in terracotta under brick labels and relieving arches (Plate 59), probably the earliest use of terracotta in England. The late 15th-century house faced inwards to the three-sided courtyard open to the W. (Plate 57). Contemporary floors and roofs remain in the side ranges (Plate 63), the roofs being originally ceiled but without attics; the roof of the main range met the roof of the cross-wings with hips, not gables.

Dissolution. The Council in the North. The Abbey of St. Mary's, York, was formally surrendered on 29 November 1539 (*L&P Hy VIII*, XIV, ii, 213, no. 603) and on the 17 December 1539 the Council of the North wrote to Cromwell asking him to thank the king for allowing them to use the house 'which of late was called St. Mary Abby without the City of York' (*L&P Hy VIII*, XIV, ii, 258, no. 698). At first little would need to be done to the abbot's house except repairs.

The accounts of Leonard Beckwith, receiver of the revenues of the lands of the dissolved monasteries in Yorkshire, show that in 1539–40 £58. 3s. 9d. were spent 'on divers buildings within the site of the monastery of St. Mary's York now called the King's Manor, which are reserved for the King's councillors in the North . . . as appears by a book of particulars made by the Bishop of Llandaff and other members of the Council' (PRO, LR 6/121/2).

More extensive alterations were called for in 1541 when Henry VIII visited York in company with Queen Katherine Howard and lived here for twelve days (Drake, 128; Knight, 366). Beckwith's account for 1540–1 shows that £8. 10s. 2d. was spent on 'dighting and cleansing' the Manor before the King's arrival, but the main outlay was entrusted to Clement Throckmorton, an official of the Court of Augmentations. It was he who was responsible for 'repairing and beautifying' such houses 'as the king intends to resort unto in his progress', including York, Leconfield and Hull. For this purpose he received at least £833. 6s. 8d. of which £400 is specifically stated to have been for works at 'St. Mary's Manor' (PRO, *L&P Hy VIII*, XVII, 135–6; LR 6/121/2). Two windows with hollow-chamfered members and set in reused ashlar may be of this date.

A survey made shortly before the death of Henry VIII in 1547 shows that many of the abbey conventual buildings still remained; it describes the church as roofless and lists as 'uncovered' most of the former monastic buildings. The abbot's house is not mentioned by name but may be represented by a block of habitable rooms listed as 'the hall, the chapell, the great chamber, the chamber over the great chamber, two litill chambers, a privie kitchen, two chambers over it, all under one roof covered with leade' and measuring 34 yards long and 14 yards broad. It is difficult to reconcile this with the present house, but the main E. range is 103 ft. long inside and the width across this range and the porch is about 42 ft. Here three fothers of lead were needed to repair the roof. One of the few buildings to be listed 'in good state' was the gatehouse by St. Olave's church, together with a chamber in or adjoining it occupied by Robert Chaloner, a member of the Council. A range of timber-framed stables 60 feet long and 24 feet wide is noted as having been 'newe amended by my Lord President's commandment' (PRO, E 101/501/17).

It has always been supposed that Henry VIII ordered a palace to be built between the abbot's house and the river and that it was ruined a few years later (Davies, 4, 5). Foundations that could belong to this building have been found in excavation (Wellbeloved, *St. Mary's Abbey*, 9, 11, 14), but the architectural features of the outer W. range, K, usually supposed to be built at this time, suggest that *c.* 1600-20 is the correct date for it.

In November 1549 the Privy Council directed the under-treasurer of the Mint at York to issue £30 to be spent on repairs by the Earl of Shrewsbury (a member of the Council in the North) and authorised the expenditure of a further £30 'in cace it shall appere unto the said Erle that the said £30 is not sufficient' (PRO, *APC*, II, 363). The money was to be paid to John Harbert, appointed Keeper of the Manor in 1543 (*L&P Hy VIII*, XVIII, i, 546).

In 1550 the Augmentations prepared to pull down the S. aisle of the abbey church, the dorter and the refectory, and were barely prevented from laying their hands on the Manor itself. As it was, 'the chief Wyndowe of the Kings Ma^ts owne bed chamber was damaged' and 'such spoyle and defacing [was] made in divers parts of his highnes said palace, that hit wold greve any man to see it' (College of Arms, Talbot Papers B 216. Letter dated April 1551). A similar protest must have been addressed to the Privy Council, for on the 15 April 1551 that body gave directions for a letter to be written to 'Mr Chancellor of th'Augmentations that the Kinge's Pallaice at York be no further defaced' (PRO, *APC*, III, 261).

When in 1562 the possibility of a meeting between Queen Elizabeth and Mary Queen of Scots was being explored, the Lord President (the Earl of Rutland) told William Cecil that the palace was not fit for the purpose 'as it has been so defaced that only one large chamber remains'. Some new building had been begun 'as will appear by the plat thereof sent by the bearer' (*CSP Foreign 1562*, nos. 215 and 218), and between 1561 and 1563 the Earl of Rutland received £380 for the work (PRO, Receivers' Accounts Yorkshire, SC6 Elizabeth I/ 1740-1). This new building, E (Plate 67), continues the line of the main range of the mediaeval house to the N. and is the earliest surviving addition.

That new buildings were being added perhaps as early as *c.* 1560 is recorded elsewhere, for on the 8 September 1570 Thomas Radcliffe, 3rd Earl of Sussex (President 1568-72), wrote to the Council: 'By order of Her Majesty I caused £200 to be received of the receiver of Yorkshire for finishing buildings begun by the Earl of Rutland [Henry Manners, 2nd Earl, President 1560/1-1563] and the Archbishop of York [Archbishop Thomas Young, President 1563-8], when Presidents, at Her Majesty's house in York, and have caused the money to be so employed. Understanding from the Council that the £200 is disbursed, that money is still due to certain workmen, and that a gallery has yet to be tiled, and with two other chambers ceiled in the roofs,— which will finish the whole work begun, and unless done before winter, the vault which is over the gallery will be in danger of falling, as the rain has already begun to pierce it. [This vault was probably in plaster.] I beseech your warrant to Sir Thomas Gargrave, Receiver of Yorkshire, and others, to take account of the £200, pay the overplus due, see what remains to be finished and disburse the premises, either of the revenues of Yorkshire or of the fines of the rebels, which being done with small charge now, Her Majesty shall have a convenient house for her President, and sufficient for a lodging for herself, if occasion shall draw her into those parts. If it not be presently done, the cost already bestowed will be lost, the like will not hereafter be done without a far greater sum, and the house will remain unfit for any nobleman that holds that place. . . . £200 or £300 at most will pay all that due and finish the rest; £150 is already due' (*CSP Dom. Add. 1566-79*, 318, no. 6). On the 10 September 1570 the Earl of Sussex told Sir William Cecil that he had written to the Council (*ibid.*, 319, no. 8). Between November 1568 and April 1570 he had spent over £400 on the house (PRO, E 101/501/16) and had been authorised to take 100 oaks from the Forest of Galtres.

The works were completed between November 1568 and April 1570 with John Hilton as 'clerk surveyor' (PRO, E 101/501/16). In April 1569 Sussex pleaded in vain for a warrant for a further 100 oaks. The Queen's Highness he was told 'layeth away her owne Buildings, by reason of the grate charges', and would not be pleased if any but essential repairs were carried out at her house in York (letter from the Marquis of Winchester to the Earl of Sussex printed by L. Howard, *A Collection of Letters of Many Princes, great Personages and Statesmen* (1753), 216-17). Eventually the Council allowed a further outlay of £260 from the fines imposed on the rebels of 1569, and the work was completed, but Sussex was still out of pocket to the extent of £89 (PRO, E 101/501/16; SP 46/14, ff. 1180-1).

In addition to the new building the mediaeval house was largely refenestrated at this time, the characteristic window having hollow-chamfered brick mullions, transom and reveals, plastered to simulate stone. The windows which Radcliffe put in on the first floor facing the court have labels joined to form a continuous string but this does not reach to the ends of the side ranges (Plate 56). The end of the string coincides with the position of a wall which closed the W. side of the court, as is shown on a drawing by Place. This wall can hardly have reached up as high as the top of the upper windows but some projecting feature on this line must be presumed to have provided a termination for the string. The enclosing wall survived till 1822 (Hargrove, III, 578–9; A. Smith's plan in Baines' *Directory*, II (1823)) but it does not appear on the plan of the YPS excavations of 1827–8. The central part of the mediaeval house was probably altered when it was refenestrated, for there is plaster of this date in the present roof space above the ceiling. At this time also a porch was built in the N.E. corner of the courtyard with reuse of sections of 13th-century plinth (Plate 57).

Earl of Huntingdon. There is little record of building work in the time of Henry Hastings, Earl of Huntingdon (President 1572–95), but he formed a Council Chamber (Plate 64) on the first floor, partly in the W. end of the N. range of the mediaeval house (C on plan) and partly over the adjoining mediaeval kitchen (D) into which he inserted the present floor. A frieze in the Council Chamber (Plate 65) contains Huntingdon's own bull's head crest within a Garter (granted 1570) and the Bear and Ragged Staff badge for his wife, Catherine (Dudley) (GEC, *Complete Peerage*, VI (1926), 657). The bay window to the room is probably also of this period. The room had been divided by 1682 but it was opened up again later. The use of the upper part of the mediaeval kitchen to form part of the Council Chamber did not mean that the kitchen could not continue in use, for it still had a fireplace with a flue.

Huntingdon also erected a building along the S. precinct wall of St. Mary's Abbey, for on the 22 June 1580 it was 'Agreed that my L Psident shall have a license to set Butteries upon the moat without Bothome barr, in the occupacon of John Farley, nere the mannor, and he is to pay therfore yearlie . . .' (YCA House Book, B 27, 243) and on the 15 July 1580 it was agreed that the Lord President should have a lease for 100 years of 'a parcell of the moate of the Walles of this Cittie nighe unto Bothome barr, to set four pillowes upon, nighe adjoyninge to a newe building in the Mannor garth, and he to pay for the same yearlie xijd' (YCR VIII, YASRS CXIX (1953), 36). In the undercroft of the S. range (J) one jamb of a window, set in a wall which bears no

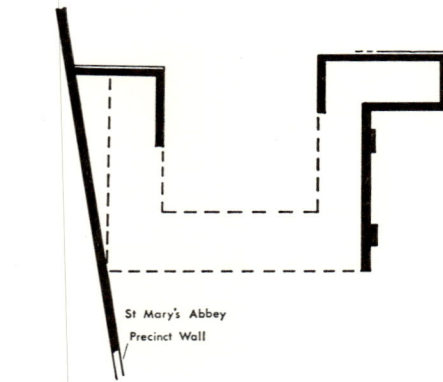

c. 1290

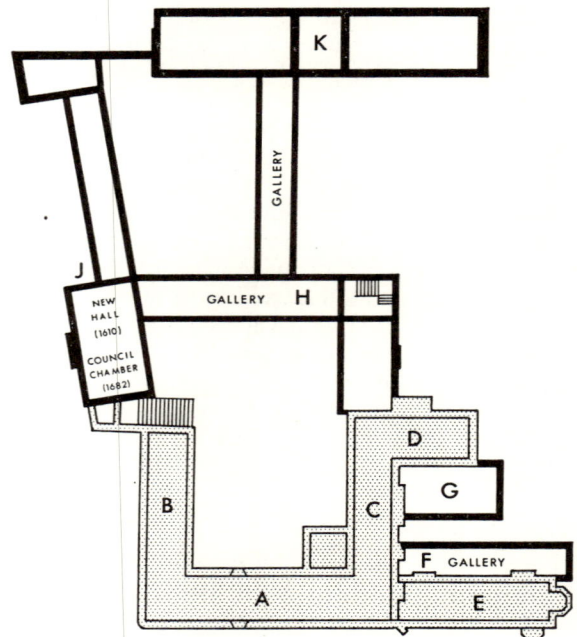

c. 1690

Fig. 30.

(11) The King's Manor. Plans showing development.

In each plan new work is shown in black outline, earlier work is stippled.

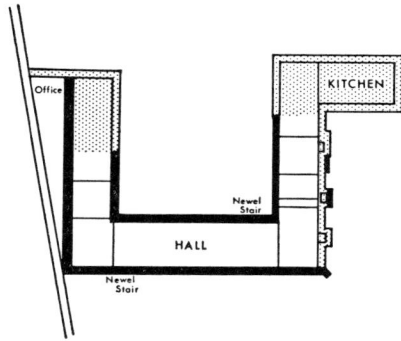

c. 1500

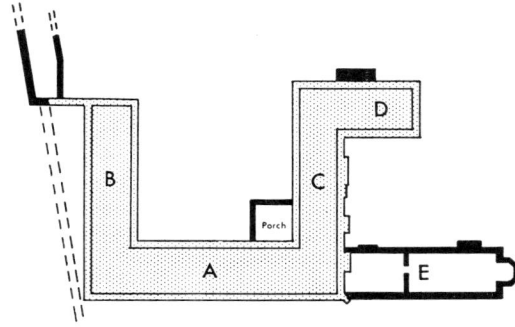

c. 1600

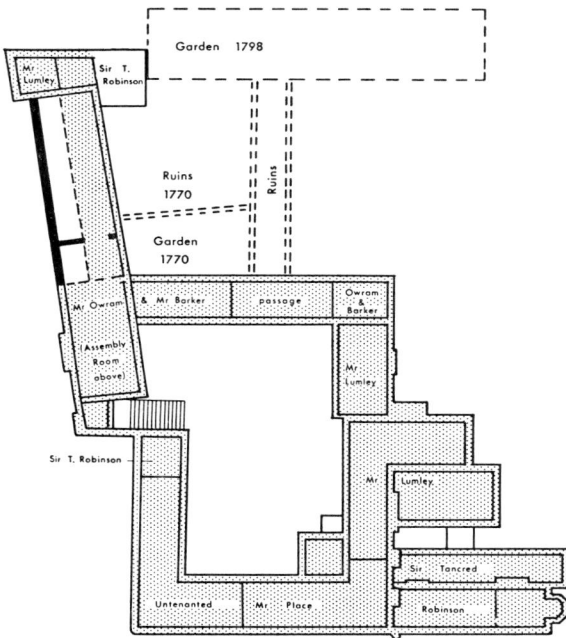

c. 1730

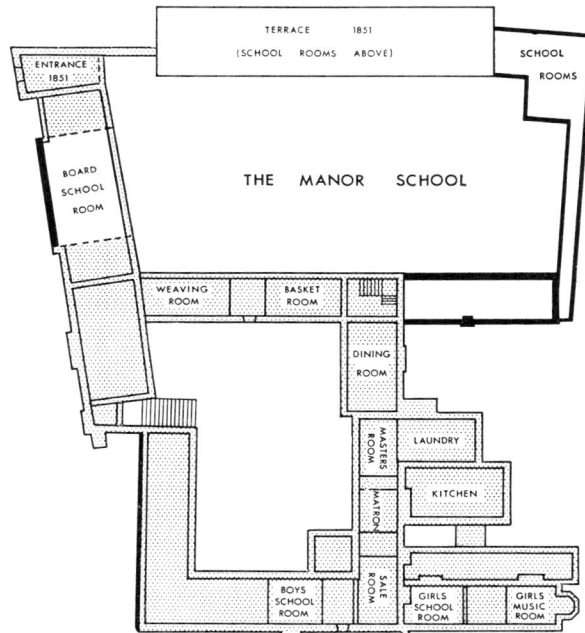

c. 1900

Fig. 31.

A. Hall block of abbot's house. *c.* 1483–1500, remodelled
    c. 1560–70 and later.
B. S. Cross-wing of abbot's house.
C. N.   „   „   „   „
D. Kitchen of abbot's house, remodelled *c.* 1575 to form the
    Huntingdon Room.

E. North Range. *c.* 1560–70.
F. Second North Range. *c.* 1610.
G. Central North Block. *c.* 1610, upper part late 17th-century.
H. Central Range. Lower part *c.* 1610, upper part 1633.
J. South Range. 16th-century and later.
K. West Range. *c.* 1610 with modern upper storeys.

relationship to the present building, is presumably a relic of Huntingdon's building. Speed's map of 1610 shows a range along the abbey wall. William Foster, Free of York in 1570/1 (SS, CII, 13) was described in 1577 as the Lord President's mason; he may have been in charge of Huntingdon's works.

Lord Sheffield. James I, on his first visit to York in 1603, is said to have ordered the house to be embellished (Drake, 574). On the 3 September 1609, Edmund, Lord Sheffield (President 1603–19), applied to the Exchequer for 500 marks a year for the repair of the King's Manor and Sheriff Hutton Castle (*CSP Dom.* 1603–1610, 534, no. 72, 541, no. 5), and the Lord Treasurer asked for an estimate. The estimate, duly presented on the 17 December 1609 (*ibid.*, 573, no. 71; Davies, 18–19), is important for it details the buildings already existing: in the following extracts an attempt is made to identify the various sites. Work is recommended to the great chamber, the dining and drawing chambers, the seven chambers above them and the passages to the chambers and half paces [the central and S. ranges of the mediaeval house (A, B) and, if this be so, it is important to note that that main rooms were on the ground floor]; the north galleries with four chambers at the east end and vaults and parlours under them [probably the N. range of 1570 (E)], the galleries next to the cloister, with four chambers at the E. end and five parlours beneath [probably the N. mediaeval range (C)]; the passages and stairs between the two galleries [not identified]; the old Hall kitchen and paistry, etc. six rooms [the W. end of the N. mediaeval range (C)]; the larders with three chambers over them, the granary, bakehouse and stables [probably removed]; the new kitchen and the building of a new Hall [the S. range (J)]; the parlours and chamber at the north end of the tennis court and the parlour and chambers next the garden [probably removed] (*cf. YAJ*, XXXVI, pt. 143 (1944), 374–8); and the gatehouse roofs and eight parlours and chambers [the main abbey gatehouse, which stood intact until 1705].

The total estimate amounted to £758. 19s. 4d. and on 28 June 1610 it was approved by Simon Basill, the Surveyor of the King's Works (*CSP Dom.* 1603–1610, 573, no. 71), but not until the 4 July 1616 did Sheffield receive a grant of £1000 towards the expenses (*CSP Dom.* 1611–1618, 379, no. 4). In 1616/17 Sheffield received £3500 for work (Receivers' Accounts Yorkshire). Craftsmen concerned were Thomas Brinsley, mason, George Wilson, free of York 1598, carpenter, Thomas Sell, bricklayer, and John Tayler, tyler.

Buildings erected for Lord Sheffield include three with a characteristic moulded plinth; these are the inner N. range (F) doubling the width of the range of 1570, the central N. block (G) (Plate 53) and the outer W. range (K). The new kitchen referred to in the estimate above was not newly built by Lord Sheffield but was so called to distinguish it from the old kitchen, and the tiles, laths, nails and lead required make it clear that it was to be reroofed. The building of a new hall must have been a more drastic reconstruction; it included £50 for stone and £50 under workmanship for rough work, windows and chimneys. It would appear to have been a remodelling of Lord Huntingdon's S. range but not extending so far W. as the present dining hall, according to a plan dated 1682 in William Salt Library, Stafford (Dartmouth MSS. D 1778/III/02). Sheffield was also responsible for linking his new hall to the original N. range by a single-storey gallery which now forms the lower story of the central range (H, Plate 69); the stone frieze now at first-floor level must originally have been at the base of the parapet and a similar frieze is seen in 17th-century drawings at the head of a lofty bow window on the W. side of Sheffield's outer W. range. It is probable that Lord Sheffield also provided the two elaborate carved stone doorcases, now both on the E. elevation but one of which was described by Hargrove (III, 578–9) as facing the courtyard. To Sheffield also can be attributed various fireplaces with jewel ornament (Plate 66). He remodelled the mediaeval house, altering the floor levels in the central block and improving the outside to present a pleasant façade to the garden for the visit of James I in 1617. There is only one mediaeval truss left, and plaster at the top of the walls within the roof space provides proof of Sheffield's alterations, which could only have been possible after the provision of a new hall.

Earl of Strafford. The last great building period in the King's Manor was during the time of Thomas Wentworth, Earl of Strafford (1628–41). On the 31 October 1628 at about 10 a.m. a fierce wind blew down seven chimney shafts on the roof of the King's Manor and the eldest son of the Vice-President, Sir Edward Osborne was killed (Benson, III, 11).

In 1633 a letter from Wentworth to the Earl of Carlisle who was staying at the King's Manor says 'the house you will find much amended since my coming to it and £1000 more to build a gallery and a chapel in that Place where you may perceive I intend it will make it very commodious' (W. Knowles (ed.), *The Earl of Strafford's Letters*, I (1739), 85). In 1634 £1712. 19s. 7d. was allowed out of distraint for knighthood in the five northern counties 'circa nova edificia de le Manner House' (PRO, E 101/668/9, m3). In 1637 the J.P.'s for the North Riding reported that their area had not contributed to the Manor repairs but had supplied timber and other materials (*CSP Dom. 1637*, 348, no. 7; *1638–9*, 99).

Charles I was in York at the King's Manor in 1633 and 1639. In a letter of August 1639 Wentworth wrote 'there is a Gloria Patri sung at St. Mary Abbey, so as the Pillars in the Kitchen now may hope to have the Honor to become the Pillars again of a Church as formerly they were' (*Strafford's Letters*, II, 381). From this it appears that Wentworth intended to remodel Sheffield's hall to form a first-floor chapel over the kitchen. At the E. end he added the external staircase and the doorway with the arms of Charles I over it (Plate 68). He put new windows in the hall itself and from the W. end of the hall he provided new access to the N. range by building a second storey upon Sheffield's gallery (H, Plate 69) across the S.W. side of the court. He inserted a new doorway with his own arms over it into Sheffield's gallery, which survives to form the central feature of the range (Plate 76).

Sir Henry Savile, who was Vice-President of the Council under Wentworth, was nephew to Henry Savile, Warden of Merton College Oxford, who introduced Halifax masons to Oxford. Possibly a similar introduction by the nephew might account for some West Riding vernacular features at the Manor.

The Civil War and Later 17th century. The Council of the North was abolished in 1641, after which there were no major additions made to the Manor until the 19th century. The place was in the forefront of action during the siege of York and at least one range, the outer one to the W. (K), was half demolished.

On the 18 June 1644 both the Earl of Manchester and Ferdinando, Lord Fairfax, wrote to the Committee of both Kingdoms at Derby House to say that the previous day the Parliamentarians had sprung a mine which blew up the corner tower of the abbey precinct wall on Bootham. Manchester's Major-General had attacked the manor house and captured 100 but he had been beaten out with the loss of 300 men (*CSP Dom. 1644*, 246).

The Manor was surrendered on the 16 July 1644 and on the 4 August 1645 Stephen Watson and two other aldermen were asked by the Committee of H.M. Revenue sitting at Westminster to make an Inventory of goods in it (*CSP Dom. 1645–1647*, 42). On the 26 October 1653 the Council of State ordered Colonel Beckwith to take care of the King's Manor which had been spoiled and wasted (*CSP Dom. 1653–4*, 217). On the 17 June 1656 the Council of State ordered that Colonel Robert Lilburne should receive £400 for repairs, and the Lord Mayor Thomas Dickenson and Ralph Rymer were to say what repairs were necessary (*CSP Dom. 1655–6*, 376, nos. 14, 15). By the 7 August

1656 Robert Lilburne's promotion to Major-General obviously required his services elsewhere, and the £400 was to be paid to Humphrey Harwood of York, who was to authorise Ralph Rymer to pay the workmen for repairs (*CSP Dom. 1656–7*, 64, no. 14). The King's Manor was thus put in order and kept habitable during the Commonwealth.

On 14 June 1662 Humphrey Harwood was still living in the Manor (*CSP Dom. Add. 1660–1685*, 69). Henry Darcy, who became keeper in May 1665, obtained a warrant for £400 for repairs in 1666 (PRO, SP 44/14, f. 59; T 48/32, f. 156v); his account survives in a damaged condition (PRO, E 101/529/5). It relates to work in 'the Presence Chamber, the Withdrawing Room, the Belcony Chamber, the King's Chamber, the Matted Chamber, the Wainscot Chamber and the Councell Chamber'.

In 1667 the Manor became the official residence of John, Lord Freschville, Governor of the City of York (Drake, 574). In 1671 and 1672 Lord Freschville was seeking money for repairs, not above £100, as he was living in part of the Manor and had 30 persons in his family (*CSP Dom. 1671–2*, 275). In 1675 the Treasury allowed £150 for repairs; old material could be used (*Cal. Treasury Books 1672–5*, 335, 840).

An important drawing by Jacob Richards (William Salt Library, Stafford, Dartmouth MSS. D 1778/III/02) gives the plan of the Manor in some detail and is associated with 'An Accompt of His Majesties Goods now Remaineing in the Manor-house at York taken by the Honourable Sir Christopher Musgrave Knight, Lieutenant General of His Majesties Ordnance ye 19th October 1682' (for more detail see RCHM, *York* II, The Defences). The plan shows that the hall built for Lord Sheffield in the S. range extended no further than the S.W. corner of the front court where it met the gallery (H); Wentworth proposed to make it into a chapel, but on this plan it is called The Councill Chamber. Access to it is from the S. end of range H and from the external staircase from the court. The previous Council Chamber (the Huntingdon Room) had been divided into two smaller rooms.

The outer W. range (K) is shown joined to the central range (H) by a gallery across the inner court, which has since been removed leaving scars on the central range (Plate 70). A plan by Archer of the same period shows also a second building connecting the central and W. ranges, along the S. side of the inner court, in continuation of the Council Chamber block. Of the outer W. range Richards says it 'never was finished within, but the roof was and covered with panntiles wch. were afterwards taken off by a certain Governour and sould for his maj^es. use but the money he kept for his owne, and left the timber to shift for it self. Under this house is a stately arched cellar the length of it' (Plate 73).

In the late 17th century when the Manor was the Governor's residence the hips at the ends of the mediaeval building were replaced by tumbled gables (Plate 55) and the central block on the N. side (G, Plate 53) was heightened, perhaps in 1682 when Sir John Reresby, Governor of York, spent about £200 on work at the Manor (J. Cartwright (ed.), *Memoirs of Sir John Reresby* (1875), 374, 378, 386–7). On the 27 October 1687 the Treasury asked John Fisher, the Deputy Surveyor General of Crown Lands, for details of the Manor so that it could be released to Francis Lawson, one of the King's chaplains, for 31 years. He was to pay 20 nobles a year to the housekeeper in place of the King. Fisher gave his report on the house and 13 acres of land and pointed out that Sir John Reresby regarded it as a perquisite of his Governorship (*Cal. Treasury Books*, VIII, 1685–9, iii, 1565/6).

On the 24 November the Treasury gave a warrant to the Clerk of the Pipe for the lease for 31 years to Francis Lawson (son of Sir John Lawson of Brough), one of His Majesty's chaplains, of the King's Manor with outhouses, stables, barns and 13 acres, now or late in the tenure of Sir John Reresby Bt., Governor of York (*Cal. Treasury Books*, VIII, 1685–9, iii, 1602). Lawson converted the Manor into a 'Popish School' and used the Hall (the Councill Chamber on Richards' plan) as a chapel.

When Father Lawson had departed, Ralph Rymer asked for the lease of the Manor at 10s. a year as granted to Father Lawson, and it was referred to William Harbord, Surveyor General of Crown Lands (*Cal. Treasury Books*, IX, 1689–92, i, 165). On the 8 June 1690 the lease was granted to Rymer by William Harbord by warrant, as Lawson had now fled the kingdom. However the house was ruinous and there is a reference to Robert Waller as housekeeper (*Cal. Treasury Books*, IX, 1689–92, ii, 711–12, 749; iii, 1416). Rymer held it until 1691/2 but on the 19 February 1691/2 the Treasury asked the Surveyor General to report as a preliminary to making over the lease to Robert Waller (*ibid.*, iv, 1502). The lease was granted to Waller on the 7 March 1691/2 (*ibid.*, 1532).

Benedict Horsley's map of 1694 shows that the back inner court was much more built up than it now is and it remained in this condition until *c.* 1800.

On the 18 May 1699 Jane Marritt, widow, petitioned for help as Robert Waller and his heirs had demolished that part of the King's Manor in which she lived (*Cal. Treasury Papers 1697–1701*, ii, no. LXI (35), 299); the demolition could only relate to the removal of the roof of the outer W. range (K).

18th century. On the 3 July 1718 Sir Tancred Robinson asked for an extension of the term granted to Robert Waller (*Cal. Treasury Books*, XXXII, 1718, ii, 434), and on the 17 July the lease was granted for 26 years from the 16 March 1722 (*ibid.*, 466). A drawing dated 19 June 1726 shows that Sir Tancred occupied the N. range (E) and its annexe (F) and also had some floor space elsewhere. Francis Place, the artist, lived in the N. part of the front range (A), Mr. Lumley and his school occupied the N. mediaeval wing (C) and (D) and adjacent buildings, and the remainder was tenanted by Mr. Owram and Mr. Barker (Leeds City Library, Newby Hall Records, N.N.2384 A/2). Ralph Thoresby brought his daughters to Mr. Lumley's boarding school in 1712.

Other drawings in the same series (A1, A3, A4) show that Sir Tancred was considering a project to develop the land to the S.W; the main designs comprised four large blocks, each consisting of conjoined symmetrically designed houses placed round a courtyard open to the river and with trees between it and the water. The design could be by the unknown architect who built similar blocks in Lendal and Micklegate. Sir Tancred Robinson Bart. was the second son of Sir William Robinson, for whom Colen Campbell built Baldersby, and Robinson's local architect there, William Etty, probably built the Red House, Duncombe Place, for Sir William in York.

Sir Tancred Robinson modernised the N. range (E) with hung-sash windows, panelling and fireplaces, and the staircase there is of his date. He probably formed the connection between ranges F and G, for he owned rooms on both sides of the yard; it was certainly there by 1770 (PRO, MPE 575).

The Banqueting Hall, which had been Lawson's chapel in the late 17th century, was 'by a strange reverse of circumstances converted into an Assembly Room and was also used by the High Sheriffs of the County, during the assizes and races, for the entertainment of their friends' (Hargrove, III, 580–1). The assemblies began in 1710 and were well attended in 1713 (VCH. York, 245). The drawing of 1726 shows the Hall enlarged to its present size.

A plan of June 1770 by R. Bewlay, Surveyor (PRO, MPE 575), shows most buildings as they are now, but most of those in the back courtyard were in ruins.

19th century. In the early 19th century the ornate door surround provided in the time of Lord Sheffield was still on the inside of the first court, the main doorway to the Banqueting Hall was blocked, and the gallery between it and the Council Chamber was occupied by Mr. Wolstenholme, carver and gilder (Hargrove, III, 580–1),

In March 1812 the York Diocesan Society and National School took over the S.W. part of the Manor (Hargrove, III, 580–1) on lease from the Crown lessee,

Lord de Grey. The Manor National School was opened in January 1813 and buildings on the S. side of the back court were partly reconstructed to form a typical school of this period. In 1818 there were 440 boys. Further accommodation was secured in 1835 and in the late 19th century school buildings in red brick with stone dressings were erected above the vault of the outer W. range (K) by J. B. & W. Atkinson. In 1922 the school was moved to Marygate (VCH, *York*, 449–50) and the Blind School, already in the other part of the Manor, occupied its premises. In 1851 the vault of the W. range had a garden terrace over it (OS).

In 1833 the Yorkshire School for the Blind was founded and in 1835 the lease of the King's Manor, less the National School, was acquired for them and the school remained there until 1956 (VCH, *York*, 459–60). For a long time a statue of Wilberforce, the founder of the school, stood in the entrance; the sculptor was Samuel Joseph, 1791–1850 (Gunnis, 222).

While the Blind School was in possession a new Headmaster's house was built in 1899, standing to the E. between the old manor buildings and the City Art Gallery, to the designs of Walter Brierley. It is in Jacobean style, treated with much sensibility. In 1958 the City of York acquired the King's Manor (VCH, *York*, 531). In 1963–4 it was restored, modernised and extended for the University of York by William Birch & Sons Ltd. of York under the direction of the architects Feilden & Mawson of Norwich in association with Robert Matthew, Johnson-Marshall & Partners.

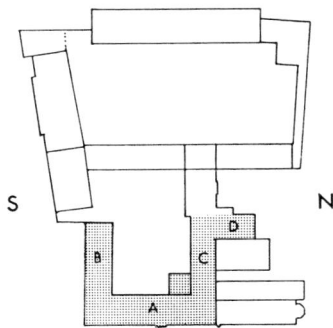

Fig. 32.

Architectural Description. The *Mediaeval House* has an E. elevation (Frontispiece, Plate 54), representing the E. side of the hall block, A, and the ends of two cross-wings, B, C; it is of *c.* 1480, but much altered. Above a stone plinth the walling is of brick with remains of a diaper pattern; it finishes at the N. end with a diagonal buttress. The positions of original windows of *c.* 1480 are indicated by the surviving brick relieving arches; the windows were constructed in terracotta (*cf.* Plate 59) of which there are remains at the extreme S. end in a small window which lit a garderobe and in the surviving fragments of small one and two-light windows by the S. chimney, which lit a staircase. Alterations to windows in the late 16th century can be identified by the remaining fragments of the plaster which was applied to the brickwork to simulate stone dressings. The upper S. window, of stone and of five transomed lights, may be as early as 1540 but other 16th-century work is probably of *c.* 1570. New work of the 17th century includes the whole of the S. chimney, the upper part of the N. chimney, the stone parapet and, *c.* 1670, the formation of gables at each end of the roof, replacing the former hips. The S. doorway, newly opened in the 17th century, has an elaborate stone surround bearing the initials IR for James I and, above, a large heraldic panel with initials CR for Charles I. Over the Royal Arms there was formerly a crown, and the niche above probably contained a bust. The N. doorway (Plate 74) is an original opening of *c.* 1480 but has a stone surround of *c.* 1610 brought from the W. elevation and reset; the S. side of it is partly restored. All the existing ground-floor windows are of the first half of the 19th century but later than Whittock's drawing of 1829. Some of the upper windows are of the 17th century; some were restored or modified in the 19th century. Two to N. of the central chimney are covered with Roman cement.

The S. elevation of the S. cross-wing, B (Plate 55), was very largely refaced *c.* 1900 in advance of the original wall face. One of the original terracotta windows of *c.* 1480 remains complete, under a brick arch similar to those on the E. front (Plate 59); it has three transomed lights with segmental heads. To the E. the gable of 1670 is finished with tumbled brickwork with two oval windows in the gable. The jambs of a 17th-century window, now blocked, remain below. All the other windows are modern. In the middle of the elevation is a small gabled projection, probably of early 18th-century date, in which is reset a fragment of a 17th-century carved stone frieze. In front of the W. part of the elevation the base of a fragment of the abbey precinct wall remains, standing only one course above the ground.

The W. end of the S. cross-wing is almost freestanding (Plate 68); the lower part is of limestone ashlar with a moulded 13th-century plinth *in situ* and contains a hollow-chamfered three-light window of the 17th century. The walling above is of brick with stone dressings rising to a gable which has been heightened, above a modern first-floor window of five transomed lights. The end wall, with the 13th-century plinth, continues S. outside the range, having previously formed the side of a small projection of which the other walls have been destroyed.

The main court was formerly divided into two parts by a wall joining the W. ends of the two wings of the mediaeval

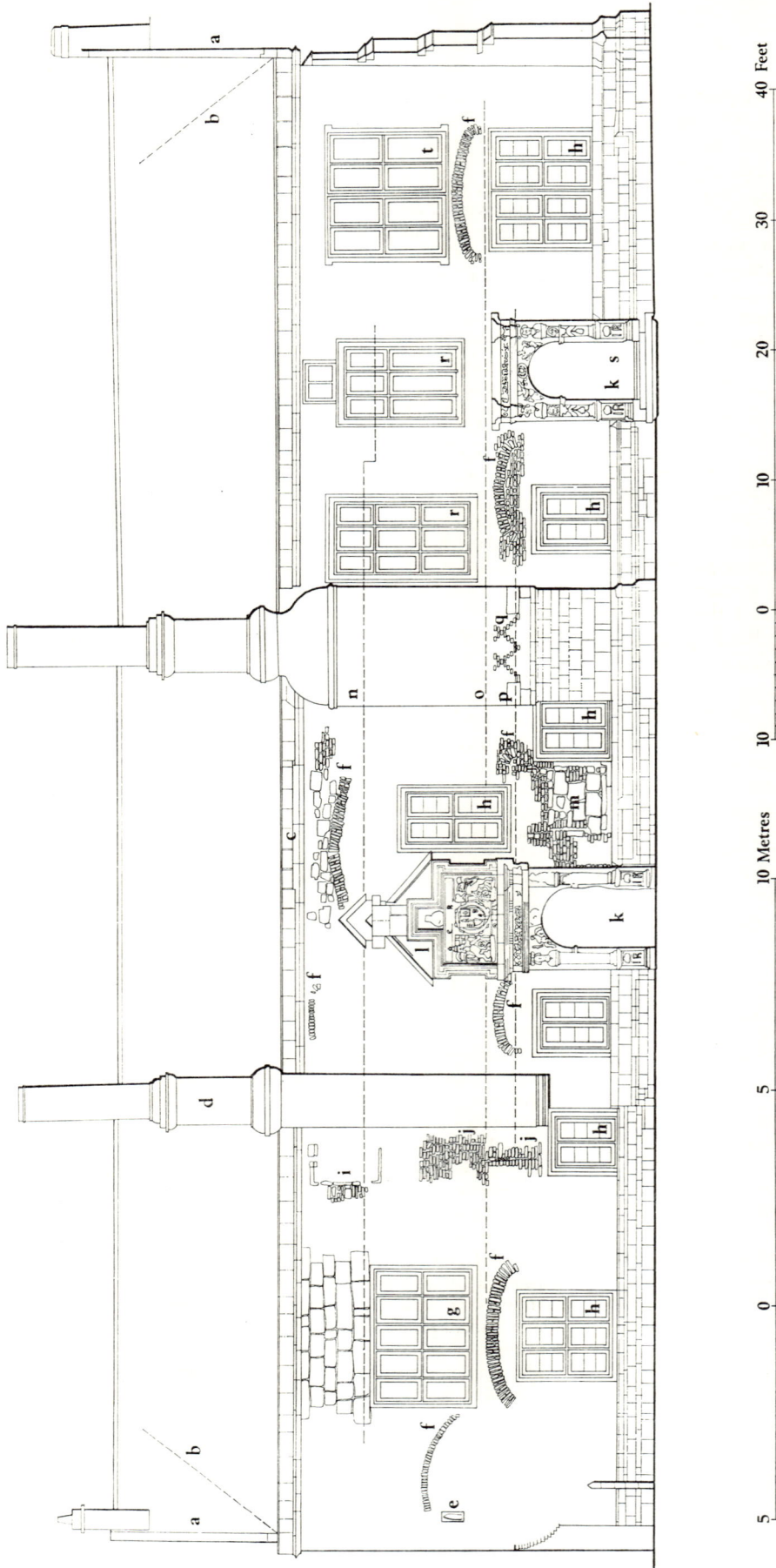

Fig. 33. (11) The King's Manor. E. Elevation.
From a photogrammetic survey by R. W. A. Dallas.

aa. 17th-century gables.
bb. line of 15th-century hips.
c. 17th-century parapet.
d. 17th-century chimney.
e. terracotta window of c. 1480.
ff. relieving arches over windows of c. 1480.
g. window of c. 1540 (?).

hh. 19th-century window.
i. 17th-century window, blocked.
jj. fragments of terracotta window.
k. doorway with initials of James I.
l. arms of Charles I.
m. window of 1570, replacing 15th-century window, now blocked.

n. level of inserted floor.
o. level of first floor c. 1580.
p. level of present first floor.
q. diaper pattern in brickwork of 15th-century chimney.
rr. windows restored, probably 17th-century.
s. entrance to former screens passage.
t. 17th-century window.

building. The E. part of the court, enclosed on three sides by buildings of 1480 formed a plain rectangle into which a gabled projection was built in the N.E. corner *c.* 1590, the lower part forming a porch in front of the screens passage as then existing (Plate 57). The walling of the mediaeval building is mainly of brick of *c.* 1480 but includes areas of stonework some of which, on the ground floor, may represent walling of the late 13th century but none has the moulded 13th-century plinth; stonework in the upper storey is reused from the 13th-century building. The lower part of the porch is of stone and on the S. side it has a 13th-century plinth but this is reset. Incomplete fragments of diaper pattern in the walls of the main range indicate that the brickwork is much disturbed and rebuilt. Few traces of mediaeval openings remain: on the W. side of the hall range are the arched heads of two first-floor openings which may have led to an external gallery; stone corbels for carrying the roof of such a gallery remain; in the N. wing on the ground floor are two reset brick niches with arched heads and on the first floor the second window from the E. is set under a 15th-century relieving arch (Plate 57).

The main doorway in the main range is quite modern and is set further N. than its predecessor. In the S. wing (Plate 56) the W. doorway is entirely modern, that near the middle of the range contains some old stonework and is flanked by a pair of windows which may be of the 16th century. In the N. range the doorway has a 17th-century head and is set in a small porch. The remaining ground-floor openings in the three principal ranges are all 19th-century or modern. In the projecting porch the openings are all of *c.* 1590, but one doorway is restored and one blocked. In the S. wing an upper window of five transomed lights set between patches of stonework may be of the mid 16th century. In the main range and the E. part of the S. wing a string-course over the upper windows is of *c.* 1570 and, by its changes in level, indicates the positions of windows at that time. They had brick jambs plastered to simulate stonework and this plaster can be seen flanking the second window from the E. end in the S. wing, which is restored but unaltered, and the N. window in the main range, where the opening is partly blocked by brickwork flanking an 18th-century window. The other upper windows were all renewed in stone in the 17th century. In the N. wing three two-light windows on the upper floor to the E. are all framed in 17th-century stonework but also retain fragments of plaster on the adjacent brickwork. The stonework of the central window appears to have been originally made for a doorway. A fourth window, further W., has four transomed lights.

The N. side of the N. wing, C, is now mostly masked by later buildings. During repair work in 1962 it was seen that this wall, of *c.* 1480, stands on the base of a 13th-century wall with its footings carried round two projecting chimneystacks corresponding to the 15th-century stacks still existing. Between the stacks is a garderobe projection built out beyond the face of the 13th-century footings, and a bay window, both part of the 15th-century building. The bay window, now partly masked, is of ashlar with four cinquefoil-headed lights at first-floor level (Plate 59). On the first floor and covered by later building is a blocked window framed in terracotta similar to that shown in Plate 59.

The mediaeval kitchen, D, projects N. from the N. wing

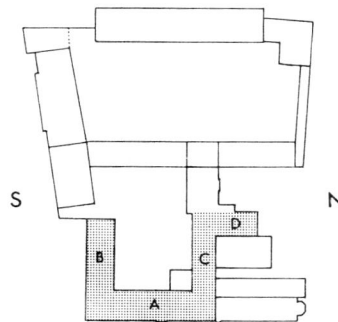

Fig. 34.

(Plate 58). The walls to the lower storey are of ashlar, partly with the remains of a 13th-century plinth, and above is 15th-century brickwork with stone quoins. In the N. wall a five-light window to the ground floor framed in ovolo-moulded stonework was inserted in the late 16th or early 17th century, replacing an earlier opening, and on the first floor a transomed window of *c.* 1610, also ovolo-moulded, cuts through plasterwork surviving from the jambs of a window of *c.* 1570. On the W. a ground-floor fireplace and its chimney have been removed, the upper part being made good in old brickwork, the lower part, now containing a doorway, in brickwork of the 19th century. Further S. is an ovolo-moulded three-light window of the late 16th century now partly converted to a doorway, flanked by masses of masonry which support a rectangular, gabled bay above with a window of four ovolo-moulded transomed lights in the main face and single-light windows in the sides, all probably of *c.* 1580 (Plate 58). The bay would then have formed a central feature in a free-standing W. elevation.

Interior of Mediaeval House. The mediaeval building was of two storeys but it has been partly converted to three storeys, the first floor of the hall range having been lowered and a second floor constructed in the 17th century. In the Hall Range, A, the present entrance hall was formed by *c.* 1610 and now contains a modern oak staircase with pine balusters probably including some of *c.* 1700 reused. At the N. end the original screens passage leads to the former porch. On the first floor there are remains of the timber-framed cross walls at each end of the range, from which it can be seen that the first floor has been lowered about 2½ ft. The present ceiling cuts across the heads of the 17th-century windows. In the W. wall are exposed the head and jambs of one of the blocked 15th-century openings visible outside (Plate 61). In the S. room is an elaborately decorated plaster ceiling of the early 17th century brought from No. 6 North Street and re-erected in 1963 (*York* III, Plate 186).

The top storey, extending partly into the roof, does not run the full length of the range; the four bays at the S. end are unused. At the N. end the upper part of the 15th-century timber-framed cross wall is exposed (Plate 60); it has a moulded head beam and plain vertical studs which at the W. end are stopped to allow for a former opening no doubt associated with a spiral staircase which came up at this point. The roof is carried on simple queen-strut trusses of late 17th-century date erected when the hall roof was extended across the wings to end in gables, replacing the mediaeval hips.

The *South Cross-wing*, B, of *c.* 1480, comprised four rooms on the ground floor. The E. room, largely refitted in the 19th century, has a deep recess in the S. wall representing a mediaeval garderobe. The fireplace has a surround of *c.* 1760 with rococo ornament enclosing an iron grate of *c.* 1820 in gothic style (Plate 66). The middle part of the wing now contains a library formed out of two rooms; the E. end has simple chamfered intersecting ceiling beams and the large W. part has moulded intersecting beams. To the W. is a timber-framed partition. At the W. end is a small room with a single chamfered beam. The whole of the upper floor now forms one large reading room, now open to the roof, but was originally subdivided, and the divisions were rearranged in the 16th century. In the S. wall is an opening to an original garderobe. There are three fireplaces: that to E. is of the late 15th century and has a segmental arched head in moulded brick (Plate 66); it was blocked in the course of 16th-century alterations when the other fireplaces were inserted but has been reopened; the middle fireplace has moulded jambs of reused stone and a flat lintel (Plate 66); the third fireplace, of *c.* 1540-50, has chamfered stone jambs and a segmental arched brick head. To the N. the timber-framing of the partition across the end of the hall range is exposed (Plate 62). The roof (Plate 63) is carried on eight simple king-post trusses with moulded tie-beams morticed for ceiling joists; most of the timbers are of *c.* 1480 but with some new members. The ridge-purlin has been removed.

The *North Cross-wing*, C, has on the ground floor a large room at the E. end with the 15th-century ceiling divided into square panels by moulded beams with mason's mitres at the intersections (Plate 60); a wider beam near the W. end originally had a partition under it. In the N. wall is a fireplace with a stone surround enriched with jewel ornament of *c.* 1610 (Plate 66). Next W. was a small compartment of one bay, with chamfered ceiling beams; it must have contained a staircase, and a doorway from it to W. led to a garderobe. In the next room to W. is a ceiling of two bays of intersecting chamfered beams but the room has been slightly enlarged. In the S. wall, between modern windows, is part of a 17th-century doorway (Plate 72). None of the partitions further W. are in original positions and the ceilings and ceiling beams in the W. part of the range are all of *c.* 1580. In the W. wall of the range is a doorway of *c.* 1610 to the added building beyond; it has moulded stone jambs and flattened four-centred head with sunk spandrels (Plate 61). The former kitchen (Plate 61), projecting to N., has a ceiling of *c.* 1580. In the E. wall is a large fireplace recess; the chamfered stone jamb to S. may be of late 13th-century date but the arched head is probably of the 15th century. The N. jamb has been reconstructed in modern times. A brick arch to S. may represent an oven. In the W. wall are recesses one of which was originally another fireplace; the chimney has been removed as described under the exterior above.

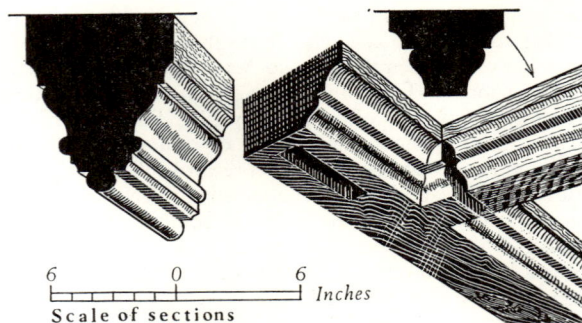

6   0   6
|—|—|—|—|—|———————| *Inches*
Scale of sections

Fig. 35. (11) The King's Manor. N. Cross-wing. Ceiling beams on ground floor. 15th-century.

On the first floor the E. room has the ceiling divided into square panels by moulded beams of the late 15th century. Towards the W. end of the room partitions have been removed and the smaller compartment formed by them must have contained a staircase giving access to garderobes projecting from the N. wall. The last bay and a half at the W. end also have the ceiling carried on late 15th-century beams but of different design and the wall-plates are embattled. The fireplace in the N. wall is adjacent to the position of a mediaeval window now blocked; further W. are the remains of the four-centred brick head of the entrance to a garderobe. In the S. wall remains of the timber-framing of the partition wall to the hall range are exposed, and further W. is the brick arched head of a former doorway to a spiral staircase.

The W. part of the range together with the upper part of the kitchen wing, D, together form the Huntingdon Room (Plate 64), L-shaped on plan with a large bay opening out in the middle of the W. side. Around the walls of the room is a plaster frieze containing three motifs (Plate 65): a *pomegranate between two wyverns,* the crest of Hastings, *a bull's head erased gorged with a ducal coronet* between two H's all within a garter under an earl's coronet, for Henry Hastings, Earl of Huntingdon, President of the Council of the North 1572-95, created Knight of the Garter 1570, and a *bear and ragged staff* for Catherine his wife, daughter of John Dudley, Duke of Northumberland. In the E. wall of the N. part of the room is a magnificent fireplace with segmental head formed of carved stone voussoirs and ornamented pilaster jambs (Plate 65). The E. end of the S. part of the room is lined with early 17th-century panelling perhaps reused; a recess E. of the fireplace represents a mediaeval bay window. The ceiling in this part of the room shows the continuation of the beam system in the adjoining room to the E. On the window glass are various schoolgirls' scratchings going back to 1745.

The roof is carried on king-post trusses; towards the E. end the last king-post has an elaborate head cut to receive the timbers of the hall range and the cross-wing, meeting at right angles (Plate 63). The later 17th-century roof covers this junction at a higher level.

Inside the porch projection between the hall range and the N. cross-wing the two main floors and the semi-attic have been

PLATE 45

(7) PARISH CHURCH OF ST. LAWRENCE. W. tower. Late 12th-century and later.

(9) PARISH CHURCH OF ST. OLAVE. W. tower. 15th-century and later.

PLATE 46

(23) INGRAM'S HOSPITAL, BOOTHAM. Archway from Holy Trinity Micklegate, reset. 12th-century.

(7) PARISH CHURCH OF ST. LAWRENCE. Doorway, reset in tower. 12th-century.

PLATE 47

Carved voussoirs.

Carved capital.

(7) PARISH CHURCH OF ST. LAWRENCE. Reset doorway. 12th-century.

PLATE 48

(7) PARISH CHURCH OF ST. LAWRENCE. Old church from S.E., 1776. © YAYAS. 12th-century and later.

(8) PARISH CHURCH OF ST. MAURICE. Window from old church, now in Yorkshire Museum. Late 12th-century.

PLATE 49

(8) PARISH CHURCH OF ST. MAURICE. Arch from old church, re-erected in church of St. James the Deacon, Acomb Moor. 12th-century.

PLATE 50

Exterior from N.

Interior from E.

(9) PARISH CHURCH OF ST. OLAVE. 15th and 18th-century.

PLATE 51

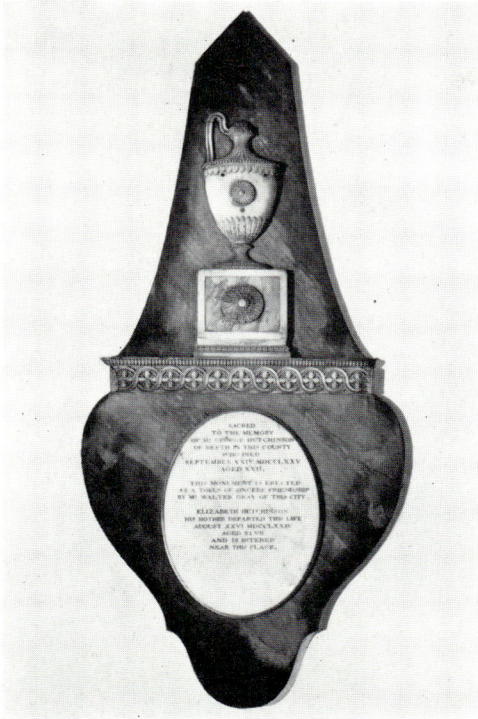

Monument to George Hutchinson. 1775.

Monument to William Thornton. 1721.

Monument to Alathea Jordan. 1741.

Royal Arms of Henry, Prince of Wales, eldest son of James I. 1610–12.

(9) PARISH CHURCH OF ST. OLAVE.

PLATE 52

(9) PARISH CHURCH OF ST. OLAVE. Panels from a stone screen, carved with angels. 14th-century.

PLATE 53 (11) THE KING'S MANOR. From N.W. 16th and 17th-century.

PLATE 54

(11) THE KING'S MANOR. E. front. c. 1480 and later.

PLATE 55

(11) THE KING'S MANOR. S. side of mediaeval house. c. 1480 and later.

PLATE 56

(11) THE KING'S MANOR. S. cross-wing, N. side. c. 1480 and later.

PLATE 57

(11) THE KING'S MANOR. Hall range and N. cross-wing from S.W. c. 1480 and later.

PLATE 58

(11) THE KING'S MANOR. Mediaeval kitchen and bay window to Huntingdon Room. c. 1480 and c. 1580.

PLATE 59

N. cross-wing. Upper part of bay window. c. 1480.

S. cross-wing. Terracotta window. c. 1480.

(11) THE KING'S MANOR.

PLATE 60

N. cross-wing. E. room on ground floor, ceiling. *c.* 1480.

Hall range. Second floor, raming at N. end. *c.* 1480.

(11) THE KING'S MANOR.

PLATE 61

Mediaeval kitchen from S.E. *c.* 1480 and later.

Hall range. First floor, blocked opening in W. wall. *c.* 1480.

N. cross-wing. Ground floor, doorway in W. wall. *c.* 1610.

(11) THE KING'S MANOR.

PLATE 62

S. cross-wing. First floor, framing at end of Hall range. *c*. 1480.

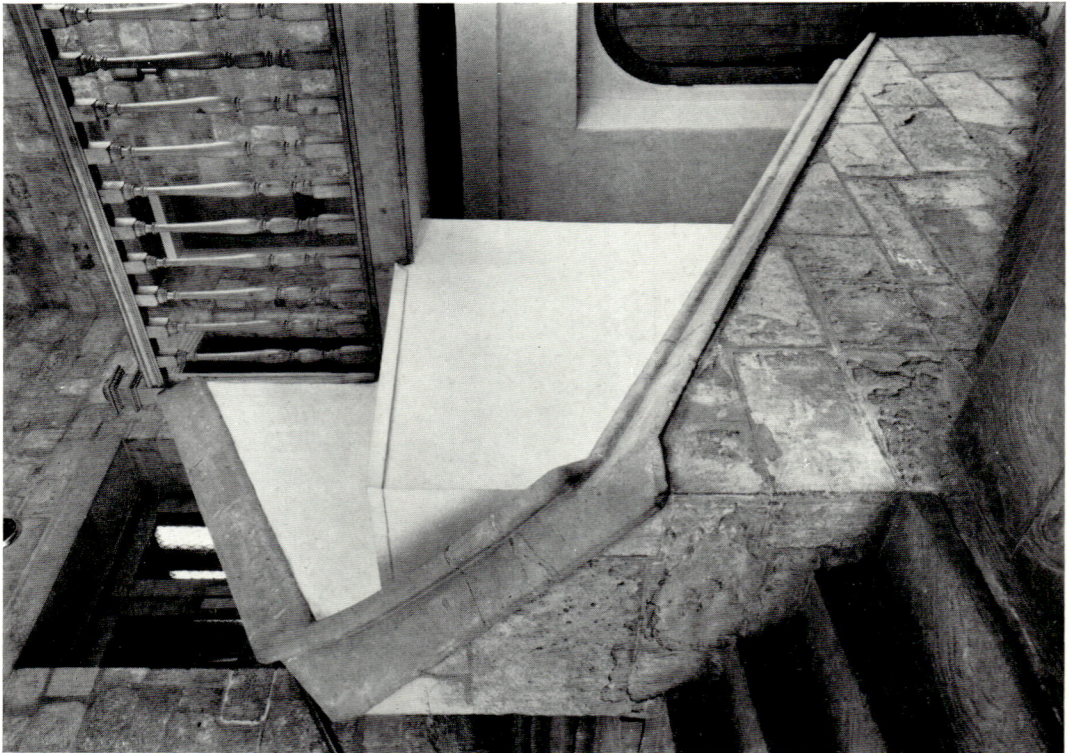

Central range. Staircase at N. end. Early 17th-century.

(11) THE KING'S MANOR.

PLATE 63

N. cross-wing. Roof, with king-post at junction with Hall roof. *c.* 1480.

S. cross-wing. Roof from E. *c.* 1480.

(11) THE KING'S MANOR.

PLATE 64

From S.W.

From W.

(11) THE KING'S MANOR. Huntingdon Room. *c.* 1580.

PLATE 65

Plaster frieze.

Detail of fireplace.

Fireplace.

11) THE KING'S MANOR. Huntingdon Room. *c.* 1580.

PLATE 66

N. cross-wing. Ground floor. c. 1610.

S. cross-wing. Ground floor. c. 1760.

S. cross-wing. First floor. c. 1480.

S. cross-wing. First floor. 17th-century, reset.

(11) THE KING'S MANOR. Fireplaces

PLATE 67

(11) THE KING'S MANOR. N. range. 1560–70.

PLATE 68

(11) THE KING'S MANOR. S.W. corner of E. courtyard. Early 17th-century.

PLATE 69

(11) THE KING'S MANOR. Central range from S.E. c. 1610 and c. 1635.

PLATE 70

W. courtyard from N.W. Early 17th-century and later.

Central range. W. side. Early 17th-century.

(11) THE KING'S MANOR.

PLATE 71

(11) THE KING'S MANOR. S. range from S.E. *c.* 1610 and later.

PLATE 72

N. cross-wing. Interior of room on ground floor. *c.* 1580 and later.

N. range. Stair hall. 16th and 18th-century.

(11) THE KING'S MANOR.

PLATE 73

Ante-room to Huntingdon Room from W. Early 17th-century.

W. range. Vaulted lower storey. *c.* 1610.

(11) THE KING'S MANOR.

PLATE 74

W. range. Doorway in lower storey. 14th-century, reset.

Central range. Doorway to stair hall. Early 17th-century.

Central range. Doorway from ante-room to Huntingdon Room. Early 17th-century.

E. front. N. doorway. Early 17th-century.

(11) THE KING'S MANOR. Doorways.

PLATE 75

Central range. Doorway from ante-room to Huntingdon Room.

Central range. Doorway to stair hall.

(11) THE KING'S MANOR. Heads of doorways. Early 17th-century

PLATE 76

11) THE KING'S MANOR. Central range. Doorway to E. courtyard. *c.* 1635.

converted to offices; the room on the first floor has an original moulded timber lintel over the window opening; in the top room the tie-beam and collar-beam of a roof truss are exposed, together with the undersides of the rafters. The room is lit on the W. by a modern dormer window below which are the remains of a window, now blocked.

The *North Range*, E (Plate 67). The line of the main hall range of the mediaeval house is continued to N., beyond the cross-wing, by a range which forms the first post-dissolution addition. It is of 16th-century date; the S. end was probably started by the Earl of Rutland in 1560 and continued by Archbishop Young after 1563, and the N. part was probably erected c. 1570 for Thomas Radcliffe, Earl of Sussex. It is of two storeys built in reused ashlar, presumably from the abbey, and brick. Stone-built dormer windows were added at the end of the 16th century. The E. wall is without plinth or string-courses, reflecting the fact that in the 16th century this was still the back of the house. On the ground floor towards the S. two original windows with later mullions flank a blocked doorway. The other windows to the main storeys are all of the early 18th century and have hung sashes; they are generally rather narrower than the 16th-century windows they replace, but there are remains of an original window, now blocked, on the upper floor. On the N. end is a large semi-octagonal bay window of two storeys (Plate 53), projecting from a stone wall with chamfered plinth and moulded strings between the storeys, and with a stuccoed gable. On the ground floor the side lights of the bay have been blocked and the mullions have been taken from other lights to allow for the insertion of hung-sash windows. In the upper storey the windows are unaltered. The W. elevation of the range is largely masked by later building; it had two projecting chimneys. Inside, in the S. room is the projection of a mediaeval chimney of the adjoining cross-wing. In the W. wall is an original fireplace, now blocked. In the middle of the range is a spacious staircase of c. 1725 with turned newels and balusters (Plate 72). An original window of c. 1570 remains in the W. wall under the stairs. The N. room, running into the bay window, is lined with 18th-century panelling and panelled pilasters flank the fireplace and overmantel. The fireplace is framed by a simple mid 18th-century moulding with a later mantelshelf above. Behind the panelling the moulded jamb of a blocked window is made up with moulded stonework of c. 1300.

On the first floor the S. room is lined with early 17th-century panelling. Behind the panelling the window openings have been narrowed by the addition of brickwork within the original stone splays. The N. wall is built largely of reused 13th-century stone and retains above the head of the present doorway the depressed four-centred head of an original doorway. The fireplace has a decorated surround of c. 1800. The room N. of the staircase is lined with 18th-century panelling. In the N. end room is an original fireplace with chamfered four-centred head and chamfered jambs. The original roof remains; it is of seven bays with simple tie-beam trusses and lap-jointed purlins passing through the principals.

The *Second North Range*, F, was added along the W. side of the main North range, probably c. 1610 (Plate 53). It is of two storeys with original attics. The walls are of stone with a moulded plinth and on the W. side rise to small gables over

8

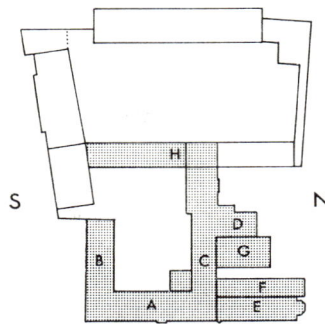

Fig. 36.

dormer windows. The W. wall abuts against a brick-built garderobe of the mediaeval N. cross-wing. The windows have ovolo-moulded stone frames and mullions.

Recesses in the S.W. corner of the S. rooms represent the garderobes of the adjoining mediaeval range. Brick vaults over the garderobes were removed in 1962, two storeys of the garderobes corresponding to the first floor of the building of 1610. A 17th-century staircase here has been taken away. Between the rooms on the first floor are doorways with moulded stone jambs and four-centred heads. In the attics is a stone fireplace with chamfered four-centred head. The roof is carried on collar-beam trusses with lapped purlins passing through the principals.

The *Central North Block*, G (Plate 53), built in the angle between the mediaeval N. cross-wing and its kitchen, is a structure of three main storeys, of which the lower part, built in coursed masonry, is of c. 1610. The upper storeys, mainly of brick, are of the late 17th century. The upper part of the N. elevation is divided into three bays by brick pilasters; above are two brick gables, rebuilt but following the late 17th-century design. The windows all have ovolo-moulded or chamfered stone frames and mullions; some are early 17th-century and some are of the second half of the century. On the N. front the central window on the first floor is set under a segmental arch. The rooms inside have been modernised but at the S. end are some remains of the mediaeval bay window; modern attics have been formed in the roof.

The *Central Range*, H (Plate 69), separates the E. and W. courtyards and at the N. end returns to meet the W. end of the mediaeval N. cross-wing (Plate 70). It was originally designed c. 1610 as a single-storey gallery with a two-storey return, to give access from Lord Sheffield's new hall to the Council Chamber devised by Lord Huntingdon and now known as the Huntingdon Room. The frieze between the storeys which forms a marked feature of the E. elevation was originally surmounted only by a stone parapet except in the return at the N. end which was always of two storeys. An upper storey was added to the range by Lord Wentworth in 1633 and an elaborate new doorway surmounted by a shield-of-arms of

Wentworth is of the same date (Plate 76). The W. courtyard was formerly subdivided, as shown by the broken lines on the plan, and another building abutted against the W. side of the central range (Plate 70).

The E. front of the central range has original windows of c. 1610 and 1633, all with ovolo-moulded members. The elaborately decorated doorway contemporary with the first-floor walling is surmounted by a panel with achievement of arms of Wentworth quartering Woodhouse, Hooton, Neville and Newmarch. Other doorways, where original, have depressed four-centred arched heads. On the W. side a gable towards the N. represents the end of Sheffield's two-storeyed return and the ashlar facing continues for one bay further to the S.; part of the walling is very disturbed and openings are blocked where the building projecting into the second court-yard has been taken away (Plate 71).

Inside, the ground-floor gallery has been subdivided. At the N. end an enriched semicircular-headed doorway (Plates 74, 75) gives access to a stairhall. The staircase is of stone (Plate 62) and was constructed c. 1633 blocking a large window in the N. wall. In the same wall is a 12th-century stone carved with pelleted interlace. A blocked doorway of Lord Sheffield's date in the upper part of the stairhall indicates a change in floor levels. The upper gallery is subdivided so that the S. end is now open to the dining hall. A blocked archway in the W. wall indicates a former projection removed; an offset in the walls indicates the top of the original parapets. From the upper part of the stairhall a modern doorway leads to the ante-room to the Huntingdon Room. At its side the original doorway, now blocked, has a moulded depressed four-centred head. The ante-room has a fireplace enriched with jewel ornament similar to that on the ground floor further N.E., shown in Plate 66. At the E. end a grand doorway to the former Council Chamber, the Huntingdon Room, has a moulded four-centred head under a stone entablature with a pediment flanked by obelisks (Plates 74, 75).

The *South Range*, J (Plates 68, 70, 71), forms the S. side of part of the E. court and of the whole of the W. court, with the middle range, which divides the courts, abutting it. It is of late 16th-century origin, but has been largely rebuilt in the 17th, 18th, 19th and 20th centuries. It comprises a first-floor porch to the hall at the E. end approached by an external stairway of the 19th century, a first-floor hall and its kitchen below built for Lord Sheffield c. 1610 replacing a building of Lord Huntingdon, an extension of the hall probably made at the beginning of the 18th century for the Assembly Room, a 19th-century Board School, and, at the W. end, a gatehouse. The N.E. corner of the porch is joined to the S.W. corner of the mediaeval house by a short length of wall with a 13th-century plinth described above.

The porch block, of three storeys, is built mainly of ashlar with some 17th-century brick in the upper part and a modern timber-framed gable to the E. A large buttress added against the E. end is near but not on the line of the abbey precinct wall. Windows are of the 17th century and modern but there is evidence of disturbance and rearrangement; a blocked window in the E. wall is represented externally by a concrete lintel but on the inside a two-light timber frame of the 18th century remains exposed. The first floor of the block, which forms an

entrance lobby to the hall, is approached from the court by a modern external stone stairway leading to a round-headed doorway flanked by tapered pilasters enriched with strap-work with entablature above and, in a moulded frame with a pediment, an achievement of Stuart royal arms, all very weathered. It is probably of c. 1635, erected by Wentworth (Plate 68).

Lord Sheffield's hall block is of two storeys, the upper storey being much taller than the lower, and now contains a dining hall on the upper floor and kitchens below. It is built of reused limestone ashlar and on the S. side it appears as five bays, but the W. bay, separated from the rest by a straight joint in the masonry, is an extension probably made at the beginning of the 18th century. The windows on the S. side are all modern on the ground floor; on the first floor are transomed windows almost completely restored but retaining a few pieces of old stone in the jambs. The elevation to the E. court has on the ground floor two large semicircular arches closed by modern windows and walling. Above are two windows of four transomed lights. Facing the W. court the extension of the hall block has a small chamfered plinth which continues W. to the gatehouse. The windows to this bay are modern and at mezzanine level are traces of a former opening now blocked.

Next W. of the hall block is a lower two-storey building reconstructed as a Board School in c. 1820–30. The N. wall is continuous with that of the hall extension with a continuous plinth as already described, but the S. wall sets forward and has a plinth finished to a very different section. The upper part of the S. wall is all 19th-century work and in the ground floor are modern windows and a modern doorway; straight joints suggest the position of two earlier doorways. Facing the court, the lower part of the N. wall shows evidence of disturbance; it contains a blocked doorway and three windows of the 17th century but with modern heads, perhaps shortened. The whole of the upper part is of the 19th century with tall mullioned and transomed windows rising into small gables projecting above the eaves.

The Board School building is continued to W. by a narrower building with the N. wall continuing in the same line as before but with the S. wall recessed. On the N. side there is a slight change in the character of the masonry in the upper floor but no clear joint. The windows are much restored and modern. At the W. end the plinth returns into the gatehall, the jamb of the gate being formed by what was a free-standing corner of the building. The gatehouse projects boldly on the S. side and the S. gateway has been formed in place of a 17th-century window the head of which remains above the arch of the gateway. Both the arches to the gatehall rise above the level of the first floor inside and low windows have been introduced in the head of the N. archway. Above the S. arch is a restored 17th-century window. The W. jamb of the N. archway is built up against a fragment of a lofty 16th-century building in which there is the moulded jamb of a gateway, with hinge-pins remaining, and the splay of a first-floor window.

Interior. In the modern kitchen under the hall, the walls are faced with stone, mostly mediaeval and reused. The E. end is divided by a length of late 16th-century walling which finishes at the W. end with an ovolo-moulded window jamb. This suggests that originally these were rooms some 12 ft. wide,

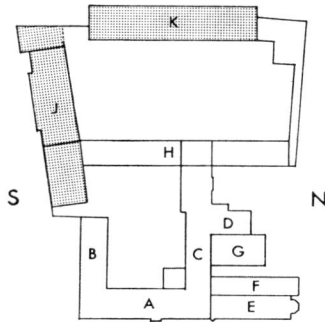

Fig. 37.

with an open loggia formed in front of them when the hall above was built. In the S. wall, between the windows, is a big 17th-century fireplace with a four-centred arched head. In the N. wall a modern doorway to the central range is a reconstruction of an older opening. On the first floor the hall now extends into the W. bay, and also includes a section of the central range. In the S. wall of the Hall a large fireplace has been restored with a modern arched head. In the middle of the hall the ceiling is carried up into an octagonal cupola within the roof, of early 18th-century date.

The lower floor of the Board School building is approached from the kitchens by a doorway with four-centred head which has a blocked window beside it, so placed as to suggest that this was the E. end wall of a building antedating the extension of the hall block. A corresponding doorway in the E. wall of the gatehouse is of different design but also with a four-centred head. It is of 16th-century date, probably reset. In the W. wall

of the gatehouse are original windows antedating the conversion to a gatehouse, now blocked. The rest of the range has been thoroughly modernised.

The outer *West Range*, K, stands in part on the site of the chapter house of St. Mary's Abbey; a fragment of rough walling near the N. end of the E. front may represent the E. wall of the chapter house. The range was erected for Lord Sheffield *c.* 1610; it included a vaulted basement, which still survives (Plate 73), containing a buttery and cellar, and lofty rooms above, lit on the W. side by semicircular bay windows. The upper storey was damaged in the siege of 1644 and removed in the 18th century. A drawing by F. Place (*c.* 1680) shows only part of the W. wall standing with a big bay window surmounted by an entablature in which the frieze matches that on the E. side of the central range. Parts of a similar entablature from a second bay window are incorporated into the walling of St. Mary's Tower as rebuilt after the siege. School buildings were erected over the basement in the late 19th century and still remain over the projection at the N. end. Over the rest of the basement new buildings for the University of York were erected in 1962 to the designs of B. M. Feilden.

The W. side of the building is completely masked by the Yorkshire Museum. On the E. side the stone wall of the 17th-century building stands some 7 ft. above ground level. It is of reused mediaeval masonry and has windows and a doorway in the plinth, the moulded capping of the plinth being stepped up over the doorway; the windows are mostly original and of two lights with ovolo-moulded stone frames and mullions. There is a similar window now blocked in the S. end. When the top of the vault was exposed during the building work of 1962 it was seen to include much reused material of the late 12th century; some of the carved voussoirs illustrated in Plates 38, 39 were found here. Inside, the basement is divided to give one long room each side of a central square compartment. This central part is covered by a brick vault but the two long rooms have segmental barrel vaults of good ashlar. A doorway at the S. end of the N. room is of the early 14th century and is reused from the abbey cloisters (Plate 74).

Fig. 38. (11) The King's Manor. Decoration from fireplace of *c.* 1760.

(12) THE YORKSHIRE MUSEUM (Plate 83) was built in 1827–9 for the Yorkshire Philosophical Society to the designs of William Wilkins R.A. with interior details by R. H. Sharp and J. P. Pritchett (VCH, *York*, 535; J. Mayhill, *Annals of Yorkshire*, 1 (1878), 332). It covers the site of parts of the E. and S. buildings round the cloister of St. Mary's Abbey. The Tempest Anderson Hall was added on the N.W. side of the Museum in 1912, covering the remains of the vestibule to the Chapter House which are preserved *in situ* in a basement below.

The Museum is in the Greek Doric style with a portico of four fluted columns in the middle of the S.W. front. The entablature returns along the front and sides and there are pilasters at the corners. The plan comprises a central space designed as a lecture room with balconies on three sides, and four ranges of display galleries. On the S.W. side the gallery is raised to the first floor over the entrance hall, curator's room, and library; on the other three sides the galleries are on the ground floor with small balconies forming an upper level. All the galleries and the central space are top lit, the roof to the latter being supported by six Corinthian columns, and beams enriched with guilloche on the soffit. Basements extend under the whole building, which is faced throughout with ashlar.

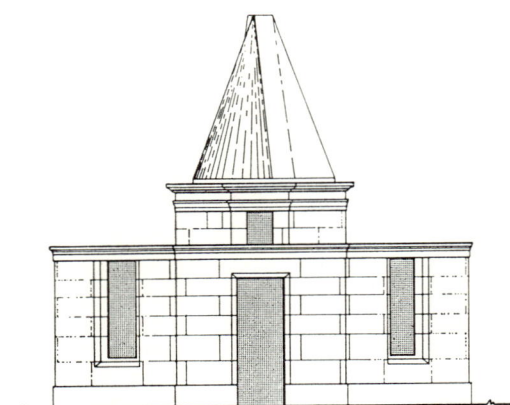

North Elevation

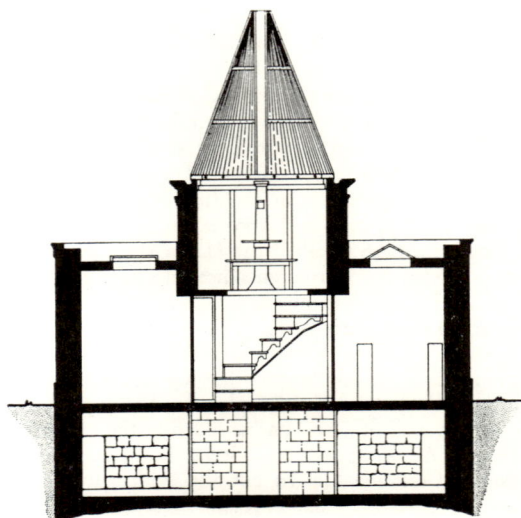

Section A-A

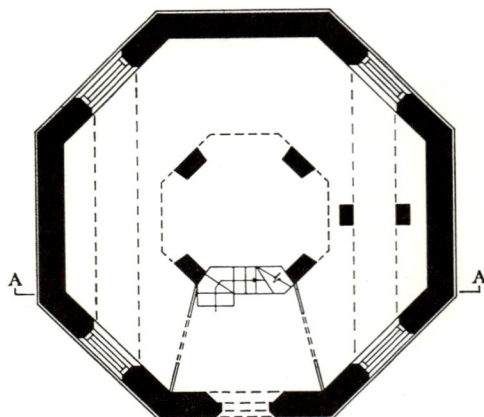

Ground Plan

N

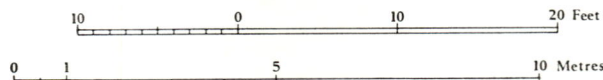

Fig. 39. (12) The Yorkshire Museum. The Observatory.

The Curator's House in the Museum Gardens, York.

Fig. 40.

An early drawing of the museum, unsigned and undated, is shown in Plate 82 (YM, Mh f. 82v). It differs in plan from the actual building in the partitions in the front range and in the columns in the original lecture room.

Mediaeval sculpture in the Museum is described on pp. xliv–xlvii and 22–24.

Curator's House, immediately S.E. of the Museum, was designed by J. B. and W. Atkinson in 1844, the original drawings being preserved by Messrs. Brierley, Leckenby and Keighley (Fig. 40). It is in Tudor style and has been little altered. The walls are of magnesian limestone mostly reused from St. Mary's Abbey.

Observatory, in the Museum Gardens, was built in 1832–3 to house the instruments offered to the YPS by Dr. Pearson, rector of South Kilworth, Leics., at the British Association meeting of 1831 (E. W. Taylor in YPS *Annual Report 1970*, 29–32). It is of stone with a lead-covered roof, and octagonal on plan. A basement contains two massive stone mountings for telescopes at ground-floor level for which there were two straight openings in the roof. In the middle of the ground floor a small stairway leads up to a central platform with telescope-mounting within a smaller octagonal upper storey surmounted by a revolving pyramidal lead-covered roof with one opening side. This roof was also given by Dr. Pearson; it was said to have been designed by the engineer John Smeaton and had served to roof a summerhouse in Dr. Pearson's garden.

(13) CITY ART GALLERY. The origins of the art gallery go back to the Fine Art and Industrial Exhibition of 1866, held on a site in Bootham Park. This was very successful, and a proposal to hold a second was approved in 1876. A site was acquired within St. Mary's Abbey precinct near the King's Manor, known as Bearpark's Garden. E. Taylor of York, who had designed the 1866 Exhibition building, was appointed architect and W. Atkinson hon. consulting architect. Work began in January 1878, and the building was opened the following year. In addition to the building which still exists there was a further wooden structure behind, to the N.W., which contained a great hall 200 ft. by 90 ft., with galleries around three sides. This part was declared unsafe and closed in 1909, but not demolished until 1941. The School of Art occupied some rooms in 1890, and the whole building was acquired by York Corporation in 1891. The permanent building was damaged in World War II, and restored and reopened in stages between 1948 and 1952. (VCH, *York*, 537; John Ingamells, 'The Evolution of York City Art Gallery' in *York Times*, I, no. 1, Summer 1961; *Report* on 2nd Yorkshire Fine Arts and Industrial Exhibition (1879).)

The building faces S.E. towards Exhibition Square. The front, of yellow sandstone, is of two nearly equal storeys. At either end is a pavilion, in the French manner, though of very slight projection. At ground-floor level the pavilions have round-arched doorways flanked by coupled Corinthian columns, between decorated panels; the whole of the centre part is screened by a vaulted loggia of quattrocento character of five bays with a similar return bay at each end; the spandrels between the arches contain portrait medallions. On the first floor each pavilion has a wide central bay divided by pairs of Corinthian pilasters from two narrower side bays. Between the pavilions are five bays of the wider type, the three middle bays being slightly advanced to carry the central pediment. The wider bays are filled with blind arcading containing, in the pavilions only, coloured pictorial panels; the narrower bays contain niches and other ornaments. Each of the pavilions is surmounted by a decorative attic.

The gallery contains an important collection of topographical prints and drawings referred to in the Sectional Preface (p. xlix). There are also two important stained glass panels by William Peckitt (1731–95) (Plate opp. p. xlix): (1) a *Self Portrait* of *c.* 1770 bequeathed to the gallery in 1952 by his direct descendant Miss M. M. Rowntree; (2) *Justice in a Triumphal Car beneath the Arms of the City*, 1753, presented in that year by Peckitt to the Corporation of York in return for his freedom to practice his art within the City; it was first installed in the Guildhall (now Committee Room 1) but was transferred to the gallery in 1951. (YAYAS, *Annual Report 1953–4*, XIV, 99, 104, Fig. 21; the panel was not destroyed as there stated.)

(14) CAVALRY BARRACKS, Fulford Road, dated 1796, are of two storeys and have brick walls with ashlar dressings and roofs covered with Welsh slates. They comprise three ranges set around three sides of a square. The buildings were designed by James Johnson and John Sanders, official architects to the Barrack Department of the War Office, as part of the barrack building programme initiated by William Pitt in 1792; considerable additions were made in 1861–5 (VCH, *York*, 541).

The officers' quarters to the E. (Plate 80) have a simple brick front in 15 bays with a pediment over the central part enclosing an achievement of Royal Arms modelled in Coade stone (Plate 81) with the maker's name and date COADE 1796. The N. and S. buildings have stables below and men's quarters above, reached by galleries carried on cast-iron columns; a second tier of columns carries the roofs above. *Demolished*

(15) MILITIA STORES DEPOT, Lowther Street, is probably contemporary with the development of Lowther Street in 1830–8. It consists of three detached buildings ranged round an open courtyard, all of two

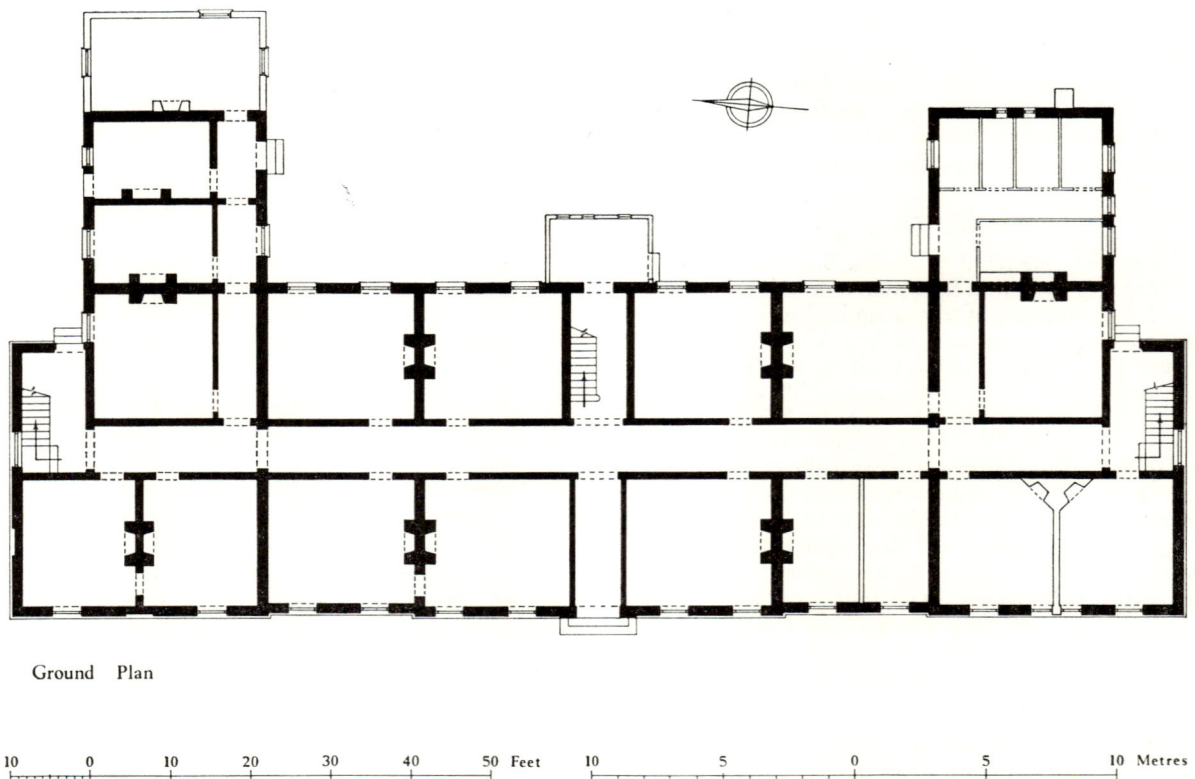

Ground Plan

Fig. 41. (14) Cavalry Barracks, Fulford Road. Officers' Quarters.

storeys with brick walls and slated roofs with widely overhanging eaves. The buildings are of plain design. The central doorway has a timber pilastered doorcase similar to those on contemporary houses; the windows have hung sashes under segmental or semicircular arches. Original iron gates and railings enclose the front of the site. To N., E. and W. there is an enclosing brick wall.

(16) CASTLE MILLS BRIDGE carrying Tower Street over the river Foss consisted of a single semicircular stone arch built *c*. 1800 by the Foss Navigation Company to replace an earlier bridge (VCH, *York*, 519–20). It was widened in 1836 when it had become 'wholly inadequate to the traffic upon it' (*YG* 16/1/1836; 14/5/1836; YCL, Evelyn Coll. plans 28, 30; Council Minute Book, vol. 1, p. 49); the work was carried out by Messrs. Craven (*YG* 27/8/1836). In 1848 G. T. Andrews reported that the bridge was in a poor state of repair (*YG* 20/5/1848). Part of the bridge was repaired in 1851 (*YG* 17/7/1852) and further repairs were carried out in the next few years (VCH, *York*, 520; Sheahan & Whellan, I, 367). *Demolished 1955.*

(17) LAYERTHORPE BRIDGE. There was a bridge over the river Foss here in 1309 and one was broken down by the defenders of the city in the siege of 1644 and restored twelve years later (VCH, *York*, 519). The present bridge was constructed in 1829 by H. Craven & Sons to a design by Peter Atkinson junior (YCA, B.50, 68, 134; K.64; and M.17A), the rebuilding necessitating the removal of Layerthorpe Postern. The single arch, spanning 35 ft., is now only visible from underneath, between later concrete additions.

(18) MONK BRIDGE carrying the road over the river Foss at the W. end of Heworth Green was built in 1794 to a design by Peter Atkinson senior (YCA, K.63; *YC* 12/6/1794). It has a single round arch of coarse yellow ashlar with large plain voussoirs, spanning 18–20 ft. An earlier bridge was broken down in 1644 (Torre, 108) and its successor was in such a bad state of repair in 1791 that an indictment was brought against the Corporation by the Crown (YCA, K.63). The Corporation bore the cost of the new bridge, aided by a contribution of £100 from the Foss Navigation Company on the understanding that the bridge should be 'fully sufficient for the purposes of the Navigation with a towing path or paths under the same'. The work was carried out by Joseph Lister, Christopher Dalton and (?) King (YCA, K.63; C.75, f. 20v). A second footpath on the E. side of the bridge was added and the approach to the bridge improved in 1844 (YCA, Council Minute Book, vol. 3, 1842–50, 103, 108). Between 1924 and 1926 the bridge was widened and the upper parts rebuilt (*YG* 18/9/1926),

(19) YEARSLEY LOCK, Foss Navigation (608536). In 1793 an Act of Parliament (33 Geo. III c.99) was passed 'for making and maintaining a navigable communication from the junction of the Foss and Ouse to Stillington Mill', 10 miles N. of York. In August of the same year J. Moon was appointed to superintend the works and by November 1794 navigation was opened up to Monk Bridge. John Rennie reported on the work in 1795 by which time four locks had been built. Two further locks were subsequently built and the canal completed to Sheriff Hutton but it was never extended to Stillington. Of the two locks within the City that at Castle Mills was rebuilt in 1859, but most of Yearsley Lock remains; the gates have been removed and a concrete weir built in place of the upper gates. The lock has sides built of brick with gritstone dressings and is 18 ft. wide.

The construction of the Foss Navigation made possible the draining and reclamation of the area of marsh which represented the fishpond formed by the damming of the Foss at Castle Mills by William I *c*. 1086, to protect the more important of his two castles.

(20) GASWORKS at junction of Monkgate and Fossbank stand on part of Piper Lane Close acquired by the York Gas Light Company from William Oldfield and Ann Tamar in 1823 (YCA, E.91, f. 175). None of the original gasworks buildings survives but the entrance to the site is flanked by a pair of lodges of the second quarter of the 19th century. The W. lodge was a house probably for the Manager; the E. lodge was offices.

The lodges are of two storeys with white brick walls and slated roofs; the principal parts of the street elevations each consist of one bay flanked by pilasters; between the pilasters each ground-floor window is or was round-headed and recessed, and small niches are recessed into the pilasters. Between the lodges are ashlar gate-piers.

(21) BOOTHAM PARK HOSPITAL was one of the first lunatic asylums to be established in Britain. The main building (Plate 77) was completed to the designs of John Carr in 1777 (*YAJ*, IV (1877), 205). Behind this a small building containing a kitchen and sitting-room for female patients was added and in 1795 an 'extensive wing' was built. This was probably the 'detached wing' which was burnt down in 1814 with the loss of the lives of several patients; it was replaced by 1817 when the present N.E. range was opened for the reception of female patients (Sheahan and Whellan, I, 609, 610), making use of fireproof floors. By 1850 two further buildings had been added to the N.W.: the first contained a wash-house, brew-house, etc. and later was

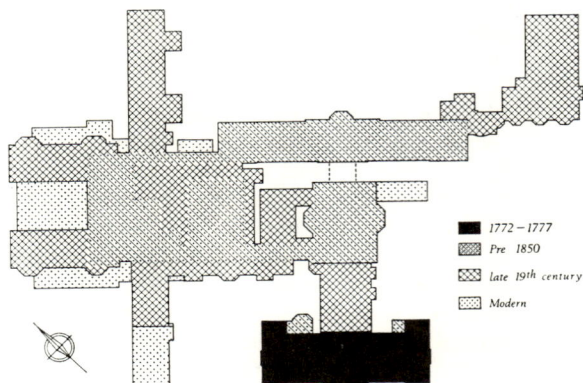

Fig. 42. (21) Bootham Park Hospital. Development plan.

converted to a recreation hall; the second contained wards for refractory patients (OS 1852). In 1886 the first three buildings were joined together to form on plan a letter I and the main staircase was moved out of the front block and a new staircase formed in a new structure immediately behind it. During the next twenty years the hospital was completely refitted internally except for the Committee Room, and recessed loggias in the back of the Female Patients range to N.E. were enclosed. Further buildings have extended the hospital to N.W. and the restrictive walls which enclosed the patients' airing yards have been removed.

The main block is of three storeys, built of red brick with stone dressings, and roofed with Westmorland slate. The front is in eleven bays, the lowest storey of the three central bays projecting and carrying four engaged Tuscan columns under a pediment to form a centre-piece (Plate 78). Above this there was formerly a circular colonnaded turret with a domed roof (VCH, *York*, pl. opp. p. 408). The end bays also project slightly. The central entrance is framed by rusticated Tuscan

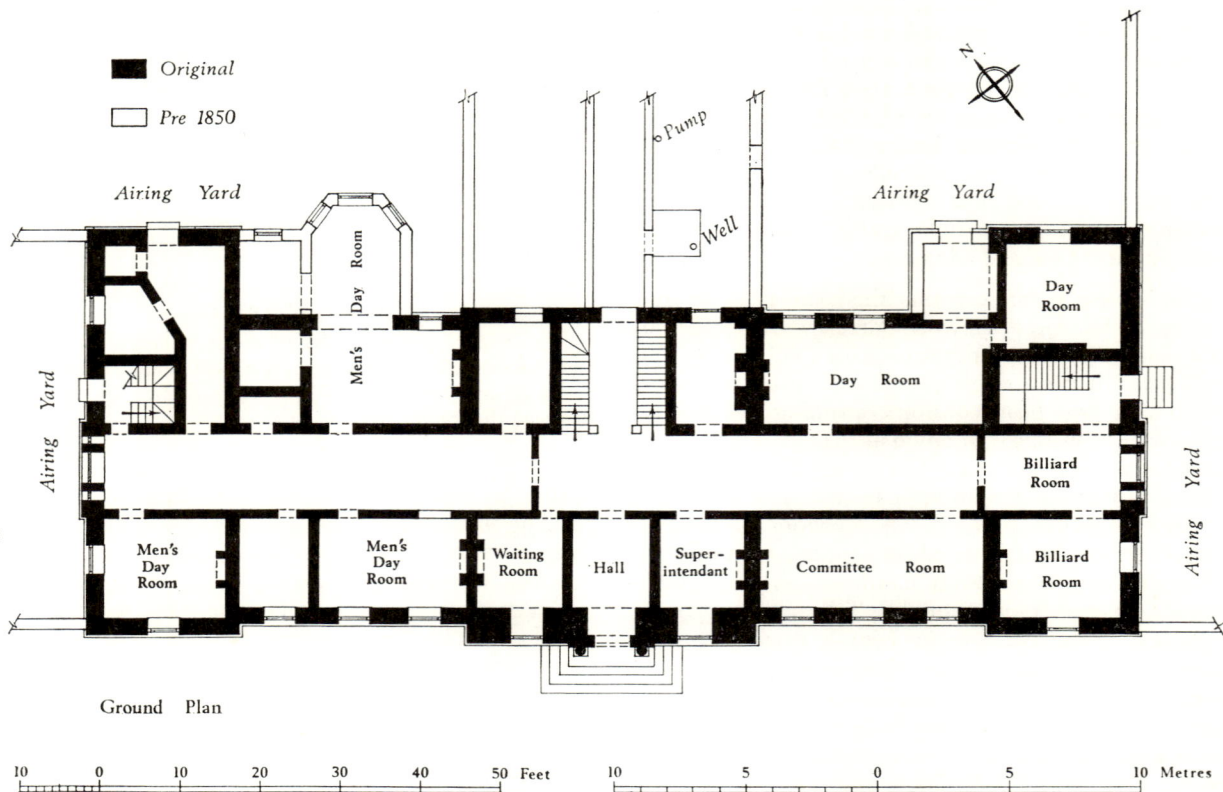

Fig. 43. (21) Bootham Park Hospital. Plan of original block as in 1850. From OS map.

columns and pediment; recessed round-headed windows give emphasis to the first floor. In the end elevations the windows lighting the ends of the main corridors are of three lights elaborated with columns and pilasters; those on the second floor are semicircular; all are placed off-centre in the four-bay elevation. At the back the end bays project more boldly; the central pediment is repeated but not the columns below it. The early 19th-century buildings are more simply designed; they have brick walls with plain sash windows and roofs covered with Welsh slates.

All the earlier buildings have been refitted, but in the original building the Committee Room on the ground floor retains the original fireplace surround and cornice, and to the walls are fixed wooden panels in frames enriched with composition (?) decoration, on which the names of subscribers are recorded. In the N.E. block of 1817 the floors of the upper wards are carried on a fireproof construction of arched brickwork spanning between iron beams.

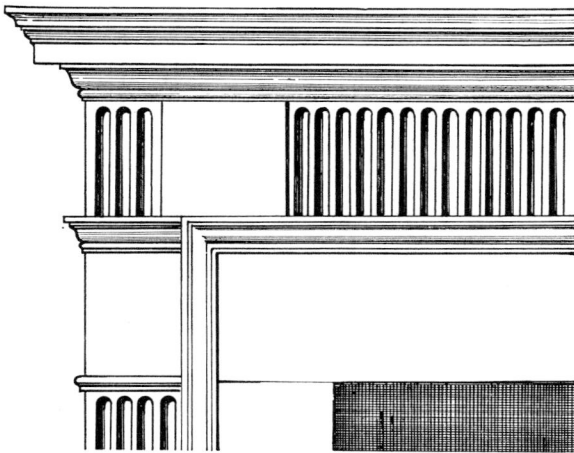

Fig. 44. (21) Bootham Park Hospital. Fireplace in Committee Room.

(22) COUNTY HOSPITAL, Monkgate, was designed by J. B. and W. Atkinson of York and opened in 1851. The original architects' drawings, dated 1849, are preserved in the office of Messrs. Brierley, Leckenby and Keighley, successors in practice to the Atkinsons. A hospital was opened in a house in Monkgate in 1740 and moved five years later into a new building which stood in front of the present building and which was pulled down in 1851 (VCH, *York*, 467–8). Additions to the new hospital include the Watt Wing opened in 1884 and a children's wing opened in 1899, both designed by Demaine and

Brierley. The same architects probably designed a Nurses' Home, added in 1905. They also designed the iron balconies which were added to the E. side of the original building in 1902 but have since been removed.

The original building, of three storeys in brick with stone dressings above a basement of massive rusticated stonework, has a slate roof and forms a rectangular block about 185 ft. long by 56 ft. wide. The main W. front is designed in fifteen bays, the centre being marked by a large entrance with rusticated stone arch and tripartite windows above. Inside, the two main staircases have stone steps and balustrades with scrolled iron standards fixed into the ends of the steps, by William Walker. In the forecourt are lamp standards also by William Walker bearing his nameplate.

(23) INGRAM'S HOSPITAL, Bootham (Plate 79), was built as almshouses by Sir Arthur Ingram of York who died in 1640. The land was acquired in February 1629/30 from Thomas Sandwith and the building must have been nearing completion in the summer of 1632 when Richard Coundall was paid for seating and stalls for the chapel, and James Ettie made the stairs (Leeds Public Library, Separate Estates 9, Temple Newsam MSS., TN/YOA 12 and TN/YOB 1). It included from the start an archway from the demolished part of Holy Trinity, Micklegate (Plate 46). The building was badly damaged in the siege of York and accounts for the repairs carried out in 1649 show that at least 5000 bricks were needed, and almost all the timberwork had to be renewed including partitions, floors, staircases and the roof. Accounts for repairs and maintenance during the later 17th and 18th centuries are preserved, and include bills for retiling part of the roof in 1674 and again in 1789 (TN/YOB 1 and 11). In 1958 drastic alterations were made to the interior and the back in conversion of the building to flats, and windows were made in the back elevation where previously there had been none.

The building is of brick with tiled roofs and comprises a central four-storey tower flanked by two-storey ranges each of which contained five dwellings. Projecting S.W. from the back of the tower, a single-storey wing contained the chapel. Set in the N.E. face of the tower is the late 12th-century archway from Holy Trinity, of two orders with a label, all enriched with nail-head ornament. The ranges to each side have stone plinths, doorways with four-centred stone heads, ground-floor windows with stone dressings, a brick band at first-floor level and upper windows with stucco dressings. At the back, the projecting chapel wing is finished with a curved Dutch gable beneath which modern brickwork covers a window of four lights with plain uncusped tracery in a two-centred head. Each wing has two projecting chimneystacks and three doorways with four-centred brick arches. The ends of the two wings are now masked by higher later buildings, but some remains of a curved Dutch gable can still be seen at one end.

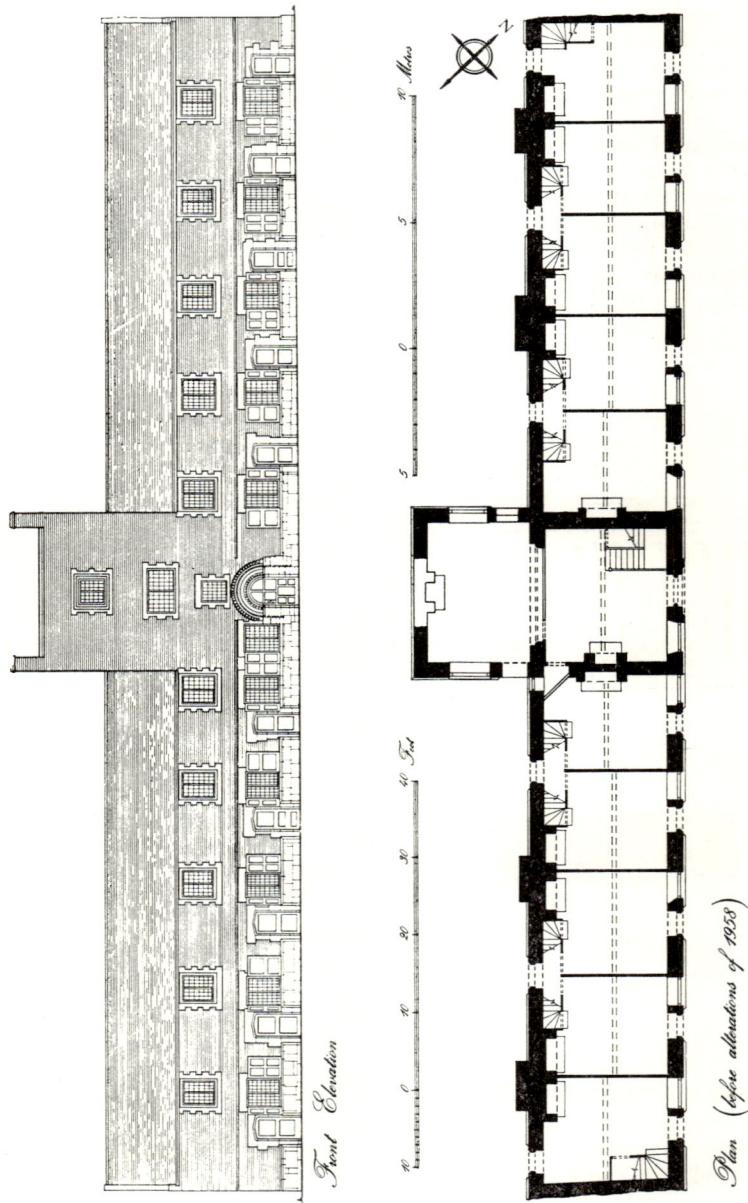

Front Elevation

Plan (before alterations of 1958)

Fig. 45. (23) Ingram's Hospital, Bootham.

Inside, the partitions between the rooms are timber-framed with brick filling between the studs housed into grooves cut in the sides of the studs. The original fireplaces had brick arches but these were covered in the late 18th or early 19th century by stone surrounds within which were placed iron ranges with ovens all enriched with a variety of raised patterns of panels with foliage, thistles, etc. The staircases were reconstructed probably in the 19th century. The roof is carried on collar-beam trusses with purlins clasped between collars and principals and receiving additional support from intermediate collars and brackets fixed to the underside of rafters.

Fig. 46. (23) Ingram's Hospital, Bootham. Timber construction.

(24) THE RETREAT (Plate 82) was established at the end of the 18th century by the Society of Friends as a mental hospital, following the death of a Friend in the York County Asylum (Bootham Hospital) in 1791 and the unsatisfactory condition of the County Asylum at that time. William Tuke with the assistance of his son Henry and his friend Lindley Murray raised money for the purchase of the site which was acquired in 1793. Subscriptions were called for from Friends in all parts of England and, in spite of considerable financial difficulties, the first buildings were opened in 1796, comprising a central three-storey block with a recessed two-storey W. wing. The following year a corresponding E. wing was erected. The designer was John Bevans of London, and construction was supervised by Peter Atkinson, of York. In the next thirty years a new block was attached to each of the four corners and a third storey was added to the original E. and W. wings. Improvements recorded in the annual report for 1843 included warming the building by hot water, drying apparatus, additional warm bathing, and the lighting of apartments, galleries

and passages with gas. Further improvements were made in the years that followed. In 1850 there were 114 patients, the highest number up to that time. Further buildings were added in the later 19th century and in the present century. An extensive programme of modernisation was carried out c. 1960. (S. Tuke, *Description of the Retreat* (1813); H. C. Hunt, *A Retired Habitation* (1932); *Annual Reports.*)

The buildings are of plain brickwork with slated roofs. The entrance retains its original pedimented doorcase but most of the windows have been refitted with modern sashes. The original sashes of which only a few now remain were of iron with iron glazing bars; in order to give security without the appearance of bars one sash filled the whole height of the window but was only glazed in the lower part, and a second, moving, sash had glazing bars which, in the closed position, came exactly behind those of the first (Plate 81). The interior has been entirely refitted.

(25) ST. MARY'S HOSPITAL and THE GRANGE WELFARE CENTRE occupy buildings begun in 1848 as the York Union Workhouse for the accommodation of 300 paupers (VCH, *York*, 280; YG 20/11/1847; 22/1/1848). A competition was held for the design and that chosen was by J. B. and W. Atkinson. The work was carried out by Thomas Linfoot and cost less than £6,000; as a contemporary newspaper report described it, 'Externally it is perfectly plain, as buildings of this class should be . . .' (YG 2/6/1849). The workhouse comprised three parallel ranges of buildings lying N. and S. on a square site within confining boundary walls. The E. block, now The Grange, housed the administrative offices; it was of nine bays but has been lengthened to eleven. The central block was the longest and has been little altered externally except for additions to the ground floor. Most of the W. range has been replaced and most of the boundary walls have been pulled down.

The buildings are of three storeys with brick walls and slated roofs, and are designed in a simple utilitarian style. The E. range has round-headed openings to the ground floor on the main E. front. The middle range has central projections to E. and W. behind which a central octagonal hall contains a staircase winding round a circular well; other symmetrically disposed projections to the E. contain lavatories.

(26) WANDESFORD HOUSE, formerly Wandesford Hospital, No. 37 Bootham, was opened for occupation by 'ten poor maiden gentlewomen' in 1743. Mary Wandesford, a spinster of York, left an endowment in her will dated 1725 and the site was purchased from William Wilberforce of Hull in 1739 (E. Brunskill, YGS, *Occasional Paper*, VII (1960), 30). Accounts for the building survive and record payment to John Terry, carpenter, and for bricklayers' work to Robert Kibblewhite, Thomas Dunn and Richard Nelstrop (Borthwick

Inst.). In the 19th century additional staircases were constructed to give access from each living-room to the bedroom above, and in 1968 further modernisation was carried out; at the same time the central entrance, previously a plain brick doorway, was given a timber doorcase with broken pediment.

The building is of two storeys with brick walls and tiled roofs. The front is designed in seven bays with the central three bays projecting under a pediment. The treatment of the walling is unusual, each bay having a round-arched recess within which the windows of both storeys are set, and a deep impost band is carried from arch to arch and across the recesses (Plate 79). At the wall-head is a heavy timber cornice and within the pediment is a plain niche containing a bust of the foundress. Two lead rainwater heads and down-pipes are original.

On the side and back elevations there are no arched recesses but the impost band from the front is continued below the sills of the upper windows. At the eaves plain oversailing courses of brickwork replace the timber cornice of the front. The windows have flat arches of gauged brick and there are two more lead rainwater heads, one dated 1739.

Front   Elevation

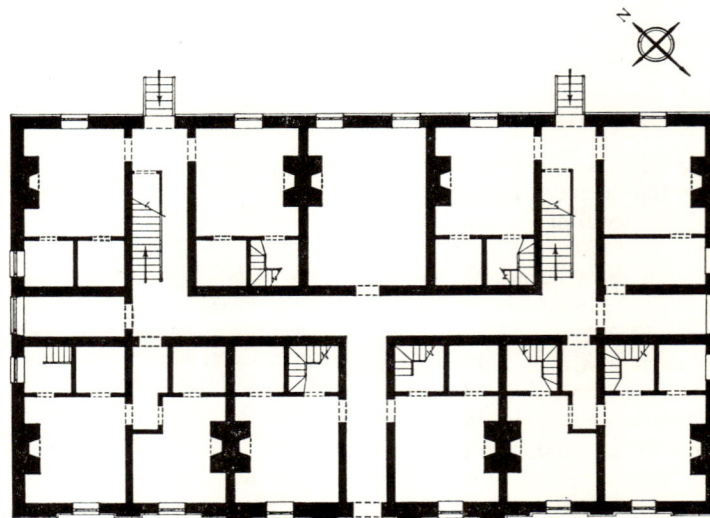

Ground   Plan

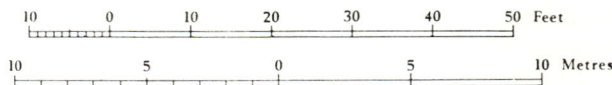

Fig. 47. (26) Wandesford House, Bootham.

The interior is very simple. The original two staircases each rise in a single flight with closed string, square newels, turned oak balusters with half-balusters against the newels, and a heavy moulded and swept handrail (Plate 124).

(27) GREY COAT SCHOOL, No. 33 Monkgate, now a Schools Clinic. The Grey Coat School was a charity school for girls opened in 1705 at No. 60 Marygate (Monument 252) in conjunction with the Blue Coat School for boys which formerly occupied St. Anthony's Hall, Peaseholme Green. In 1784 the girls' school was moved to new premises in Monkgate part of which survives as the back wing of No. 33, the front part having been rebuilt soon after 1850.

The surviving building is a two-storey brick range with slated roof; the principal elevation to the N.E. is in six bays with tall ground-floor windows set in shallow round-arched recesses. The OS map of 1852 shows that the range was originally of seven bays but the end bay to the S.E. is now enclosed within the later building. The ground floor contained a spinning-room and a sewing-room and above was a large lodging-room reached by a flight of stone steps (Hargrove, II, 569–70); the existing stone staircase is not the original one.

(28) ST. JOHN'S COLLEGE, Lord Mayor's Walk, occupies two buildings erected between 1840 and 1850 to which many later additions have been made. To the S.E. stands the building opened in 1845 as the York and Ripon Diocesan Training College for schoolmasters; for the previous four years the College had been accommodated at premises in Monkgate previously occupied by the Manchester College, and subsequently by the York and Ripon Diocesan Training College for schoolmistresses until the removal of that college to Ripon in 1861. To the N.W. is a building erected by the Diocesan Boards of Education and opened in 1846 as the Yeoman School, described as a 'middle-class' boarding school providing a practising school for the college. It was amalgamated with Archbishop Holgate's school in 1858 (VCH, York, 452, 458).

Both buildings are of two storeys and have walls of brick with stone dressings and are designed in Tudor style (Plate 80). The original Training College building is H-shaped on plan with two long parallel wings joined by a main range in the middle of which is the entrance and centrepiece, elaborated with gables and octagonal turrets. The rest of the building is plain, with stone mullioned windows with four-centred heads to the lights.
    The School building consists of a long main range with short cross-wings, of unequal length, at the ends. The front (S.W.) end of the N.W. cross-wing is masked by an addition of the later 19th century which contains the principal entrance. The main range has had modern windows inserted between the original three-light windows on the ground floor; the upper floor is lit by single-light windows.

(29) ST. PETER'S SCHOOL buildings (Plate 84) were erected in 1838 for a Proprietary School started by a company formed by leading York citizens; the architect was John Harper (Colvin, 266). In 1844 the Dean and Chapter bought the buildings to accommodate St. Peter's, the school attached to the Minster since its foundation probably in the eighth century, which at that time was accommodated in a building in the Minster Yard. The two schools were thus amalgamated and the original buildings now form only the nucleus of a much larger complex (A. Raine, History of St. Peter's School (1926)).

The central block which contains the main entrance and the Assembly Hall is of two storeys with an elaborate turretted elevation to the N.E. faced with stone. To each side is a low wing with three traceried, square-headed windows lighting a classroom and with a passage behind, and beyond is a two-storey pavilion with octagonal corner turrets flanking a lofty mullioned and transomed window lighting a classroom on each floor. The other elevations are of brick with stone dressings but are largely masked by additions; the N.W. end was enlarged to S.W. soon after 1850, the chapel, designed by Messrs. J. B. and W. Atkinson, was added in 1861 and other additions are dated 1905.
    In the N.W. pavilion the original staircase remains with iron balustrade incorporating a simple fleur-de-lis design. The staircase in the S.E. pavilion has been removed.
    At right angles to the school buildings and projecting N.E., stands the schoolhouse. This was originally joined to the school building by a curved screen wall, now removed. The schoolhouse is an irregular building of two storeys in brick with stone dressings, having small octagonal turrets at the prominent angles and mullioned bay windows to the principal elevations. Inside, the staircase is of the same design as that in the school building. Further N.E. stands the porter's lodge, a plainer building, of one storey with brick walls and stone mullioned windows.

(30) PIKEING WELL, New Walk (Fig. 48), is a stone structure designed by John Carr. It was commissioned by the City Corporation in 1752 (VCH, York, 208) to form a decorative well-head feature. It is a simple rectangular structure with a round-headed doorway in the side facing the river; the coping on this side is made up with reused stones including three 12th-century capitals placed as finials.

(31) ARCADE from the Theatre Royal. At the front of No. 73 Fulford Road are the remains of an arcade of 1834–5 which originally formed part of the Piazza erected 'in the Elizabethan style' in front of the Theatre Royal to plans by John Harper, architect (YCA, M17/A; Hudson, 168v). In 1879 the Theatre was again altered (Ben Johnson, Practical Guide (1886), 101) and the arches moved to their present position.

The arcade, built in magnesian limestone, has been much mutilated. Four complete bays remain, the arches four-centred, of a single chamfered order; three are glazed and closed in with wood to form workshops.

(32) BURTON STONE, at the junction of Clifton and Burton Stone Lane, is a large square base for a cross. In addition to the central hole for the cross-shaft there are four cup-like depressions. It is probably the Clifton Stone mentioned in 1575 (YCA, E126). Mother Shipton's Stone, which formerly stood at the corner of Rawcliffe Lane, may have been part of the cross (T. P. Cooper, Miscellaneous Notes, 24, MS. in YCL).

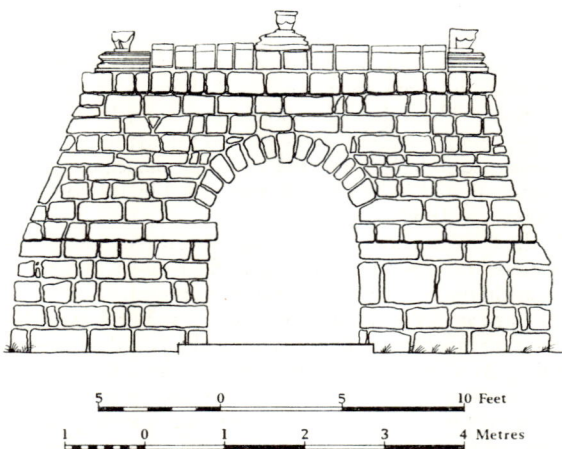

Fig. 48. (30) Pikeing Well, New Walk.

(33) FULFORD CROSS, an octagonal stone shaft raised up on three steps, stands on the W. side of Fulford Road opposite Imphal Barracks (608501). It is no doubt the cross ordered to be set up by an award of 1484 between the City and St. Mary's Abbey (Drake, 597).

(34) WHITESTONE CROSS, on the Haxby road now at 60735367 but moved from its original position, is a large irregular stone tapering from 6½ ft. to 3½ ft. wide by 5½ ft. long. It shows no sinking for housing an upright stone. It is said to be the stone referred to as 'the Whitestone Cross above Astell Brigg' in 1374 and the 'stone cross that is written upon, above Astyl Brigg' in the award of 1484 referred to under (33) above (Raine, 258).

BOOTHAM (Monuments 35–57)[1]

Bootham is the main road out of York to the N.W. (Plate 85). There were houses here in the Middle Ages but none now remaining is earlier than the 17th century. Nos. 39 to 61 (odd), on the N.E. side, are mainly of the 18th century and form one of the most distinguished groups of houses in the City. On the S.W. side there is only one house earlier than the 19th century, built on the narrow strip of ground between the road and the precinct wall of St. Mary's Abbey, where in the Middle Ages was the ditch outside the wall. The houses in this street are of three storeys unless otherwise described.

N.E. side:

(35) HOUSE, Nos. 15, 17, is of two builds. The S.E. part was built as a two-storey house probably in the early 18th century; in 1790 it was acquired by Thomas Wolstenholme (see p. lvi) who recorded in his will, dated 1800, that in 1799 he built a smaller house to the N.W. and added a third storey to the earlier house. Both parts of the building have been much altered and now comprise a shop with a single maisonette above.

Both houses were small, having frontages of about 17 ft. and 10 ft. respectively. The street front has been much altered. Internally many of the fittings have been renewed but in the N.W. part the upper flights of the original staircase of 1799 remain, and in the S.E. part are two fireplaces of the same date; one of these, on the first floor, has composition decoration of anthemion pattern between profile heads, almost certainly by Wolstenholme himself, and very similar to work in his house at No. 3 Gillygate (117). His composition decoration also appears on the architrave to one of the doors.

(36) EXHIBITION HOTEL, No. 19, was built in the late 18th century as a fairly large house, which was occupied in the last decade of the century by James Fenton and after 1800 by Lady Royds (Rate Books); it had certainly become a hotel by 1872 (Benson, III, 166) and the name probably commemorates the Fine Art and Industrial Exhibition of 1866 (see (13)). The street front has been remodelled in modern times but its earlier state can be seen in watercolours by R. Dighton (c. 1815) and C. Dillon (c. 1840) (Evelyn Collection, YCL); the interior has been altered.

The house is of three storeys with attics with a brick gable to the rear above storeys defined by brick bands. The interior has been stripped of original fittings but the main lines of the plan can be recognised; the entrance was to one side, leading to a staircase placed transversely in the middle of the house, and at the back two rooms flanked a central passage.

---

[1] Information as to owners and builders of houses in Bootham is largely based on the researches of Mr. J. H. Harvey.

(37) HOUSES, Nos. 21, 23, were built in the late 17th century as a single dwelling on an **L**-shaped plan. In the 18th century the re-entrant angle was filled in. In the 19th century the ground floor was converted to shops, the upper windows enlarged and the front stuccoed.

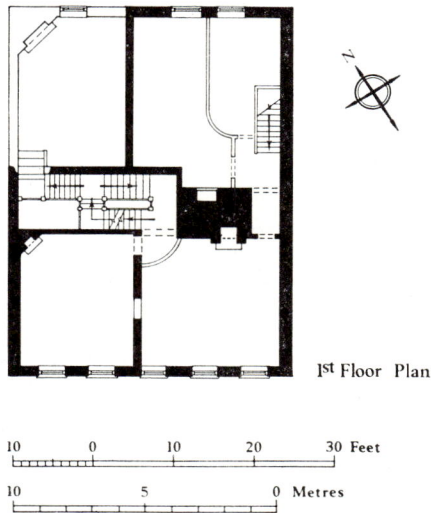

1st Floor Plan

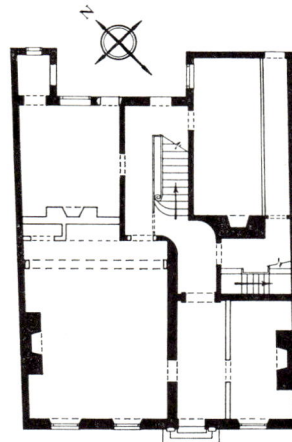

Fig. 49. (37) Nos. 21, 23 Bootham.

The building is of two storeys with attics. The original central entrance remains, with moulded surround and key-block. There are five windows to the first floor with similar keyblocks to the heads, fitted with 19th-century sashes. The gabled back wing also has windows refitted with hung sashes. Inside there is a fine staircase with close strings, heavy turned balusters and square newels (Plate 124). One of the front rooms on the first floor has the plaster on the walls painted with a diaper pattern in pale blue and black, probably original; it was preserved under late wallpaper but has now been exposed.

(38) No. 25 was erected in 1766 and has a rainwater head (Plate 106) dated 1768 with the initials TG for Thomas Gilbank who purchased the house in that year from Mrs. Woodyear (Rate Books; YCA, M13, M14). After 1850 the house was enlarged, increasing the front from five bays to seven, and the ground floor was converted to shops.

The original house was a substantial three-storey building with a central entrance. The late 19th-century shop front extends across the full width of the enlarged building incorporating the original entrance into the unified design. The shop front is no longer complete but is of interest as one of the few remaining shop fronts in the town which includes cast-iron construction. The maker's nameplate is almost illegible but the words Micklegate, York can still be made out. The brickwork above is in Flemish bond with windows under arches of gauged brick and fitted with 19th-century sashes.

At the eaves is a timber cornice with modillions and dentils. The back is partly masked by modern commercial premises; a polygonal bay projects for the full height of the house. Inside, many original cornices, dado rails and window architraves remain but no fireplaces. The elegant staircase (Plate 125) has turned balusters, clustered round a turned newel at the foot.

(39) HOUSE, No. 33, was erected by Robert Clough, builder, between 1752 and 1755 (YCA, E93, f. 293, E94, f. 4) (*see also* Monument (128)).

The principal elevation to Bootham is of four bays, with cement-rendered plinth, painted stone bands at both upper

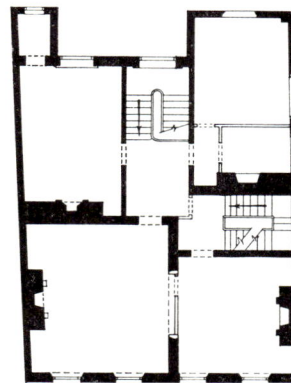

1st Floor Plan

Ground Plan

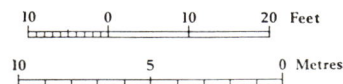

Fig. 50. (39) No. 33 Bootham.

The rear elevation (Plate 93), with walling of common bricks, has good red brick dressings to the window arches and bands. Wings of full height project from both end bays, that to the N.W. being narrower; it probably contained privy closets. The brick plinth, and the plat-bands to both upper floors are continued round the wings. The windows have flat arches of gauged brick. Above the central rear entrance is a round-arched window lighting the staircase. The roof is in two spans parallel to the street.

Inside the house some alterations have been made to the fittings. The ground floor has four rooms, with the entrance hall passage made by partitioning the original hall, and the staircase hall situated behind, between two smaller back rooms. A secondary servants' staircase is located in the middle of the house against the S.E. side wall. The main staircase (Plate 126; Fig. 8, p. li) has three balusters to each tread, swept at the foot of the stair in a cluster round the newel. The staircase window is flanked by Roman Doric pilasters on pedestals, with moulding and enrichment to the arch above, under swags of plaster drapery with ribbons and tassels.

The servants' staircase has close strings, turned balusters and square newel posts, with attached half-balusters. The moulded handrail is carried vertically up the face of the newel to reach the level of the next stair flight, an arrangement made necessary by the confined space and steep ascent (Fig. 51).

(40) HOUSE, No. 35, now of three full storeys with cellars, was built as a two-storey house before 1668 (Deeds), and presumably after the siege of York in 1644. In the middle of the 18th century the third storey was added and the house was extended at the back. Further remodelling in the early 19th century included the addition of a small wing on the N.W. side and the fitting of the present doorcase to the entrance.

On the street front the storeys are separated by brick bands. The doorway is flanked by reeded half-columns and curved brackets carrying a pediment over a semicircular fanlight. The windows are fitted with 19th-century sashes; at the wall head a timber eaves cornice carries a concealed gutter discharging to a lead rainwater head on the side elevation. The plan originally provided three rooms, all with fireplaces, one behind the other. The original staircase was at the side of the chimney between the front and middle rooms; this has been replaced by an 18th-century staircase with turned balusters, and the original staircase has been reset at the back of the house (see Fig. 8, p. li). Other 17th-century fittings remaining consist of a bolection-moulded fireplace surround and a door with two panels set in bolection-moulded framing. Some fittings survive from the 18th and early 19th centuries, including rococo plasterwork to the ceiling over the main staircase and a short stairway to the added wing with cast-iron balusters of quatrefoil section.

(41) HOUSES, Nos. 39–45 (odd) (Plate 86; Figs. 52, 53), were built as a terrace in 1748 by Thomas Griffith; the site of Nos. 39–43 had been a garden and had a frontage of only 55 ft.; No. 45 replaced an earlier building giving a wider frontage. Griffith occupied

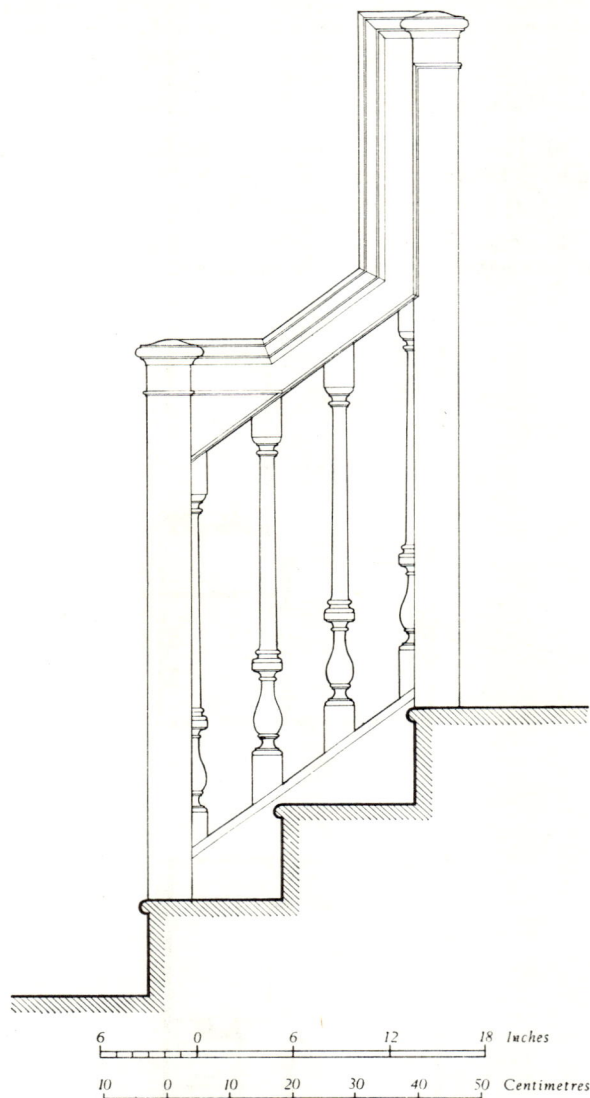

Fig. 51. (39) No. 33 Bootham. Back staircase.

floor levels, and continuous sill bands to the ground and first-floor storeys. The sash windows have flat arches of gauged brick; those to the two lower storeys have recessed frames and painted rendered reveals, those to the top storey flush frames which are replacements probably of 19th-century date, as is the moulded timber cornice. The front door, approached by a flight of four stone steps, has a simple surround of the early 19th century, with sunk panels to the pilasters and a triangular pediment; the door is original and has eight fielded panels.

PLATE 77 (21) BOOTHAM PARK HOSPITAL. S.W. front. John Carr, 1777.

PLATE 78

(21) BOOTHAM PARK HOSPITAL. S.W. front, centrepiece. John Carr, 1777.

PLATE 79

(23) INGRAM'S HOSPITAL, BOOTHAM. *c.* 1632 and *c.* 1649.

26) WANDESFORD HOUSE, BOOTHAM. 1739–43.

PLATE 80

(14) CAVALRY BARRACKS, FULFORD ROAD. Officers' Quarters.
James Johnson and John Sanders, 1796.

(28) ST. JOHN'S COLLEGE, LORD MAYOR'S WALK. S.E. building, centrepiece. 1845.

PLATE 81

(14) CAVALRY BARRACKS, FULFORD ROAD. Officers' Quarters. Pediment with Royal Arms by Coade. 1796.

(24) THE RETREAT, HESLINGTON ROAD. Original barred window. 1796.

(10) THE CEMETERY, CEMETERY ROAD. Stone pier. 1837.

PLATE 82

(24) THE RETREAT, HESLINGTON ROAD. 1796 and later. View by H. Brown, early 19th-century.

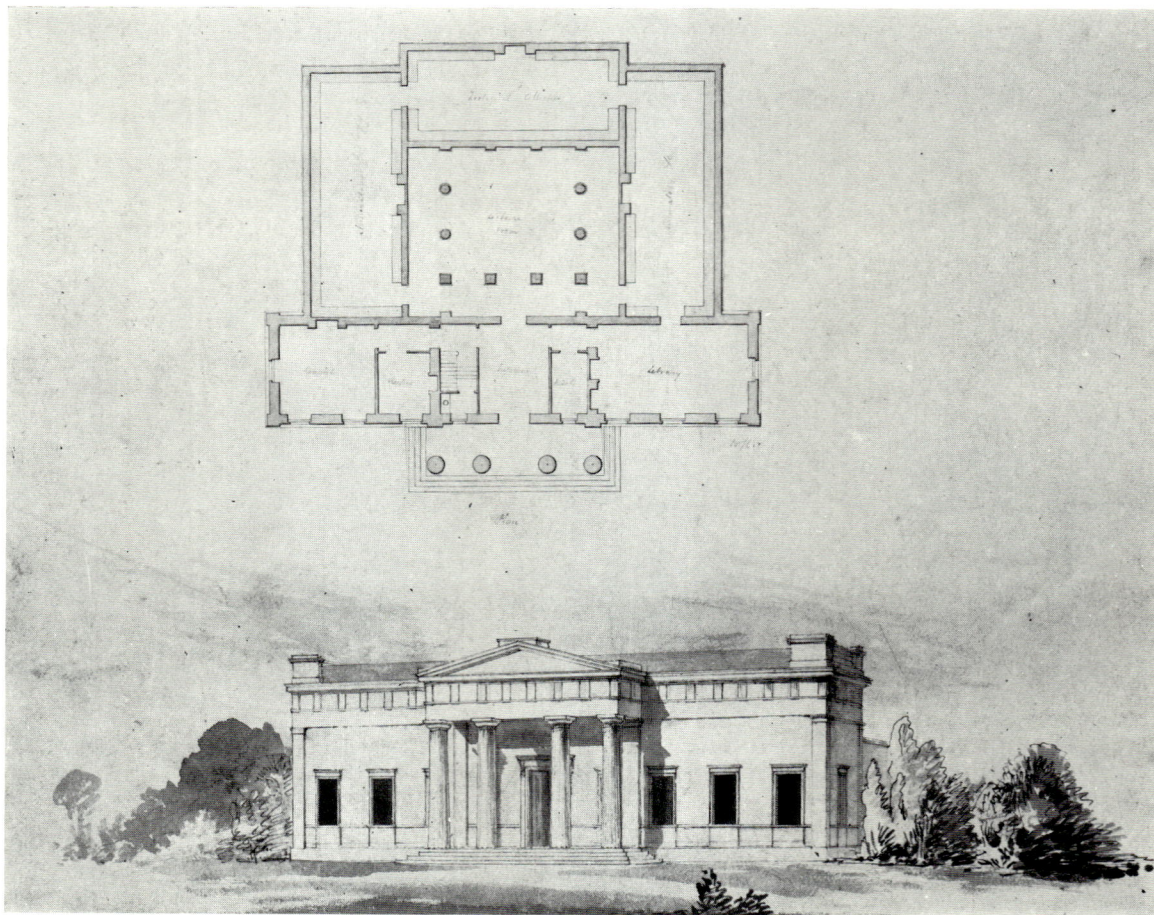

(12) THE YORKSHIRE MUSEUM. Unsigned drawing in the Museum, early 19th-century.

PLATE 83

(10) THE CEMETERY, CEMETERY ROAD. Mortuary Chapel. James Pritchett, 1837.

(12) THE YORKSHIRE MUSEUM. William Wilkins, 1827-9.

PLATE 84

(29) ST. PETER'S SCHOOL, CLIFTON. N.E. front, centrepiece. John Harper, 1838.

Fig. 52. (41) Nos. 39–45 Bootham. Front Elevation (reconstruction).

No. 45 himself; the other houses were let (YCA, E93, 195, 213, 241). At the back of No. 45 a substantial block was added *c.* 1800 and late in the 19th century Griffith's house was pulled down and replaced by a new, four-storey house, retaining the structure of 1800 behind it. Nos. 41 and 43 are now combined as one property and divided into flats.

The street front, of three storeys above a basement, is terminated at each end by stone quoins and the storeys are divided by stone bands; the quoins comprise small stones flush with the brickwork alternating with larger projecting stones. At the eaves is a bold timber cornice, with modillions, carrying a concealed gutter; part has been restored. The design is now interrupted by the four-storey front of the later No. 45 but the quoins of the original N.W. angle remain in position. The doorcase to the entrance to No. 45 (Plate 107) is however of mid 18th-century date, and is probably one of the original doorcases reused. For Nos. 39–43 there are now only two doorways, one with a semicircular fanlight in an open pediment of *c.* 1800 (Plate 108) and one with a late 19th-century doorcase. The fenestration has been altered so that only No. 41

9

now has windows all of the original size; they have flat arches of gauged brick with keystones; the sashes have lost the original glazing bars and early 19th-century wrought-iron guards have been added to the first-floor windows. The S.E. end has a double gable with parapets concealing the end of the double-span roof. The back has been much altered. Projecting behind No. 45 is the four-storey block of *c.* 1800 with two very large hung-sash windows to the first floor; these and the other smaller windows all have flat arches of gauged rubbed brick, very markedly splayed.

Nos. 41 and 43 were joined into one in *c.* 1800 and, together with No. 39, were extensively refitted, but on the top floor No. 39 retains an original fireplace and panelling. No. 41 retains an original panelled dado in the front room on the ground floor and a simple cornice in the room behind; in No. 43 the front room on the ground floor and the saloon above are both lined with original fielded panels above plain dados. The staircase of No. 41 has been removed but most of the original staircases remain in Nos. 39 and 43, with open strings and turned balusters. Otherwise fittings are mostly early 19th-century with free use of reeding (Plate 111) and mouldings of symmetrical section combined with composition decoration almost certainly by Thomas Wolstenholme (Plate 113).

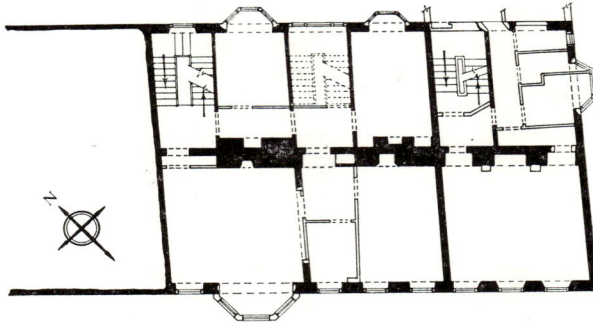

1st Floor   Plan

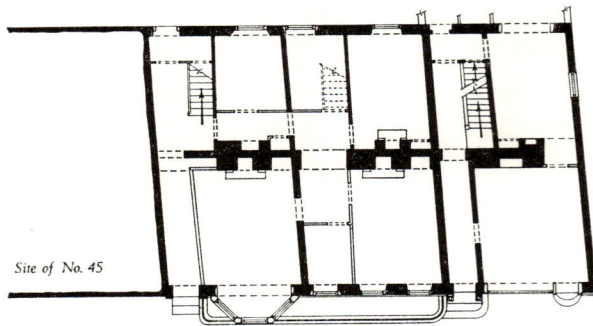

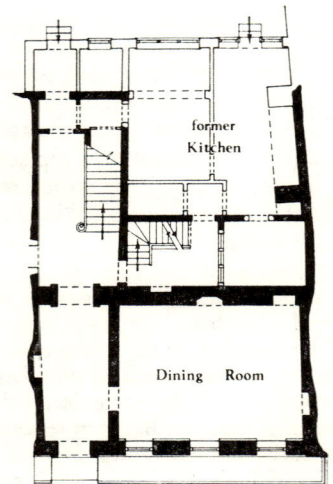

Site of No. 45

Ground   Plan

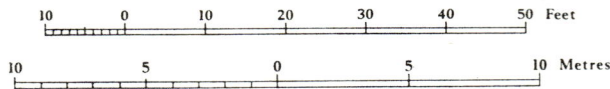

Fig. 53. (41) Nos. 39, 41, 43 Bootham.

where there is a large round-headed window to the staircase and a projecting three-sided brick bay with sash windows, originally of two storeys but heightened to three storeys in the 19th century.

Internally the house is well fitted and has been little altered. The principal rooms have moulded ceiling cornices and enriched architraves to door and window openings. On the ground floor the former dining-room (Plate 117) has the walls panelled above a dado rail; over the door is an entablature with decorated frieze (Plate 110) and over the enriched fireplace surround is an eared overmantel with carved pendants at the sides. The main staircase (Plate 123) rises only to the

Saloon

1st Floor   Plan

(42) HOUSE, No. 47, was built in 1753 by John Carr for Mrs. Mary Thompson, widow of Edward Thompson, M.P., of Oswaldkirk. Mrs. Thompson died in 1784 and the house was subsequently owned by Leonard Pickard (1798) and the Rev. Pickard (1855). (YCA, E93, 288; Rate Books of St. Michael-le-Belfrey in YML; Deeds; APS, *Dictionary of Architecture*, II, 36.) It is now owned by Bootham School.

The house is of three storeys with attics and basement, and has a front of four bays. The most notable features of the elevation (Plate 86) are the double bands to the two lower storeys and the lengths of stone cornice over the window arches, unsupported by an entablature or brackets; the window cornices of the ground floor are aligned with the cornice over the architrave to the lofty entrance doorway, which has supporting brackets to each side. At the eaves is a heavy timber cornice carrying a concealed gutter. A rainwater head has a monogram of initials probably of Mary Thompson (Plate 106). The eaves cornice is continued round to the back

former Kitchen

Dining   Room

Ground   Plan

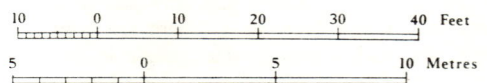

Fig. 54. (42) No. 47 Bootham.

first floor and has turned balusters alternately plain, fluted, and twisted, and newels in the form of fluted columns. Over the staircase the ceiling is enriched with plaster panels and foliage (Plate 118). On the first floor the Saloon occupies three bays of the front and has an ornate fireplace and overmantel enriched with fruit, flowers and foliage (Plate 116). The secondary staircase rises through the full height of the house; up to the first floor, where it was for servants' use only, it has a close string and square newels, but above the first floor it has an open string, turned newels as well as turned balusters and a more delicate handrail. The second floor was partly refitted in the 19th century. In the attics the queen-post construction of the roof trusses is exposed.

(43) HOUSE, No. 49, was built in the late 17th century as two dwellings of two storeys with attics. In *c.* 1738 the two houses were converted into one (Deeds); the front part was heightened to three full storeys and remodelled with the addition of rusticated quoins (Plate 87). The back elevation was completely rebuilt in 1965 but the previous disposition is shown in the accompanying figure and in Plate 92.

The main front to the street is built in Flemish bond with a brick plinth, a plain band at first-floor level, and at the second floor a moulded stone string which may have formed the eaves cornice to the original houses; at each end are rusticated quoins in applied stucco. At the eaves a timber cornice with modillions projects boldly to carry a concealed gutter. The entrance has a late 18th-century doorcase with timber pilasters carrying an open pediment. At the S.E. end is an arch for an opening now blocked which may have been an original entrance. The rear elevation, before rebuilding, had two 17th-century gables and projections which probably housed staircases and closets. None of the original windows remained, all having been replaced by sash windows of various dates.

Surviving 17th-century work includes a moulded plaster

Back Elevation

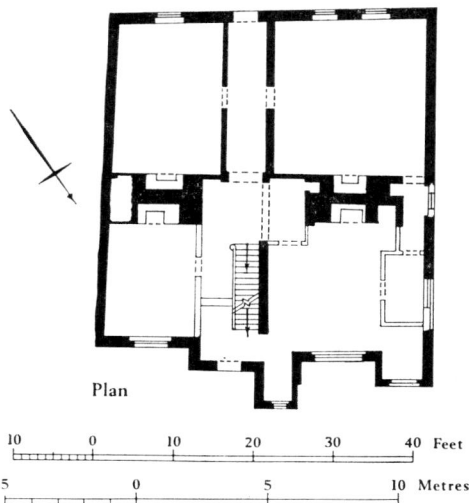

Plan

Fig. 56. (43) No. 49 Bootham before reconstruction.

cornice, stop-chamfered and stop-moulded ceiling beams and, in the kitchen, moulded joists. Bolection-moulded panelling forming a dado in the entrance hall must also antedate the alterations of 1738. The two front rooms on the ground floor are lined with ovolo-moulded and fielded panelling and have fireplaces with moulded surrounds all of *c.* 1738. Irregularities in the ceiling of the S.E. room suggest the removal of a partition enclosing an original entrance passage at the S.E. side. The staircase has an open string, turned balusters, and a heavy handrail terminating at the foot in a volute over a turned newel and clustered balusters; the lower flight is built up on a heavy inner string reused from an earlier staircase. On the first floor one of the back rooms has a surround to the fireplace, architraves to door and window and an overdoor (Plate 110) all moulded and enriched and of *c.* 1738. Other rooms on this floor have simpler 18th-century fittings. The top floor, appearing as a full storey on the elevation, comprises attic rooms, since the 17th-century roofs were modified but not destroyed when the front was heightened. The trusses are of simple collar-beam form and the purlins are staggered and held by tusk tenons projecting through the principal rafters.

Fig. 55. (43) No. 49 Bootham. Construction of upper floor.

(44) HOUSE, No. 51 (Plate 87), was built for Sir Richard Vanden Bempde Johnstone, Bart., who bought No. 49 and the ground adjoining before 1800. The house was designed by Peter Atkinson senior and was nearing completion in 1804 (Deeds; APS, *Dictionary of Architecture*, I, 119). The Yorkshire Gazette (22 March 1834, 14 May 1842) shows that it was formerly known as Bootham House. It was acquired for Bootham School in 1846 and the back wing of the house was remodelled and a second back wing added; this last was destroyed by fire in 1899 and has been rebuilt since.

The house is of three storeys, built in brick with stone dressings. The front, of five bays, has a balcony outside the first-floor windows with cast-iron lattice railings. The central porch, in the Roman Doric order, has a pair of columns to each side, the spaces between the columns corresponding to small windows between pilasters flanking the doorway. On the upper floors this rhythm is repeated in triple windows, that on the first floor being framed by Ionic pilasters and entablature with a curved pediment. The variations in size and decoration of the windows between one floor and another are shown in the figure opposite p. 55.

The disposition of the front part of the house has not been altered. The imposing staircase (Plate 117), approached between fluted columns, is of stone with cast-iron balusters supporting a slender mahogany rail. Original fittings that remain include doors and doorcases (Plate 110) and a number of fireplaces, some with composition enrichments (Plate 115), probably by Wolstenholme. Other fireplaces have reeded marble surrounds carved with realistic floral sprays (Plate 114).

(45) HOUSES, Nos. 53, 55, form one building with a symmetrical five-bay façade (Plate 88). In a trust deed of 1773 (YCA, E94, f. 144) the houses are referred to as two new built messuages, one occupied by John Lund and the other by Mrs. Robinson, but it seems probable that the whole building originally formed the one house on which Lund was paying rates in 1766 (Parish of St. Giles, YCA, M13, M14), erected *c.* 1765 and divided into two *c.* 1770–1. The style of the front suggests that John Carr may possibly have been the architect. An extra bay was added to the N.W. end early in the 19th century. The building was acquired by Bootham School in 1923 and is now used as school offices.

The street front, facing S.W., has stone bands over the basement and ground-floor windows and joining the first-floor sills. Over each of the first-floor window arches is a cornice without architrave or frieze, similar to the cornices at No. 47 (monument 42) by John Carr. At the eaves a timber cornice carries a concealed gutter. In the centre two doorways are sheltered by a Doric porch added in the early 19th century. On the first floor the central window is blind and retains the original glazing bars but the bars have been removed from the other windows of the two lower storeys. The N.W. addition breaks forward and repeats the first-floor bands of the original

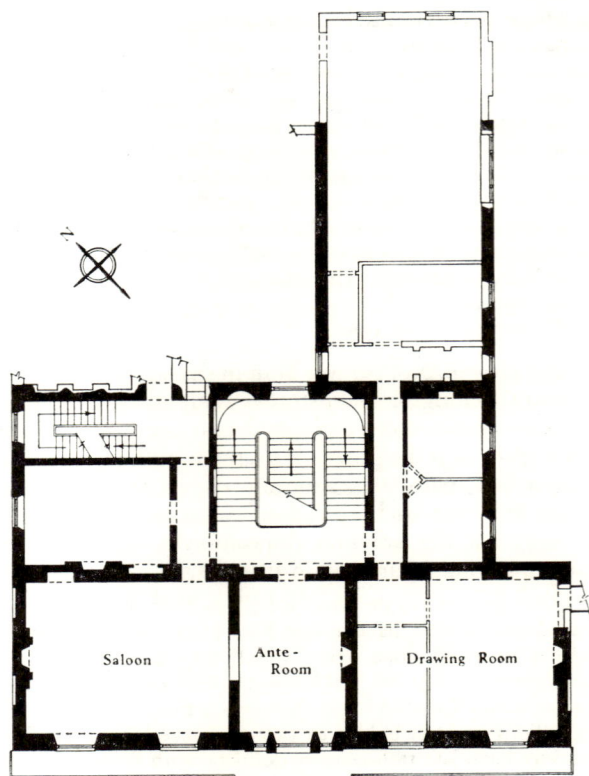

1st Floor Plan

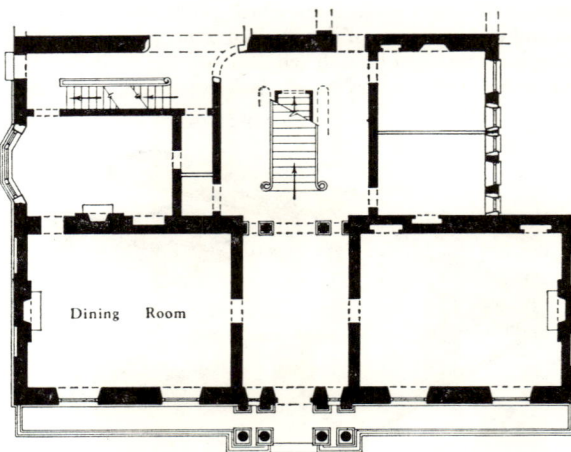

Ground Plan

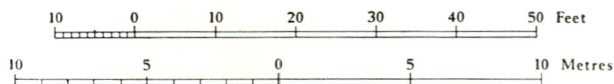

Fig. 57. (44) No. 51 Bootham.

building. The back has a small three-storey closet projection at the S.E. end.

Each house has the very common plan of an entrance passage widening at the back to accommodate a staircase, one front room and one back room; the staircase has been removed from No. 53 and some alterations have been made to the internal partitions. Throughout the building most of the rooms have moulded cornices; only one 18th-century fireplace surround remains, with simple timber mouldings. The staircase has an open string and turned balusters clustered at the foot of the stair round a turned newel.

(46) House, No. 57, is referred to as 'new erected' in a conveyance of 1759 and soon afterwards became the residence of Dr. William Burgh who lived there for nearly forty years till his death in 1808 (R. Davies, *A Memoir of the York Press* (1868), 271–3, 299; R. H. Skaife in *YAJ*, I, iv (1870), 322). The main part of the house assumed its present appearance *c.* 1830, when it was heightened and stuccoed at front and back. An extension to the N.W. is mostly modern above the bottom storey.

The stuccoed front is shown in the drawing opposite p. 55; the back has a small projection at the E. corner for closets on two floors and in the centre is a projection probably added *c.* 1830 providing closets on three floors opening off a back porch on the ground floor and off extensions of the half landings of the staircase above; round-headed windows giving light to the staircase are paired with round-headed blind recesses in the closet walls. The staircase, with open string and turned balusters, is of the 18th century but the rest of the fittings are mostly of the 19th century, including those to the two intercommunicating reception rooms which occupy the whole of the front of the house on the first floor.

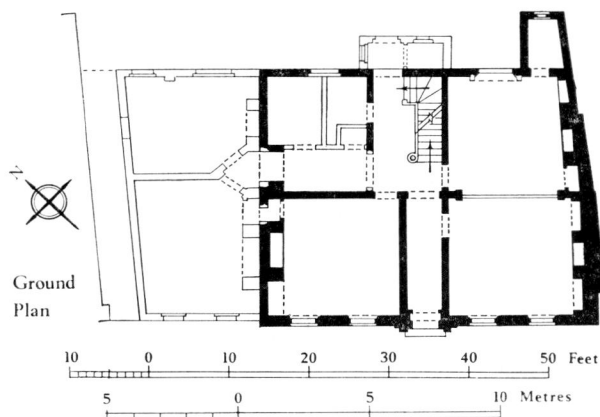

Ground Plan

Fig. 58. (46) No. 57 Bootham.

(47) House, No. 59, was built in the 18th century on an **L**-shaped plan probably having a back porch with a closet over it in the re-entrant angle but the lower part of this projection has been surrounded by later additions. The property was purchased in 1764 by Henry Teasdale, butter factor (Deeds), and was still occupied by Mr. Teasdale in 1791.

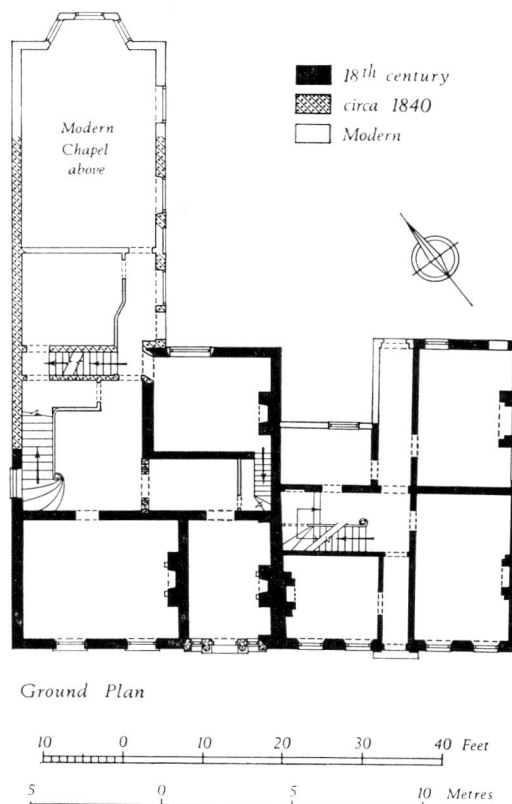

Modern Chapel above

Ground Plan

Fig. 59. (47, 48) Nos. 59, 61 Bootham.

The street front has a stucco band at first-floor level and a timber gutter cornice. The central entrance has been re-modelled with a 19th-century doorcase and the lower windows have been reglazed without bars. The back wing has a double-pitch mansard roof, and bay windows have been added to the first and second floors. The interior fittings are generally of the late 18th century and of a modest character. The staircase has good turned balusters, three to a tread, and the handrail is swept from one flight to the next without newels.

(48) House, No. 61, was built towards the end of the 18th century on an **L**-shaped plan, three storeys high with cellars and attics (Plate 88). In *c.* 1840 a long back wing was added, with a spacious reception room on the first floor approached by a new staircase in an enlarged stair hall. At the same time the original back wing was heightened to four storeys, the front entrance was re-modelled and iron balconies were added to the first-floor windows, uniform with those of 1843 at No. 29 Penley's Grove Street (297). Fittings of this period are generally very similar to those in Nos. 57 and 65 Bootham (46, 49). In modern times the first-floor room

in the back wing has been adapted for use as a chapel and the whole wing extended to provide a sanctuary.

The elevation to Bootham is shown in the drawing opposite p. 55: at first-floor level is a painted band of applied timber; the sills of the first-floor windows were probably lowered when the balconies were added; the top windows have been refitted with 19th-century sashes. On the N.W. side an 18th-century window to the staircase with a round-arched head was reglazed in the 19th century with marginal panes.

The internal fittings are mostly of c. 1840. The main staircase rises the full height of the house under a lantern. The stair hall is entered through a screen with Ionic columns between pilasters echoing the motif of the front entrance. The staircase has cantilevered stone treads and iron balusters alternately of elaborate arabesque design and of symmetrical design about a central roundel (Plate 131). Some of the rooms have elaborate friezes (Plate 122) and centrepieces to the ceilings (Plate 120). At the top of the house some of the rooms retain original doors and architraves.

(49) RECORD HOUSE, No. 65 (Plate 98), was built c. 1827, the rating assessment being more than doubled in 1828. The owner was then Mrs. Barbara Ashton Nelson and she built the present house further back from the road than the previous house on the site. From 1838 the owner was R. W. Riddle and the occupier from that year until c. 1854 was John Clough, banker. The house is now (1971) used as government offices.

The house, of two storeys above a basement, is built of ashlar and white brick with a slated roof carried down to widely projecting eaves; it is built on a nearly square plan with the rooms arranged around a central stair hall lit by a glazed lantern which rises above the roof with a chimney at each of its four corners. Two small wings project N.E. at the back. The S.W. front towards Bootham is in fine ashlar with a central curved bay projecting forward with three round-headed windows to the ground floor. The entrance is on the N.W. side which is of white brick with stone dressings; a *porte-cochère* with two columns supporting a simple entablature stands in front of the doorway. The S.E. side is also of white brick with stone dressings, each floor having five windows with moulded stone architraves. The back, to the N.E., is of common red brick.

The central staircase has cantilevered stone treads and cast-iron balustrading enriched with honeysuckle ornament; below the lantern lights is a band of richly moulded plaster-work with cornucopiae and foliage (Plate 122). On the ground floor the middle of the S.W. front is occupied by the drawing-room, extending into the curved bay where the windows have panelled pilastered jambs with enriched imposts. A plaster border to the ceiling and the jambs of the marble fireplace are both enriched with a flowing leaf pattern. The rest of the house is more simply fitted, without decoration.

The grounds are separated from the road by iron railings set between stone piers carved with honeysuckle ornament (Plate 106). The railings were made by John Walker of Walmgate for John Clough.

(50) HOUSES, Nos. 67, 69, 71, 73, two pairs forming a three-storey terrace, were built in the first half of the 19th century. No. 71 retains the original reeded door-case to the entrance and a shallow bow window on the ground floor; the others have been converted to shops. All have shallow segmental bows to the first floor. Each house has one front room and one back room to each floor; in Nos. 67, 69 the staircases are placed transversely between the rooms.

(51) HOUSES, Nos. 75, 77, have the date 1770 on a rainwater head (Plate 106), and there is no reason to doubt that this is the date of construction. Both houses had wings added to the rear in the 19th century; they have been much altered in conversion to shops in the present century.

The S.W. front to Bootham is in five bays with continuous bands joining the sills of ground and first-floor windows and a timber cornice at the eaves. The doorways are set centrally, side by side. In No. 75 the entrance passage widens out at the back to accommodate the staircase which retains the original turned newels and balusters.

S.W. side:

(52) HOUSES, Nos. 8, 10, were built probably as one dwelling in the mid 18th century. The building is of two storeys with a plat-band at first-floor level; the ground floor has been converted to shops.

(53) HOUSE, No. 34, was built in the early 19th century. It was of three storeys with a tiled roof. *Demolished.*

(54) HOUSES, Nos. 40, 42, are of three storeys; No. 40, built in the early 19th century, retains an original doorway with attached fluted shafts and patterned frieze but No. 42, of the mid 19th century, has been much altered.

(55) HOUSE, No. 52, of three storeys, was built in the early 19th century. In the middle of the century the front was altered: the wall was heightened some 3 ft.; a new entrance doorway was made with small windows to each side; some of the upper windows were altered; and a small three-storey wing was added at the back. Later some further extensions were made at the back.

The front is faced with red brick and is three bays wide, with hung-sash windows under flat brick arches. The back is in common brick. The interior has a dining-room at the front, occupying two bays, and an entrance hall occupying the third bay which seems originally to have contained a separate little room now opened out into the hall. Behind, the central stair-case was rebuilt c. 1850 and is placed between the kitchen to the N.W. and the study which runs into a later addition. On the first floor a saloon at the front occupies two bays and communicates through a wide opening with a smaller room in the third bay. Few of the original fittings survive.

(56) HOUSE, No. 54, is a substantial three-storey house of *c.* 1840 (Plate 89).

The symmetrical five-bay front, built in white brick, has the central bay and pilaster strips at each end projecting forward; brackets under the eaves cornice are set against a white plaster frieze which is matched by a deep white band under the second-floor windows. The front door is protected by an open porch and the window above is emphasised by a bold plaster surround. The back is built in red brick and has a central projection which on the upper floors contains alcoves leading off the half-landings of the staircase and giving access to water-closets also in the projection. The plan, apart from the projection at the back, is square with the common arrangement of four rooms disposed two on each side of a central hall. This central hall has the entrance hall divided from the stair-hall by Corinthian pilasters carrying an enriched entablature, the doorways to principal rooms have enriched architraves and overdoors and the alcove off the half-landing is reached between fluted columns with foliated capitals, but the detail of these decorative features is all coarsely designed.

(57) BOOTHAM LODGE, No. 56, now offices, was built by Thomas Walker, solicitor, between 1840 and 1845 (Deeds) (Plate 90). A kitchen wing at the back appears to be a very early addition, built to replace an original kitchen in the basement. The design is developed from the usual square house with central hallway by addition of side wings; that to the N.W. accommodates the main staircase and additional rooms, one of the front rooms in the main block being opened to the entrance to make a spacious hall leading to the main staircase; the S.E. wing consists only of one first-floor room over a carriage entrance.

The house is of two storeys with walls of pale-coloured brick. On the street front the windows are emphasised by painted surrounds, those on the ground floor and the central window above having side pilasters with entablature above. The central porch has Greek Doric columns and forms a balcony to the window above with smaller balconies formed on the cornices of the flanking windows, all with cast-iron balustrades.

Inside, the entrance passage opens to the hall between Doric columns. The main staircase is beyond the hall; the secondary staircase opens off the entrance passage. Both the staircases have cast-iron balustrades with mahogany handrails (Plates 130, 131). The principal room on the ground floor, at the back, has a richly decorated ceiling cornice (Plate 122) and over the moulded architraves to the doors are overdoors decorated with urns and festoons in low relief by Wolstenholme (Plate 112). On the first floor a large saloon occupies the front of the house, originally with a fireplace at each end and two doorways symmetrically disposed. One fireplace has been removed and the position of one doorway has been altered. Both the fireplace and overdoors are decorated with figure subjects (Plates 113, 115) and the enrichment is in composition by Wolstenholme. The ceiling cornice is decorated with flowers and leaves.

CLARENCE STREET (Monuments 58–65)

Clarence Street was constructed between 1830 and 1834 when Thomas Snowdon, tailor, is listed as living there (Directories); there were houses being built in 1832 (*YG* 17/3/1832) and later in the 1840's when five newly-erected houses were offered for sale in *YG* 22/5/1841. The even numbers on the E. side of the street were built on some of the 20 building plots advertised as forming the S. side of Clarence Street in *YG* 22/2/1845. Unless otherwise described the houses in this street are of two storeys.

W. side:

(58) HOUSES, Nos. 3–11 (odd), were built *c.* 1840; Nos. 5, 7, 9 are of one build and Nos. 3 and 11 are separately constructed. No. 3 has a carriageway to the rear on the ground floor. Most of the decorative features were added *c.* 1860. *Demolished.*

(59) HOUSES, Nos. 13–23 (odd), date from *c.* 1838. Nos. 15–19 have bay windows to the ground floor; all first-floor windows have segmental arches of common bricks. *Demolished.*

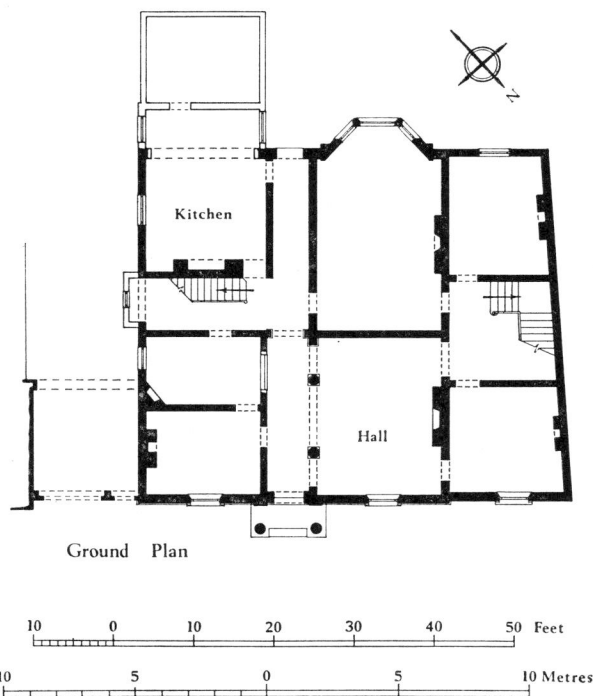

Fig. 60. (57) Bootham Lodge, No. 56 Bootham.

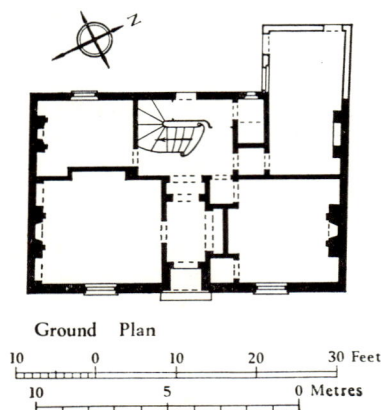

Ground Plan

```
10    0    10    20    30 Feet
10         5         0 Metres
```

Fig. 61. (60) No. 47 Clarence Street.

(60) No. 47, is a symmetrical double-fronted house which was built during the 1840's. An extra bay with a carriageway was later added to the S. The front doorway has a semicircular head and stucco architrave with five raised voussoirs. The plan shows an ingenious attempt at providing architectural interest in a confined space (Fig. 61). The staircase is lit by a round-headed window in the rear wall.

(61) Nos. 49–75, 79–85, 91–95 (odd), are small terraced houses of *c.* 1840, generally uniform in character but of different builds and with variations in detail and size. Each house has a timber doorcase with a rectangular fanlight over the door. The doors are generally of six moulded and fielded panels. Most houses have a projecting bay window to the ground floor and a single hung-sash window under a segmental arch to the upper floor, but not all the bays are original. Nos. 91 and 95 are of three storeys. *Nos. 49–73 demolished.*

E. side:

(62) CLARENCE HOUSE is a detached house of three storeys and attics built to an irregular plan in *c.* 1830. It was called Clarence Cottage in 1838, when it was occupied by Mrs. Eleanor Wilson, and in 1843 and 1846, when it was occupied by Thomas Meynell Esq., Secretary to the Philosophical Society (Directories), and in 1850 (OS).

The front elevation is of three bays, with continuous sill-bands to the first and second-floor windows; the central window on the second floor is a dummy. The central doorway has side pilasters and a small hood supported on coarse console brackets. The eaves project boldly. *Demolished.*

(63) HOUSES, Nos. 12–24 (even), were built *c.* 1840. *Demolished.*

(64) HOUSES, Nos. 36–74 (even), were built during or

after 1845. They are similar to those on the W. side of the street but more extensively altered. Nos. 48 and 50 are of three storeys.

(65) HOUSES, Nos. 84–86 (even) and No. 167 Lowther Street, were built shortly before 1850.

CLIFTON (Monuments 66–98)

Until the early 19th century Clifton was a rural village largely dependent on dairy farming. A new era was introduced in 1836 when Earl de Grey offered the greater part of the village for sale. Papers relating to this sale among Clifton Manor papers in The Guildhall, York, together with a large amount of other documentary evidence, provide information as to owners and occupants of houses (*see* Barbara Hutton, *Clifton and its People* (YPS 1969)). Unless otherwise described the monuments in this area are of two storeys.

BURTON STONE LANE

(66) HOUSES, Nos. 19–37 (odd), were built on land bought by William Bellerby, joiner, at the 1836 sale of the de Grey estate, almost certainly by Bellerby himself.

The houses form a terrace, apparently of one build but with slight variations in size and detail. There is a carriageway between Nos. 27 and 31 which gives access to the rear, where William Bellerby had his yard, and to No. 29.

CLIFTON

The road forms the continuation of Bootham northwestward, leading to the village of Clifton and continuing as the main road to Thirsk. *See also* p. 94.

N.E. side:

(67) Nos. 2, 4, a pair of large three-bay town houses of three storeys and attics, were built soon after 1825 on land known as Bootham Garth or Coates' Closes, purchased by the Acaster Malbis Charity Trust from Christopher Coates in 1658. In 1825 the site passed to Thomas Bell, bricklayer, and William Bellerby, joiner, on a 99-year lease, one of the conditions of which was that they should erect buildings of at least three storeys and to a cost of £2,000 within 17 years (documents in possession of owner). No. 2 retains its original central doorcase with slender three-quarter round fluted columns (Plate 109).

(68) HOUSE, No. 8, was built between 1782 and 1784 for Joseph Goodlad of Harrogate, innholder, (Deeds) and has a wing at the back added *c.* 1830. (Fig. 63.)

The house, of three storeys, is four bays wide (Plate 89) and has a stone band and continuous stone sills in the idiom used by John Carr; at the eaves is a dentilled timber cornice. The back wing has large hung-sash windows set in a segmental bay and provides an additional spacious living-room. Inside, the staircase retains the original turned balusters and the front rooms have the original enriched and decorated surrounds to the fireplaces (Plate 114).

(69) THE WHITE HOUSE, No. 10, was built in the 18th century, possibly before 1731, when William Roberts of York, mercer, bought the property which then included the site of No. 8 (YCA, E93, f. 63; Deeds of No. 8). It was originally only two storeys high but later in the 18th century it was heightened to give a symmetrical three-storey front of five bays (Plate 91). From 1828 to *c.* 1853 it was the home of the Rev. D. R. Currer, a cleric without clerical office but an influential magistrate (Hutton, 20), and in his time the house was extended to each side, the additions being of two storeys only but equal in height to the original building, and the whole front was stuccoed and given a new eaves cornice; a two-storey bay window to the upper floors is a later addition. When the wings were added, the house was also extended at the back, the original eaves cornice remaining above the additions. In modern times the house has been converted to offices with extensive wings added at the back.

The front elevation has a good timber Doric doorcase. Inside, the curved staircase, contemporary with the heightening in the later 18th century, has stone treads and plain iron balusters. Some of the fireplaces retain the original Adamesque surrounds. A large first-floor room in the N.W. addition (now divided) has an elaborate foliated cornice.

(70) HOUSES, Nos. 14, 16, were built probably *c.* 1800 and form a symmetrical pair with a wide pedimental gable over the front (Plate 91) in a style favoured by the architect Thomas Atkinson (d. 1798). For a time during

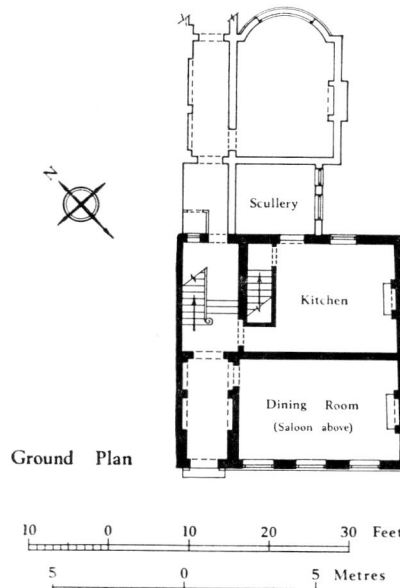

Fig. 63. (68) No. 8 Clifton.

the 1840s No. 14 was occupied by David Russell, solicitor, son of David Russell of Clifton Grove, and No. 16 by his friend William Whytehead, a solicitor active in local government as a Commissioner under the Act of 1825 (VCH, *York,* 264–5).

The house is of three storeys and the front has stone bands at the first floor and at first-floor sill level. The entrance doorways have timber Doric doorcases with open pediments.

Inside No. 14, the staircase has slender turned balusters and two fireplaces have surrounds decorated in the Wolstenholme manner. In No. 16 the staircase has 19th-century cast-iron balusters.

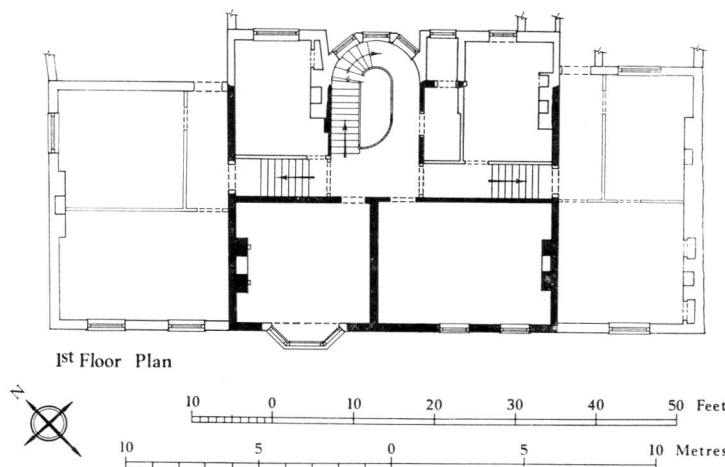

Fig. 62. (69) The White House, No. 10 Clifton.

(71) HOUSE, No. 18, was built shortly after the sale of the de Grey estate in 1836 on a site purchased by Mr. Henry Elsworth. It is detached and double-fronted and executed in white brick. Both front and rear elevations were originally symmetrical. An original doorcase of the Tuscan order remains at the rear.

(72) TERRACE, Nos. 26–32, of three storeys with semi-basements and attics, was built on a plot bought by William Bellerby, joiner, at the sale of the de Grey estate in 1836. The houses are probably those referred to in Bowman, 11–12, in connection with the finding of a Samian bowl in 1841 'when excavations were made for some houses on a piece of land called Chapel Close opposite the Proprietary School at Clifton'. They were called Burton Place in 1850 (OS).

Each house is of three bays with a doorway at the N.W. end. The windows have stone lintels with simulated voussoirs and key blocks. The terrace has a heavy moulded cornice.

(73) HOUSE, No. 36, was built probably in the second quarter of the 18th century as a small three-storey town house with front and back rooms on one side only of the entrance passage. Late in the 18th century or early in the 19th the house was widened by two bays to give a double-fronted house with rooms on both sides of the entrance.

The street front has projecting bands dividing the storeys and continuous stone sill-bands to the two lower storeys. The front door is deeply recessed within a semicircular-headed opening. The interior has been converted to form flats and a new staircase built in a new position. Few of the original fittings survive.

(74) No. 40, is a three-bay town house of three storeys and attics, built in the early 19th century.

The principal feature of the house is the front elevation to Clifton which has a stone plinth, continuous sill band to the first-floor windows and round-arched heads of fine red gauged brick to the ground-floor windows and the doorway. The first and second-floor windows have flat arches of gauged red brick (Plate 90). It has a conventional plan, two rooms deep, with the staircase at the rear opposite the front entrance. The saloon is on the first floor occupying the whole width of the frontage. The interior has good Regency fittings.

(75) CLIFTON VIEW, Nos. 42, 44, dates from the second half of the 18th century but incorporates an earlier chimney which may have formed part of a cottage on the site first mentioned in 1696 (Deeds); it was first insured with the Sun Insurance Company in 1786, the owner then being Mrs. Dorothy Elston and the occupier Mr. Ellis. The house was built on a wide frontage giving an elevation of five bays, the position of the early chimney being reflected in the unequal spacing of the windows (Plate 90); on plan the house is only one room deep and the entrance passage leads to a staircase projecting at the rear. In the middle of the 19th century the house was divided into two and a second entrance doorway made between the two S.E. windows. At the same time a new timber cornice was made at the eaves.

The house, of three storeys, is divided at the front into three stages by projecting brick bands. The front entrance has a timber surround with an open pediment over a semicircular fanlight, and the windows are set under segmental arches of common brick including blue headers. Inside, the staircase has slender turned balusters; two late 18th-century fireplace surrounds survive on the upper floors, and a number of iron firegrates, one marked Carron.

(76) HOUSE, Nos. 64, 66, (Plate 85) of two storeys and attics with brick walls and tiled roof, was built in the late 17th century incorporating parts of an earlier timber-framed structure, probably of the 16th century. The house was modernised and partly rebuilt in 1962.

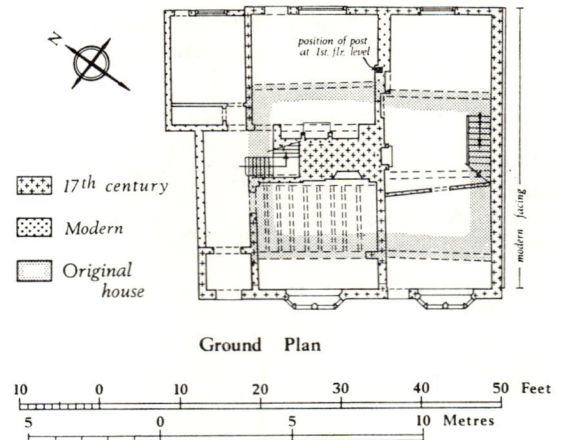

Fig. 64. (76) Nos. 64, 66 Clifton.

The S.W. elevation comprises two main bays and a projection for a porch on the N.W. side. The ground floor is of plain brickwork; the upper floors are fully rusticated and rise to curved gables surmounted by finials. Brick string-courses at the floor levels are cemented over. The entrance has a four-centred head, and jambs and head are cement-rendered to simulate stone dressings. Under the main gables two-storey bay windows are carried on moulded brick corbelling; the mullions were all originally of brick, probably plastered, but one window has been replaced in timber and the others are rendered in modern cement. The S.E. wall has been rebuilt. The back is mostly of 17th-century brickwork with modern window openings; the back gables are plain and rebuilt in modern brick. On the N.W. are modern additions. Inside, the exposed beams and ceiling joists, indicated by broken lines on the plan, are mostly of 16th-century date and on the first floor is a timber post with a big curved brace rising to a tie-beam, also part of the earlier structure. Its position indicates that the original N.E. elevation was jettied. The entrance door, of pinewood, which has been moved from its original position, is of 17th-century style and panelled to form a geometrical pattern. The central chimney has 17th-century brick fireplaces with arched heads. One of the rooms has an 18th-century fitted corner cupboard with scrolled brackets under the shelves.

(77) HOUSES, Nos. 74, 76, are a three-storey pair built in the early 19th century. No. 76 has a pantiled roof.

(78) SHOP, No. 88, is a small building of one storey and attic which was built probably in the 17th century as a single-storey timber-framed structure of two unequal bays. It may originally have been for farm use rather than domestic, but had been converted into two cottages before it became a shop. In the sale map of de Grey property in Clifton (1836) it appears as the property of Oswald Barker. The walls have been re-built in brick leaving only some of the main posts, from which braces rose to tie-beams and wall-plates. The present roof, with queen-struts carrying side purlins, appears not to be original.

(79) BARKER'S TERRACE, Nos. 96–106 (even), was built after the sale of the De Grey estate on a site bought by Oswald Barker, who in 1843 was living in one of the houses (Directory). The houses form a range of six small and plain tenements with simple plans. Nos. 104 and 106 were built slightly later than Nos. 96–102.

S.W. side:

(80) BOOTHAM GRANGE was built as a pair of large semi-detached houses in the second quarter of the 19th century and converted to flats in the 20th.

The street front is six bays wide and three storeys high and terminated by giant brick pilasters with Corinthian capitals. The window sills are joined to form continuous bands across the upper storeys. The two doorways, one now converted to a window, have semicircular fanlights set within continuous moulded architraves without imposts. Inside, only one of the original staircases remains; it occupies a large open well, top-lit by a large glazed lantern, and has stone steps with elaborate cast-iron balustrades in which decorative roundels alternate with standards entwined with vines (Plate 131).

(81) ST. CATHERINE'S, No. 11, is a detached house of c. 1840 built in white brick. It has a front of four bays giving good-sized rooms with two windows to one side of the entrance and small rooms with one window to the other. Beyond these last is a single-storey bay containing domestic offices. Bay windows and closets have been added at the back. The front elevation is simple, with a pedimented porch and hung-sash

windows. Inside, the main staircase has undulating iron balusters of plain square section. The secondary staircase is all of wood. At the front of the house are simple iron railings by Walker of Walmgate.

(82) BURTON COTTAGE, No. 27, is a villa of c. 1830–1840, of two storeys, built of white brick (Plate 100). The house is very compactly arranged with the staircase in the front entrance hall between the original kitchen and a small study; the principal rooms are at the back facing S.W. and looking onto the garden. At the front the arched entrance is set in a slight projection under a pediment. At the back the lower windows have small side lights flanking the main lights and on the upper floor the same proportions are maintained by sliding shutters flanking the windows, under fretted pelmets. Inside, the staircase has slender turned balusters; the ceilings of the principal rooms have elaborate foliated centre-pieces (Plate 121), one having cast-iron foliage conceal-ing ventilation ducts.

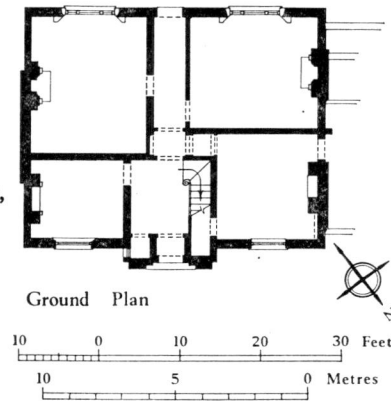

Fig. 65.
(82) Burton Cottage, No. 27 Clifton.

Ground   Plan

(83) HOUSE, No. 29, was built in the second quarter of the 19th century. It is of two storeys and built of white brick. The front entrance is set between projecting bay windows rising through both storeys and finished with gables.

(84) HOUSE, now St. Olave's School, was built for David Russell, solicitor, between 1813 and 1818 (Hargrove, I, 288; YC, 5 April 1813), forming a sub-stantial residence standing in its own grounds. It is of two storeys with stuccoed walls and slated roof and comprises a main block, almost square, and a lower wing for domestic offices to the N. The main front is symmetrical with a central porch; the roof is low-pitched with a deep projection of the eaves. Many of the original interior fittings have been removed, but the original staircase with slender turned mahogany balusters remains.

(85) HOUSES, Nos. 51–57 (odd), were constructed after the sale of the de Grey estate in 1836 in *c.* 1840.

No. 51 retains a doorcase with reeded pilaster jambs and a shallow bow window with two-panelled shutters. Nos. 53–57 have been converted into shops.

THE GREEN

(86) THE OLD GREY MARE, p.h., was built in the late 17th century. It is said to stand on the site of the Maypole Inn which was burnt down in 1648; in 1820 it was known as the Grey Horse. A gabled cross-wing to the N. was added in the late 19th century. The building is of two storeys with walls of colour-washed brickwork; it has been very much altered and a hood to the front door, of late 17th-century character, was added by W. G. Penty in the late 19th century (*Builders' Journal and Architectural Record*, 7 Nov. 1900).

(87) Nos. 9, 10, 11, are a range of three cottages which were built *c.* 1835. No. 11 has a symmetrical front elevation.

(88) Nos. 14, 15, were built *c.* 1840 as a symmetrical double-fronted house (No. 14) with an outbuilding at the W. end. This last was later converted into a small house to form No. 15. The window arches are segmental and the roof tiled. *Demolished.*

(89) No. 16 is a gardener's cottage in the Gothic style (Plate 101), built after 1836 for the Roper family, owners of Clifton Croft (*see* (98), p. 69). Roper arms are carved on the W. wall. The weathervane has on it the initials JR for John Roper who owned the property with his brother Edmund from 1826 until his death in 1875.

(90) HOUSES, Nos. 22, 23, (Plate 101) were built on a site bought by Seth Agar, grocer, at the sale of the De Grey estate in 1836, by George and Eli Horsfall, joiners, and completed in 1839 when the roof truss of one was signed.

The houses are a pair of detached dwellings of one storey with dormered attics, built in a Gothic cottage style in stucco-rendered brick. The doorways have four-centred arches and doors with four trefoil-headed lights to the upper parts. No. 23 retains its original fenestration at the front but two bay windows have been added to No. 22.

WATER END

Water End leads from Clifton Green down to the river where, until 1960, there was a ferry.

(91) Nos. 2, 4, 6, Ellison Terrace, are a symmetrical terrace of two single-fronted houses flanking a double-fronted three-bay house, two of which were standing by 1819 (Deeds). The doorways have slender reeded pilasters and reeded entablatures; the doors have six

fielded panels and semicircular fanlights above. The windows to the ground floor have louvred shutters and those to the first floor slightly segmental arches of gauged brick. (Plate 102.)

(92) No. 8 (Plate 102) was built, probably as a small farmhouse, in the late 17th century. It is of two storeys and L-shaped on plan with tumbled brickwork to the gable at the back. In the second half of the 18th century cottages were built adjoining on the W. and No. 8 was refronted to give a uniform elevation to the whole range. In the 19th century the back wing was made into a separate tenement and the interior was much altered. The original staircase, with heavy bulbous balusters (Plate 124), remains but it has been moved from its original position. Modern alterations include the construction of a fireplace opening of thin 15th-century bricks from the filling of the timber-framing of the Fox Inn, formerly in Petergate, the introduction of a late 18th-century fire-surround from No. 64 North Street (*York* III, 107), and a late 18th-century grate (Plate 132).

(93) COTTAGES, Nos. 10, 12, were built in the second half of the 18th century and may originally have formed a single dwelling (Plate 102).

(94) GREEN TREE COTTAGE, No. 28, now a private house, was built in the early 19th century. It was called the Green Tree in 1830 and kept by George Holgate and in 1836 the Sycamore Inn, kept by Alice Holgate (Directories; sale map of de Grey Estate). In 1850 it appears as the Sycamore Cottage p.h. (OS).

The house is detached and has symmetrical, stucco-rendered front elevation with two gables. The front entrance is in Tudor style.

(95) HAVERFORD, formerly Cliff House, was built as the residence of William Catton, woollen draper of High Ousegate, in *c.* 1842. The house stands in its own grounds and follows the general pattern of St. Olave's School (84) and Clifton Croft (98): all three have a square main block containing the principal rooms and a lower wing containing the domestic offices and servants' rooms, but Haverford is architecturally rather more ambitious, having stone dressings to the white brickwork and an elaborate stone porch with three arched openings between Tuscan pilasters; the wide eaves are supported by paired brackets. The side and back elevations have bay windows. The roof is low-pitched, hipped, and slated. The plan shows four rooms in the main block, flanking a central hallway, with the main staircase to the rear and a secondary staircase to one side.

(96) Nos. 39, 41, 43, are a range of three houses built

in *c.* 1849, possibly to house the domestic staff of Government House (post 1850). All three have windows with slightly arched heads of gauged brick. No. 43 is more elaborate than the other two and has a heavy doorcase and a large single-storey stucco-dressed bay window to the ground floor and continuous sill band in stucco to the first-floor windows. *Demolished.*

(97) St. Hilda's Garth, formerly Clifton Holme, at the end of Ousecliffe Gardens, was the house of a solicitor, Joseph Munby, for whom it was built in 1848 (Hutton, 24; Directory; OS).

It is a large two-storey house in white brick with stone dressings (Plate 98). The main block is rectangular with a porch of three bays in the front and a segmental bay projecting at the back; a lower servants' wing projects to N.E. The porch with three stilted segmental arches carried on Tuscan columns is the first real departure from the Georgian tradition in this area, and is stylistically more closely allied to the buildings of the third quarter of the century than to its predecessors.

(98) Clifton Croft, a substantial house standing in its own grounds and facing Greencliffe Drive, was built *c.* 1830 for John Roper, wine merchant; it is of two storeys and comprises a square main block in white brick, and a lower wing partly in red brick; the roofs are covered with Westmorland slates. The main front, to N.E., is symmetrical with an open Tuscan porch in the centre.

De Grey Street

De Grey Street was 'newly set out' in 1847 (*YG* 26/6/1847); it does not appear in the 1846 Directory.

(99) Houses, Nos. 5–11, 25–28, were built *c.* 1849. They are of two storeys with basements and attics. The doorcases have pilaster jambs with sunk panels, and simple entablatures with modillioned cornices. *Demolished.*

Eldon Street

(100) Nos. 39–51 (odd) and 92–102 (even) Lowther Street are 13 two-storey houses of one build erected in the 1840s (Plate 103).

Fawcett Street

(101) The Woolpack, p.h., is a three-storey house built *c.* 1845. The front elevation is of two bays with a simple entrance and a large three-light sash window beneath an arch of gauged brick to the ground floor.

Fishergate (Monuments 102–105)

Fishergate is a continuation of Piccadilly southwards outside the walls. The site of the Augustinian Priory of St. Andrew (5), which stood on the W. side of Fishergate, is now covered by factory premises. Redevelop-

ment on a large scale began *c.* 1830; modern houses were being advertised for sale in 1831 (*YG* 19/2/1831).

E. side:

(102) House, No. 29, until 1972 the nunnery of the Sisters of St. Vincent, was built in the late 18th century. It was acquired by the Sisters in the late 19th century and in 1902 a large wing was added to the S. for a day nursery; this is now disused.

The house (Plate 94) is of two storeys with a five-bay front, the middle three bays projecting slightly under a pediment. At the first floor is a stuccoed band and at the eaves a timber cornice with dentils. The central doorway is flanked by reeded columns carrying an entablature over a semicircular fanlight. The windows have cemented surrounds, not original, and in the pediment is an oval dummy window. To the N. a single-storey projection with a blind window was built probably as a screen wall only; a balancing projection to the S. has been removed but original openings in the S. wall indicate that the wing of 1902 replaced an earlier structure.

The house is built on a simple plan with front and back rooms each side of a central hall, the back part of the hall containing the staircase. This has an open string and iron balusters of square section, alternately straight and undulating, carrying a shallow, veneered handrail. Inside the staircase window is mounted a second casement glazed with a collection of coloured quarries, flashed and patterned, of *c.* 1830, probably by one of the Barnett family of York glaziers, and made up at some later date to fit its present position.

(103) No. 33 is a two-storey detached house with an asymmetrical four-bay main elevation built in the first quarter of the 19th century. The windows and the fanlight over the entrance have triangular heads and glazing bars forming geometrical patterns (Plate 101).

W. side:

(104) Houses, Nos. 16–40 (even), all date from *c.* 1830. Nos. 30 and 32 are of three storeys and the others of two only. Nos. 16–28 have doorways with reeded columns or flat pilasters. Nos. 18 and 20 form a pair and Nos. 22–28 a range of four with an open passageway through the centre. Nos. 34–40 are a range of four; their upper windows have painted lintels simulating arches, with markedly bowed soffits.

(105) Fishergate House (Plates 99, 128) was built in 1837 for Thomas Laycock Esq. to the designs of J. B. and W. Atkinson; it cost £4,500. Drawings preserved in the office of Messrs. Brierley, Leckenby and Keighley include a plan dated 1837, not exactly as built, another plan of the ground floor also dated 1837 marked 'now erecting', corresponding almost exactly to the existing house, and a plan for the first floor dated 1840 marked 'erected'. The house forms a solid rectangular block mostly of two storeys but partly of three, faced with

Fishergate House, York.

Redrawn from original plans
dated 1837 & 1840
by
J.B. & Wm. ATKINSON
Architects, York.

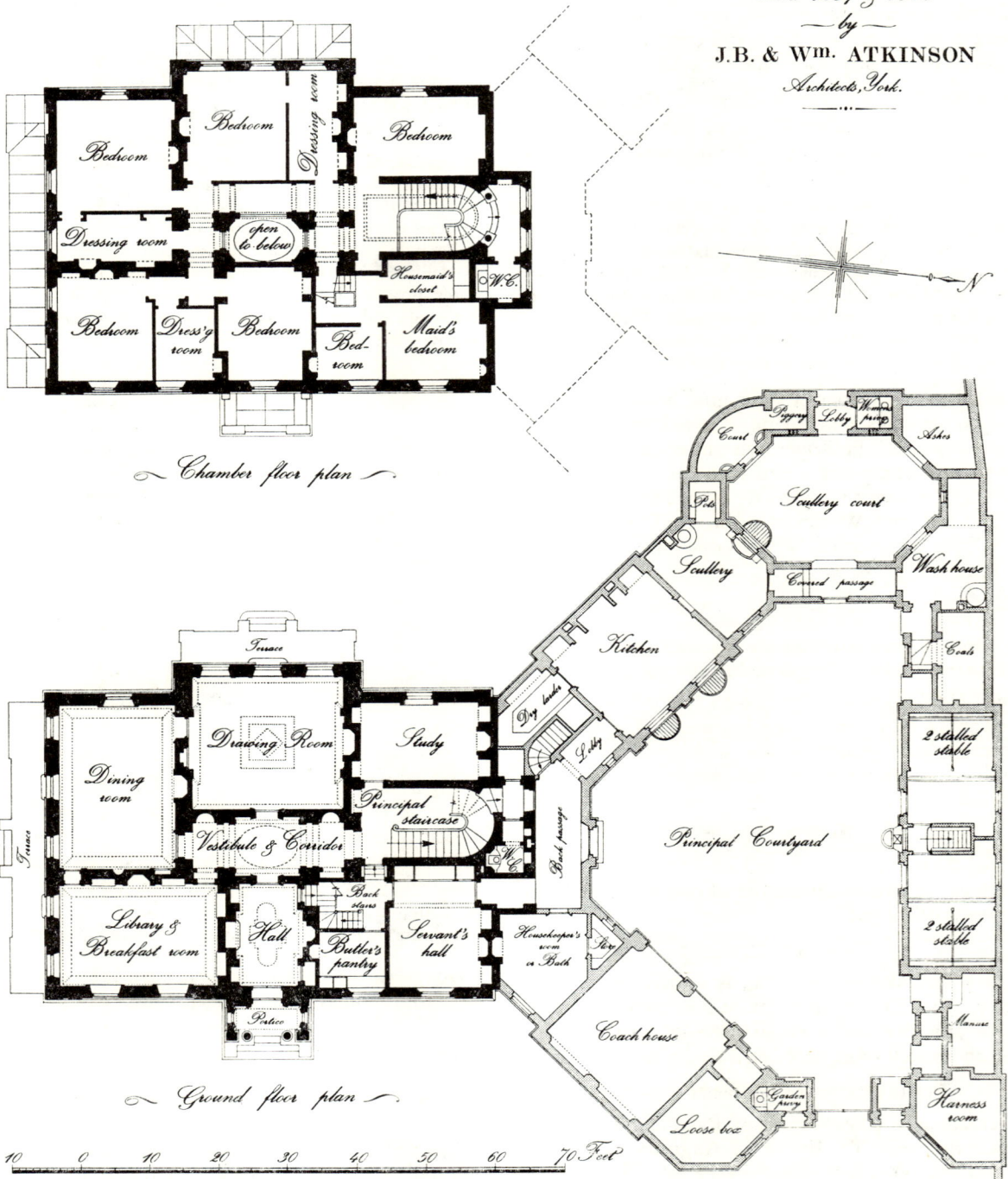

Chamber floor plan

Ground floor plan

10   0   10   20   30   40   50   60   70 Feet

Fig. 66.

white brickwork, with two low wings to N.E. and N.W. enclosing a service courtyard. The elaborate design of the central light-well is remarkable. The plan appears to be derived from Sir John Soane's plan for Tyringham (Stroud, Plate 97).

The E. front is divided into three unequal bays by brick pilasters with simple stone capitals. The central porch has Ionic columns *in antis*. The fenestration was originally all in two storeys although the staff rooms inside to the N. were arranged in three storeys. The windows have been altered to correspond with the internal arrangement. The S. side has the lower storey masked by a modern addition, only one round-headed window being visible; on the first floor three flat-headed windows are flanked by round-headed recesses. The W. elevation is divided by pilasters matching those to the E., with the central part projecting. Some additional windows have been put in the ground floor. On the S. side three round-headed windows in a central projection light the staircase. The roof rises from widely projecting eaves to a central flat surrounded by a stone balustrade with a chimneystack at each corner, and enclosing a lantern light. The service wings, which include stables and coach-house, are built of red brick; the coach-house has two segmental-arched openings, now blocked, and the windows in the range opposite are set in corresponding segmental-arched recesses.

The entrance hall has arched recesses in the side walls, some containing doorways, and a ceiling decorated with raised mouldings and rosettes (Plate 128). The inner hall has in the centre an oval opening in the ceiling to admit top light from the lantern above (Plate 128); to N. and S. are vaulted spaces with niches in the W. wall. The principal rooms on the ground floor are subdivided by modern partitions and only one fireplace remains, having a mantelshelf carried on shaped and carved brackets. The staircase to the N. has thin canti-levered treads and iron balustrading; it rises in two flights with a segmental landing, opening between columns with scrolled and foliated capitals to a further small landing lit by round-headed windows (Plate 128).

On the first floor there is an arcaded gallery (Plate 129) around the central light-well, which is surrounded by modern iron balustrading. Eight bedrooms and dressing-rooms have moulded plaster cornices but fireplace surrounds have all been removed.

The service quarters in the N.E. part of the main block are in three storeys served by a back staircase rising round an open well and having close strings, square balusters and turned newels. The two wings enclosing the courtyard have been much altered internally in conversion to offices.

Foss Bank

(106) HOUSES, Nos. 4, 5, are a two-storey pair built *c.* 1840. The window openings have segmental arches.

Fulford Road (Monuments 107–116)

Unless otherwise described Monuments 107–116 are of two storeys.

(107) FULFORD CONSERVATIVE CLUB, built *c.* 1810, is a detached house with symmetrical front elevation. The doorway has fluted pilasters and an open pediment.

(108) Nos. 196, 198, are a pair of three-storey terrace houses built in the second quarter of the 19th century. The window openings have cambered arches.

Grange Garth:

(109) FULFORD GRANGE, off the Fulford Road and 200 yds. S. of (105), is a house mainly of *c.* 1830–40 but which incorporates a building of the late 18th century. It has been added to in modern times and divided into three parts, known as The Grange, No. 37 Grange Garth, and The Croft.

The front part of the house, to the E., is of two storeys built in white brick. The E. façade, of five bays, has a central door-way with an Ionic porch; to the S. is a semicircular projecting bay. The staircase in the middle of this part has an iron balus-trade (Plate 131) and many of the openings have symmetrically moulded architraves butting against square blocks at the angles (Plate 111), typical of the period. The middle part of the building, with its principal elevation to the S., is of three storeys, built in red brick of the late 18th century. It may have formed the service wing of an earlier house; it has been much altered but retains some late 18th-century fittings. On the S. front is an entrance doorway with a good 18th-century timber pilastered doorcase with fluted architrave (Plate 107), probably not in its original position. Further W. is a lower extension contemporary with the E. part of the house, with a modern addition behind.

(110) No. 2, was built *c.* 1835 as a lodge to The Grange (109). It is of one storey and built in white brick with ashlar dressings (Plate 101).

(111) No. 32, is a cottage with Gothic details built *c.* 1840 on part of the Grange estate. The door and window openings have two-centred arched heads and the glazing bars are in the form of tracery (Plate 101).

Love Lane:

(112) LILAC HOUSE has been formed from a narrow range of three cottages, of which two are perhaps pre-1850.

(113) NEW WALK ORCHARD, cottage at 60394978, is a two-storey building incorporating a small earlier building, probably of the late 18th century, of one storey with brick walls and a low-pitched single-slope roof.

St. Oswald's Road:

(114) No. 3, is a two-bay detached house built *c.* 1840–50. The shorter side faces the road and additions have been made to the W. and E. The windows have rusticated arched stone lintels.

(115) HOUSE, No. 10, built c. 1840–50, is similar to No. 3, with a two-bay front facing the road and a gabled wing at the rear projecting to the E., but is all of one build.

(116) HOUSE, No. 20, also built c. 1840–50, is detached and double-fronted but with the original central doorway now blocked.

GILLYGATE (Monuments 117–140)

Gillygate lies along the outside of the City Wall to the N.W. It takes its name from the church of St. Giles which stood at the N. end of the street. The church had disappeared by the 17th century, though the churchyard was then still in use for burials. Houses were being built in Gillygate in the 12th century.

N.W. side:

(117) HOUSES, Nos. 3, 5, form a symmetrical pair four storeys high, built in 1797 by Thomas Wolstenholme, carver (1759–1812), whose decoration modelled in a plastic composition is to be found on fireplaces and doorways, etc. in many York houses (YGS *Report* 1969, 37–45). Wolstenholme occupied No. 3 himself and the property remained in the hands of the Wolstenholme family until 1887.

The original, very elegant, front elevation has been sadly mutilated; it was symmetrical with the two front doors together and central blind windows above. To each side was a matching tier of elaborate windows shown complete in a photograph of c. 1880. On the ground floor, where there are now shop windows, were shallow segmental bays each of three lights under round arches. The segmental form was continued to the first floor where square-headed windows were divided by attached columns with a shallow entablature above and a balustraded apron below. The second-floor windows did not project; each was of three round-headed lights with cornices above. The top storey has semicircular windows divided into three lights by timber mullions. At the eaves No. 3 retains the original timber cornice with coupled brackets. The back has plain hung-sash windows, many of which have been altered.

Inside, the ground floor has been stripped of its original fittings. The staircases have open strings with shaped ends to the treads with leaf decoration; in section the balusters form hollow-sided squares.

The fittings throughout the upper part of the house are mostly original and enriched with applied decoration. Bay windows are framed by pilasters enriched with reeding and garlands, with a frieze above enriched with anthemion ornament. Principal doorways have side pilasters with floral trails between bands of reeding and overdoors enriched with urns and garlands (Plate 110). Decorated segmental panels are incorporated into these overdoors or used elsewhere as isolated units over doors (Plate 112). In other places the architrave is enlarged to form similar panels over doorheads.

Similar segmental panels also appear over fireplaces. Throughout both houses the surviving fireplace surrounds are enriched with reeding, foliage, urns and garlands (Plate 115), even the plainest, in the top back bedroom, having enriched shelves and modelled masks on the frieze. Many of the fireplaces retain the original iron grates.

Front Elevation (as built)

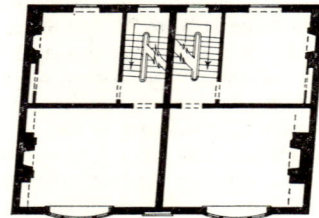

1st Floor Plan (as built)

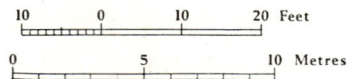

10   0   10   20 Feet

0   5   10 Metres

Fig. 67. (117) Nos. 3, 5 Gillygate.

(118) No. 9 is a three-storey single-fronted town house with two windows on both upper floors, built c. 1800. The ground floor was converted to shop premises c. 1900. The roof is pantiled.

(119) HOUSES, Nos. 11, 13, are of the early 19th century. They are both of three storeys and have pantiled roofs. No. 13 has been much altered.

PLATE 85

BOOTHAM. General view from W.

76) Nos. 64, 66 CLIFTON. Late 17th-century.

PLATE 86

(41) Nos. 39–45. 1748 and late 19th-century.

(41, 42) Nos. 45, 47. No. 47 by John Carr. Late 19th-century and 1753.

BOOTHAM.

PLATE 87

(43) No. 49. Late 17th-century and later.

(44) No. 51. Peter Atkinson senior, c. 1804.

BOOTHAM.

PLATE 88

(48) No. 61. Late 18th-century and later.

(45) Nos. 53, 55. c. 1765.
BOOTHAM.

PLATE 89

(68) No. 8 CLIFTON. 1782-4.

(56) No. 54 BOOTHAM. c. 1840.

PLATE 90

(75) Nos. 42, 44 CLIFTON. 18th-century and later.

(57) Bootham Lodge, No. 56 BOOTHAM. 1840–5.

PLATE 91

(69) The White House, No. 10. 18th-century and later.

(70) Nos. 14, 16. *c.* 1800.

CLIFTON.

PLATE 92

BACKS OF HOUSES

(43) No. 49 BOOTHAM. Late 17th-century.

(137) No. 70 GILLYGATE. *c.* 1770.

PLATE 93

BACKS OF HOUSES

(39) No. 33 BOOTHAM. 1753–5.

128) No. 28 GILLYGATE. Robert Clough, builder, 1769.

PLATE 94

(205) Rockingham House, No. 25 JEWBURY. 1792.

(102) No. 29 FISHERGATE. Late 18th-century.

PLATE 95

(241) No. 29. Late 18th-century.

(252) No. 60. Early 18th-century.

MARYGATE.

PLATE 96

(128) Nos. 26, 28. Robert Clough, builder, 1769.

(137) No. 70. *c.* 1770.

GILLYGATE.

PLATE 97

(279, 280, 282, 283) Nos. 40–48. Early 18th-century and later.

(278) Middleton House, No. 38. *c.* 1700 and later.

MONKGATE.

PLATE 98

(97) St. Hilda's Garth, CLIFTON. 1846–50.

(49) Record House, No. 65 BOOTHAM. *c.* 1827.

PLATE 99

(105) Fishergate House, FISHERGATE. J. B. and W. Atkinson, 1837.

(164) Heworth Croft, No. 19 HEWORTH GREEN. 1842.

PLATE 100

Front elevation.

Rear elevation.

(82) Burton Cottage, No. 27 CLIFTON. *c.* 1835.

(120) HOUSES, Nos. 19, 21, are of the early 19th century. They are of three storeys and have a modern uniform elevational treatment. No. 19 has been much altered on ground and first floors and has a pantiled roof. No. 21 is only one bay wide.

(121) HOUSES, Nos. 23, 25, of two storeys with stuccoed walls and pantiled roof, were built in the 18th century as a single dwelling. The building was re-modelled *c.* 1800; bay windows were added, new entrances with reeded columns and open pediments to the doorcases were formed (Plate 108), and a second staircase was put in.

(122) Nos. 59, 61, of two storeys may have been built as one house but are now two properties with two shop fronts to the ground floor. No. 59 retains an original doorway. Early 19th-century.

(123) HOUSE, No. 65, of two storeys, has a moulded doorcase with paterae in the angles and a pantiled roof. Early 19th-century.

(124) Nos. 67, 69, are a pair of two-storey single-fronted houses designed symmetrically with a single central chimneystack. The doorcases have reeded architraves. Early 19th-century.

(125) HOUSES, Nos. 71, 73, built *c.* 1835, are a three-storey pair forming a symmetrical composition with paired entrances, with decorated frieze and reeded jambs, in the centre. The ground-floor windows have segmental arches.

The entrance to each house is to a through passage widened at the back to accommodate a staircase which has slender turned balusters and newels. The fireplaces have reeded surrounds.

S.E. side:

(126) HOUSE, No. 12, three storeys high and five bays wide, was built in the first half of the 18th century but was largely refitted late in the same century. At the beginning of the 19th century the front was modified when the windows of the top storey were reduced in number from five to three. The ground floor is now occupied by modern shops.

The front is built in red brick with a projecting band of five courses over the first-floor windows. The central doorway retains the original moulded timber architrave and dentilled cornice carried on console brackets. The house is L-shaped on plan with a central entrance and staircase between two front rooms, and a back wing containing two rooms.

(127) HOUSE, Nos. 16, 18, 20, was built in the early 18th century, two storeys high on an L-shaped plan with a long seven-bay range facing the street and a rear wing at the S.W. end. Early in the 19th century a third

10

storey was added so that the roof is now continuous with that of No. 12, the N.E. part was enlarged at the back, and the whole was divided into two dwellings. Joseph Halfpenny, the York artist, lived here from 1803 till his death in 1811 (YCA, E96, f. 15v; Deeds; J. W. Knowles, *York Artists*, I, p. 199). On the front modern shops occupy the whole of the bottom storey. The first floor is of 18th-century brick with seven hung-sash windows under 19th-century lintels.

Inside, some plaster cornices and rehung doors are the only 18th-century fittings. Two 19th-century staircases have open strings and slender turned balusters (Plate 127).

1st Floor Plan

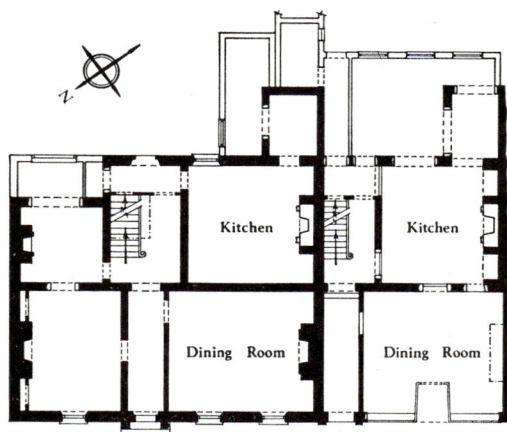

Ground Plan

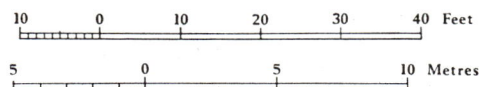

Fig. 68. (128) Nos. 26, 28 Gillygate.

(128) HOUSES, Nos. 26, 28 (Plate 96), were built in 1769 by Robert Clough, bricklayer and master builder of York, and the first occupants, in 1770, were Francis Smyth Esq. and Col. Robert Prescott. The houses were very well finished; notable among the fittings is the ceiling to the first-floor saloon in No. 28, probably executed by Robert Clough III, plasterer and son of the builder, baptised 1736. The front windows have been reglazed with large plate-glass panes and a shop front has been inserted in No. 26. The houses are not of equal size. No. 26 occupies three bays and No. 28 four bays. The original entrance to No. 28 remains (Plate 107), and at the side of it is a contemporary extinguisher. At the back round-headed windows light the staircases of both houses, and both have a small projecting closet wing, three storeys high (Plate 93).

Inside, No. 26 has lost many of its ground-floor fittings in conversion to a shop but the original staircase remains. On the first floor many of the fittings were replaced in the mid 19th century but an original fireplace remains. Throughout the house there are good 18th-century ceiling cornices. In No. 28 one of the front rooms on the ground floor and the saloon above have ceilings decorated with rococo plasterwork (Plate 119). Over the staircase the ceiling has a roundel with Gothic cusping (Plate 120). Amongst the fireplaces (Plate 114) are later insertions by Thomas Wolstenholme; one includes a figure panel which also appears in overdoors at Bootham Lodge (57) and Garrow Hill (147). Some of the rooms have enriched ceiling cornices (Plate 122). The staircase is original and has turned balusters with large plain umbrella-shaped knops (Plate 125). *See also* figs. 8f, 9b.

(129) HOUSE, Nos. 38, 40, was built in the late 18th century, probably *c.* 1787, this date having appeared on a pump formerly in the back yard. It has been altered by the conversion of the ground floor to shops. The house is of two storeys, rectangular on plan with four rooms disposed about a central entrance passage widening to a staircase at the back. Some of the rooms retain original cornices, simply moulded.

(130) HOUSES, Nos. 42, 44, are a three-storey pair with pantiled roof, built in the early 19th century.

(131) Nos. 50, 52 were built as a two-storey house in the 18th century. A third storey was added later and the whole has been very much altered.

(132) HOUSE, No. 58, is a very small two-storey building now converted to a shop; it is probably of 18th-century date.

(133) No. 62, is a house of the late 19th century but it has a small two-storey wing at the back of late 18th-century date, much altered.

(134) No. 64 was built at the end of the 17th or the beginning of the 18th century as a small two-storey house which probably had two rooms on each floor and a small projection at the back for a staircase. In the late 18th century the depth of the house was increased by the addition of rooms at the back under a second span roof parallel to the original. Later the two roofs were incorporated into one and the central valley eliminated. The front has been stuccoed and completely modernised but retains a rainwater head dated 1770. The wooden frame of a staircase window remains in the original back projection.

(135) No. 66 was built probably in the third quarter of the 18th century as a two-storey house with front and back rooms on one side of a through passage. A third storey was added in the early 19th century together with a back wing, and the doorcase to the entrance, with reeded pilasters, is also of this date.

(136) HOUSE, Nos. 68, 68A, was built in the late 18th century; it is of two storeys and was originally four bays wide but the front has been completely altered with shop fronts below and new windows above. The original plan was L-shaped but the re-entrant angle has been filled in, and the interior much altered.

(137) HOUSE, No. 70, was built *c.* 1770, of two storeys with a four-bay elevation to the front (Plate 96). It is generally similar to (136) above, but has not been so drastically altered. The original L-shaped plan provided two front rooms flanking the entrance passage and one small room and the staircase projecting at the back (Plate 92).

(138) HOUSE, No. 82, was built in the late 18th century. It is of three storeys, two bays wide, and with a projecting wing at the back. The ground floor has been converted to a shop.

(139) Nos. 84, 86, a pair of small houses of three storeys and attics, were built before 1823 by which time No. 84 was occupied by John Ware, yeoman, who remained there as John Ware, gent., until 1832 or later (Directories; title deeds of No. 84, York Town Clerk's Office, File 2604/3).

The front elevation of the pair was originally symmetrical. At the centre of the composition and covering the dividing line between the houses there is a blind recess to each of the upper floors; these and the upper floor windows have segmental arches. No. 84 has a pantiled roof.

(140) HOUSE, Nos. 88, 90, was built *c.* 1735. John Carr, yeoman, purchased the property in 1733 and his will dated 1738 refers to his 'newly erected dwelling house'. The house was drastically remodelled *c.* 1795. A

kitchen was added at the back in 1829 and further alterations took place in the second quarter of the 19th century when part was converted to a shop and the entrance was refitted to match the new shop front.

The house has two storeys, cellars and attics and the original plan comprised a single room on each side of the entrance hall which contained the staircase. The staircase is of the late 18th century with close string, turned balusters and square newels. *Demolished 1958.*

HAXBY ROAD (Monuments 141–143)

Monuments 141–143 are of two storeys.

(141) THE PUNCH BOWL HOTEL and HOUSES, Nos. 2–12 (even), were called Clarence Place in 1850 (OS) and some or all of them were built by 1838 (Directory).

The Punch Bowl Hotel is possibly a remodelling of two of the houses in Clarence Place on the 1852 OS map but has been altered out of recognition. Nos. 2 and 4 form a pair, but No. 2 i now incorporated into the hotel. The doorways together form a unified composition with three fluted pilasters with Ionic capitals and a frieze with paterae and incised fret ornament. No. 2 has a segmental bow window with similar frieze to the ground floor. Nos. 6, 8, 10 have bay windows with canted sides which are probably additions. No. 12 is a double-fronted terraced house of three and a half bays, with the main doorway in the middle and a smaller service door at the N.E. end. There are bay windows with canted sides on the ground floor.

(142) NO. 44 is a double-fronted house of the early 19th century but all the openings have been modernised. It was called Stray Cottage on the 1852 OS map.

(143) ASYLUM COTTAGE is a small double-fronted house of the early 19th century. The central doorway has pilasters with sunk panels and the window openings plastered arches simulating rusticated ashlar.

HESLINGTON ROAD (Monuments 144–147)

(144) NOS. 11–45, 49–51A (odd) and 1 Apollo Street, are small two-storey houses built in pairs and threes to form an irregular line of terrace development, dating from *c.* 1845. Nos. 11–27 form a straight range of houses. Some pairs have central passageways or carriageways to the rear and some have modern shop fronts or added bay windows. Nos. 29 and 31 have pantiled roofs.

(145) BELLE VUE HOUSE, a villa decorated in Gothic style, was built in the 19th century before 1838 (Plate 104). It was the home of William Abbey Plows, sculptor and stone and marble merchant who had works at Foss Bridge in Walmgate. Plows bought the site in 1833 and it is probable that the house was built for him shortly after that date and that the carved stone decoration came from his own workshop. Plows sold the house

in 1852 and it changed hands several times thereafter before it was acquired in 1879 for The Retreat to form a part of the villa system of treatment of patients in separate houses. The upper part of the house was pulled down in 1935 and the S. wing has been removed but the lower parts of the walls of the main part have been retained to enclose a swimming pool. The upper storey was castellated (*YC* 25/7/1839) and appears faintly on N. Whittock's view of the City of *c.* 1858; photographs of the house before alteration are preserved at The Retreat. A lodge, similar in style to the main house, was demolished in 1971 (Directories; Deeds and *Annual Reports*, York Retreat).

The main part of the house consisted of a rectangular block with octagonal turrets terminating the front elevation which is built in white brick with stone dressings. In the middle of this front is a porch with rectangular stone piers carrying stone lintels carved with triple ogee arches cusped and enriched with ivy trails. The piers have capitals carved with figure subjects. The arched decoration is repeated over the windows, and a string course is enriched with dog-tooth ornament.

Plows offered the house for sale in 1846 when it was described as having a basement providing dry cellarage; on the ground floor, breakfast room, dining room, drawing room opening into a vinery, and two kitchens; a stone staircase with bronzed iron balustrade of the vine pattern and a secondary servants' staircase; upstairs, four principal bedrooms, two dressing rooms and three servants' bedrooms. The dressing rooms were supplied with soft water, and a water closet and a bath were provided. The grounds were formally decorated with urns and figure sculpture; the only garden ornaments now remaining are three carved stone capitals 'removed from Carlton Palace [Carlton House, London] on the demolition of that splendid building' (*YC*, 25/7/1839; *YG*, 15/9/1846; OS 1852).

(146) COTTAGE, No. 103, was built in the last decade before 1850; it is marked on the OS map of 1852 as The Herdsman's Cottage. It is a small single-storey dwelling in plain brick with hung-sash windows, all much restored.

(147) GARROW HILL, Thief Lane, now a hostel for nurses of The Retreat, was built as a large private house in the early 19th century; it is probably the house occupied by Henry Bland, banker and partner in the firm of Messrs. Swann, Clough & Co., from 1828 (Directory) till his death in 1835 (*YG* 14/2/1835). In 1836 Thomas Barstow was living there, having moved from Blossom Street, and the house remained in his family till 1927. The house has been somewhat altered externally with the addition of a large bay window on the E. side and the alteration of some of the windows to take modern casements in place of the original hung-sashes. Internally some of the rooms have been subdivided but many of the original high quality fittings remain.

The house is of two storeys with walls of white brick and low-pitched roofs covered with slates. The main part of the house is a large rectangular block built round a central hall, lit from a lantern projecting above the main roofs (Plate 129). The entrance on the N. front has an added porch with stone Tuscan columns and entablature. The eaves on this side are of very slight projection. On the S. side and to the E., where there are two gables, the roofs have a wide overhang supported by paired brackets. To the W. two projecting wings enclosed a courtyard. The N. wing contained the kitchen and is little altered; the courtyard has been partly roofed over to form a dining hall and the S. wing, which formerly comprised store-rooms only, has been drastically altered. The staircase in the entrance hall has elaborate iron balusters (Plate 130). A number of doorcases are original and of unusual design; panels in the overdoors can be identified as being from the moulds of Thomas Wolstenholme (Plates 111, 113).

development of Heworth Grange Crown Estate on the N.W. side of Heworth Green (PRO, MPE 1006). These houses were never built but they are typical of the good quality houses which were erected in Heworth Green from this period onwards. In the streets behind Heworth Green the standard is much lower, and in some instances definitely mean.

Monuments 148–198, unless otherwise described, are of two storeys and were built in the first half of the 19th century.

BULL LANE

(148) GLENFIELD is a detached double-fronted house built *c.* 1840–50 on an **L**-shaped plan and later extended.

(149) THE GLEN RESIDENTIAL NURSERY was built as

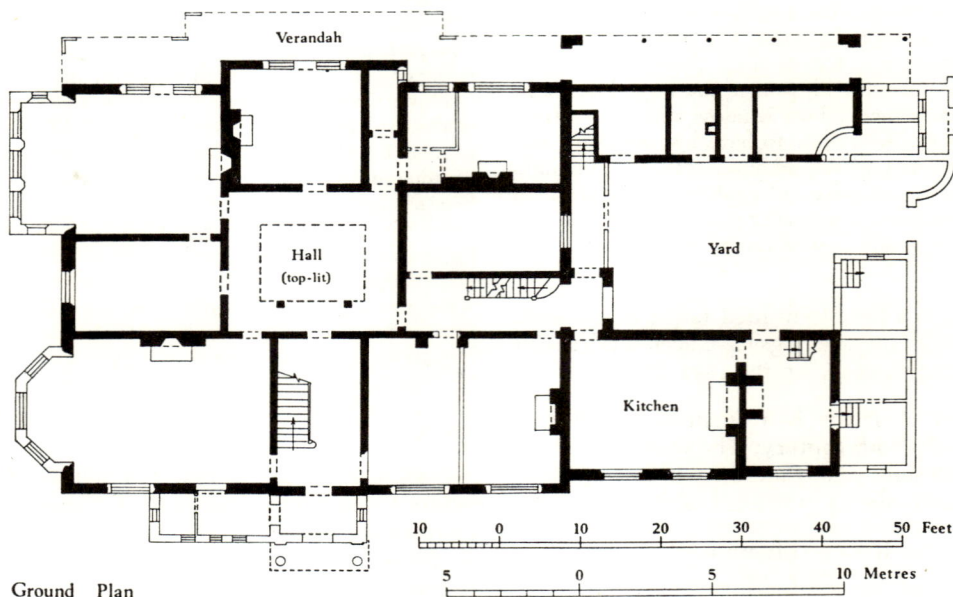

Fig. 69. (147) Garrow Hill House, Heslington Road.

HEWORTH (Monuments 148–198)

Heworth is an area to the N.E. of the City, part of which has always been in the City and part of which was included by extensions of the City boundary in 1884 and later. The plan of a small mediaeval village street with long tofts is traceable on the OS map of 1852 in the road now called Heworth, but most of the area was moor or common until the rapid expansion of the City in this direction after *c.* 1825. At this date plans were drawn up by Peter Atkinson II and R. H. Sharp for the

an isolated detached house in the last decade before 1850. On the OS map of 1852 it is called Glen Heworth and is shown surrounded by extensive grounds. Later in the 19th century it was enlarged to double its original size. The house was designed in a picturesque Tudor style featuring many gables with boldly projecting barge boards, the principal ones shaped and pierced, and chimneystacks of various designs; one of the stacks had three separate octagonal shafts with moulded brick bases and tops. *Demolished.*

EAST PARADE

Some of the houses are described as modern built in *YG* 20/5/1848 but others date from earlier in the century. The street is architecturally undistinguished and many of the houses have been altered.

(150) Nos. 1–15 (odd), are small terraced houses of uniform design but with breaks in the brickwork between Nos. 7 and 9 and 9 and 11 indicating different builds. They date from *c.* 1845. The doorways, where unaltered, have simple pilastered doorcases with panelled reveals.

(151) Nos. 49–55 (odd), 59, 65, 67, 71, 73, 137, 139, 141, 153, 155, 157, are terraced houses, all broadly similar but with variations in detail. Nos. 49–55 form a range of four; Nos. 65 and 67 are a pair. No. 141 is double-fronted and a more ambitious house than the others. The gutters are carried on single or paired brackets but No. 139 has a moulded cornice and No. 141 a bracket cornice. Some of the houses have tiled roofs.

(152) HOUSES, Nos. 79–83 (odd), form a terrace which was mentioned as occupied in the Directory of 1846 and called Eastern Pavilion on the 1852 OS map. Nos. 81 and 83 have entrances with engaged shafts to the jambs and bold entablatures with modillions, and three-sided bay windows with similar entablatures.

(153) DANBY TERRACE, Nos. 117–133, was built *c.* 1835. Nos. 123 and 125 are double-fronted.

EASTERN TERRACE

(154) Nos. 1–10, form a terrace of ten double-fronted houses with a datestone of 1846. Nos. 2–9 are of three bays with a central doorway; Nos. 1 and 10 have a carriageway with a low elliptical arch of shaped bricks in place of the lower rooms on one side.

HAWTHORN GROVE

(155) HOUSE, No. 31, is a detached villa of three storeys with semi-basement and attic, gabled to the N.E. and S.W., which was built *c.* 1830.

There is a two-storey porch to the S.E. and a large semicircular bay, possibly added, with modillioned and dentilled cornice at the S.W. end. The ground level varies between the front and back of the house and the entrance hall is therefore divided between two floors. The main part of the hall has a plaster panel above each of the three doorways leading off it, that over the door to the middle room with griffins and a central urn very similar to that in No. 61 Bootham (48). The middle room itself has a frieze with Classical figures on both of the long walls. The staircase rises in the hall between the two front rooms; it has very slender turned balusters, two to a tread, and a mahogany handrail, and is lit by a tall window with marginal panes of ruby and white glass, stretching up over two landings.

HEWORTH

(156) HOUSES, Nos. 49–55. No. 49 is a double-fronted house, the rear of which is attached to the terrace formed by Nos. 51–55.

(157) THE WHITE HOUSE is double-fronted and has a pantiled roof.

(158) HOUSE, No. 56, of two storeys, was built in the second half of the 18th century on a simple rectangular plan with front and back rooms each side of the entrance hall.

(159) SPARROW COTTAGE, No. 58, a small double-fronted house, was built *c.* 1810–20. The central entrance has reeded panels to the jambs and entablature. The window heads have segmental arches. The roof is pantiled.

(160) HOUSES, Nos. 60, 62, are a symmetrical pair. The doorways have reeded pilasters and friezes; the roofs are pantiled.

(161) TRENTHOLME, No. 68, is a double-fronted house and has a central entrance with reeded pilasters, and semicircular fanlight. The tiled roof has two gables to the W. end.

(162) HOUSE, No. 86, double-fronted and of three bays, was built *c.* 1840. The door has six fielded panels and a rectangular fanlight with marginal panes and the windows hinged casements and plastered lintels simulating flat arches of ashlar. The roof is pantiled.

(163) HOUSE, No. 97, of three storeys, has a lead rainwater head with initials and date S.E. 1794, probably for Samuel Ella, who acquired property in Heworth after 1790 (N. Riding Registry of Deeds, CI 334 488, CI 336 490). A later porch has been added to the symmetrical front. On plan, two front rooms flank the central entrance hall, and there are smaller back rooms each side of the staircase behind.

HEWORTH GREEN

(164) HEWORTH CROFT, No. 19, is a detached villa of two storeys and attics (Plate 99) standing, with a coach-house and other outbuildings, in its own grounds. It is first mentioned in the Directory of 1843, under its former name, Queen's Villa, when it was occupied by the Reverend John Acaster, incumbent of St. Helen's, Stonegate. There is a large house on this site on Robert Cooper's map of 1832 but Heworth Croft appears to be later than this in style. An advertisement in the Yorkshire Gazette of 5 August 1854, shortly after John Acaster's death, states that he built the house for himself and that he obtained a lease for 99 years from the Crown in 1842. It is therefore likely on documentary as well as stylistic grounds that the house was begun in 1842.

The house is built of white brick in an Italianate style and roofed in slate. There are raised brick pilasters to the corners of the elevations. The main garden front elevation, of three bays, has a central pilastered porch with round-arched windows to the sides, a plain ashlar band at first-floor level and a slight projection to the central bay of the first floor. In the angle formed by the main rectangular block of the house and a wing to the N.W. is a low tower which further emphasises the Italianate style in which the house is built. The wing is connected to an original coach-house building now converted to other uses. The interior has a staircase with stone treads and cast-iron balustrade, and the principal ceilings are decorated with moulded plaster in various patterns.

(165) No. 26 is a detached house dating from c. 1835. The symmetrical front elevation has a central entrance with reeded half-columns and a fanlight with geometrical glazing bars (Plate 109). The windows have slightly segmental cement arches with key blocks. The roof is hipped.

(166) St. Maurice's House, No. 36, is a white brick detached villa dating from c. 1849. The front elevation is symmetrical and has neo-Norman details. The interior is plain and much altered.

(167) No. 44 and No. 1 Mill Lane are two three-storey houses (Plate 102), perhaps originally built as one house with a service wing, of c. 1835–40. It is shown as one building on the 1852 OS map. The main elevation of No. 44 is of four bays with a continuous sill-band to the first floor and a heavy moulded cornice. The entrance has a shallow hood supported on brackets. The first-floor windows have flat pediments supported by similar brackets; those to the ground and second floors and to all floors of No. 1 Mill Lane have segmental cement arches with key blocks and simulated voussoirs. No. 1 Mill Lane has a bold entrance with rounded arch with key block and simple pilastered jambs.

(168) House, No. 46, (Plate 102) of three storeys, was built c. 1840. It has a simple entrance with Tuscan pilaster jambs. The windows have segmental arches of cement with key blocks and simulated voussoirs; those on the first floor have a continuous sill-band.

(169) Houses, Nos. 48, 50, (Plate 102) are a pair, of three storeys with basements, built c. 1845–50. They have heavy doorcases and angular bay windows to the ground floor and a continuous sill-band to the first-floor windows; the window openings have segmental arches.

(170) Houses, Nos. 52, 54, (Plate 102) both of three storeys, were built c. 1845–50. No. 52 has an entrance similar to that of No. 50 and a modillioned cornice. No. 54 has a heavy doorcase with bold brackets and an angular bay window; the upper-floor windows have segmental arches.

(171) Heworth Moor House, No. 56, is a detached house of three storeys built c. 1849. The front elevation is symmetrical and has a heavy central porch with pedimental top flanked by two-storey bay windows with modillioned cornices; the other windows have segmental arches.

(172) Houses, Nos, 58, 60, are a three-storey pair of c. 1840 with conjoined entrances with a continuous entablature and Doric half-columns. The windows have cement arches with fluted key blocks and simulated voussoirs.

(173) No. 62 is a detached house built in white brick in the second quarter of the 19th century. The symmetrical front elevation has a central entrance with heavy Tuscan doorcase flanked by brick and timber bay windows; these three features have a continuous modillion cornice. The eaves cornice is of moulded timber beneath a parapet with stone coping.

(174) Shoulder of Mutton Hotel, No. 64, of two storeys with attics, was built as a substantial private house c. 1840. It was called Heworth Green on the 1852 OS map. The symmetrical front elevation has a central open Tuscan porch flanked by bay windows. The three first-floor windows have moulded architraves and an entablature and cornice supported on console brackets. The tiled and hipped roof has a central dormer window.

(175) The Lodge, No. 66, a detached house of white brick, dates from c. 1845. The symmetrical front elevation has a central doorway beneath a fanlight with segmental arch, flanked by plain bay windows with modillion cornices. The upper windows have cambered arches and the eaves cornice is made up of two courses of moulded corbels to carry cast-iron gutters.

(176) House, No. 72, of two storeys and attic in brown brick with white dressings, was built c. 1849. There is a bay window with enriched modillion cornice to the ground floor. The first-floor windows are set beneath semicircular arches with blind tympana springing from nook-shafts with foliated capitals. The gutter is carried on decorated bearers.

(177) House, No. 74, called the Shoulder of Mutton p.h. on the 1852 OS map and kept by John Poulter in 1843 (Directory), was built c. 1840. It is almost square on plan and has a symmetrical front elevation in painted stucco. The central entrance has a heavy hood supported on large consoles; it is flanked by angular bay windows. The two first-floor windows have segmental arches; two first-floor windows in the E. wall retain their original glazing. There is a single-storey addition at the E. end.

(178) SCARBOROUGH PARADE, Nos. 76–94, was built by 1830 when eight houses are entered as occupied in the Directory. They are terraced houses of two and three storeys of several builds and with variations in detail. The entrances have reeded half-columns or pilasters and rectangular fanlights with geometrical glazing.

(179) HOUSE, No. 96, was originally all of brick but was refaced in stone at the end of the 19th century.

(180) HOUSES, Nos. 98, 100, built c. 1840, form a pair, of three storeys with semi-basements. The entrances have fluted attached columns and doors of six panels beneath rectangular fanlights with geometrical glazing bars. The windows of the three main floors have segmental arches of cement with key blocks and simulated voussoirs.

(181) HOUSES, Nos. 104, 106, built between 1817 and 1832, formed a symmetrical building marked on the OS map of 1852 as Cupola House. No. 104 has been enlarged by the addition of a W. wing; No. 106 has been heightened; both have been completely refronted, but the Greek Doric columns and entablature framing the entrance to No. 104 are probably original, reset.

Each house was originally entered from the side, the front door leading to an entrance hall, containing a staircase, behind the front room. Two further rooms lay behind the hall.

(182) WYNSTAY, No. 108, a small villa in cement-rendered brick with a symmetrical front elevation of three bays, was built in the second quarter of the 19th century. It was called Arlington Cottage on the 1852 OS map. The central entrance is framed by panelled pilasters and an entablature with a pediment; there are bay windows on either side. The roof is hipped.

(183) HEWORTH VILLA, No. 110, a detached three-bay villa in white brick with rusticated quoins, was built in the second quarter of the 19th century. It was called Heworth Green Cottage on the 1852 OS map. It is attached to an older 19th-century house which remains at the rear.

The symmetrical main front has a central entrance with a hood with a modillion cornice supported on large brackets resting on stone corbels, flanked by bay windows with similar cornices. The first-floor windows have segmental arches. The roof is hipped. The staircase has turned balusters, two to a tread, a mahogany handrail and shaped cheekpieces. It rises between the two ground-floor front rooms to a large landing lit by a window with a two-centred arched head and arched glazing bars in a Gothic pattern.

(184) THE LIMES, No. 112, a detached three-bay villa-type house in white brick with a stone plinth, was

called Terrace Cottage on the 1852 OS map. The central entrance has fluted attached Roman Doric columns and a canted bay window to one side. Inside, the rooms have reeded cornices and the doors have four sunk panels. The staircase has plain balusters of square section, two to a tread, a slender mahogany handrail and elaborately-shaped cheekpieces.

HEWORTH ROAD

(185) HOUSES, Nos. 28–36 (even), form a range of five small dwellings with plain entrances built c. 1845.

(186) No. 38 is a small stucco-rendered house with a hipped roof. The main elevation of four bays is articulated by giant pilasters.

(187) Nos. 42–52 (even) are a range of small single-fronted houses with pantiled roofs. No. 50 was called Letter Receiving Office on the 1852 OS map.

(188) THE NAG'S HEAD, p.h., was built in the second quarter of the 19th century. It had the same name in 1850 (OS) and in 1843 when it was occupied by Charles Vaux (Directory). It is a double-fronted house of three bays. The details have been modernised.

(189) HOUSES, Nos. 58, 60, built c. 1820, have entrances with reeded half-columns to the jambs. The upper windows have segmental arches.

(190) HOUSE, No. 62, has a doorway with a reeded frieze over a rectangular fanlight flanked by deep brackets. There is a segmental bow window with hung sashes to the ground floor. The roof is pantiled.

(191) HOUSES, Nos 77, 81, 80, 82 have been considerably altered. No. 81 has been widened to be double-fronted but retains the original doorcase to the entrance, with attached reeded columns. Nos. 80 and 82 form a symmetrical pair, with similar doorcases.

(192) No. 83 is a detached double-fronted house originally of three bays. It has a modern pantiled roof.

JOHN STREET

(193) HOUSES, Nos. 4, 6, form part of a small terrace, dating from 1840–50. The openings have been modernised. *No. 4 demolished.*

MILL LANE

(194) HOUSES, Nos. 13–15, form a small terrace similar to that in John Street but retain their simple pilastered doorcases and cambered arches to the windows. *Demolished.*

STOCKTON LANE

(195) THE MANOR HOUSE, No. 1, was formerly called the New Manor House (OS). It was built before

1830 when William Hornby Esq. occupied it and it appears, with its name, on Robert Cooper's map of 1832.

The house is partly of two, partly of three, storeys, above a high semi-basement and is of unusual, nearly cubical, shape. A bay window was added later in the 19th century.

(196) THE COTTAGE, No. 11, dates from c. 1800 and appears to be on Robert Cooper's map of 1832. It was called Belle Vue Cottage in 1834, when it was occupied by Mr. John Scott, in 1846, when it was occupied by Henry Janson, gent., and in 1850 (Directories; OS).

It is a double-fronted cottage with rendered walls and a hipped roof and was extended c. 1840. It has an enriched doorcase of mid 18th-century date, brought from elsewhere.

(197) ROSE VILLA, No. 34, was called Heworth Villa in 1850 (OS). It was built in the early 19th century and may be the house which appears on the site on Robert Cooper's map of 1832. It is a double-fronted house with later extensions.

WOOD STREET

(198) HOUSES, Nos. 2, 4, 12, 14, 20, 22, 26, 34, 36. and 1, 7, 9, 23 are simple small terraced houses of different builds and with variations in detail. They all date from c. 1849. No. 9 is double-fronted.

HULL ROAD (Monuments 199–200)

(199) No. 147 is a small early 19th-century cottage with a pantiled roof.

(200) MILL FIELD HOUSE, N.G. 62615136, of two storeys and a basement, has walls of white brick and a low-pitched slated roof with wide projecting eaves. It was built c. 1830 as a substantial four-square villa. It was occupied by Richard Faircloth Esq., Receiver of Taxes for the County, during the 1840s (Directories).

On the W. the ground floor has been extended and the E. side is masked by later building. To the S. is an open entrance porch with four Tuscan columns; the windows have stone lintels cut to simulate arches with enriched key-stones. The original staircase has been removed but many of the doors and fireplaces retain their original surrounds.

HUNTINGTON ROAD (Monuments 201–202)

(201) PARK PLACE, Nos. 9–13, a terrace of two storeys and basements with variations in details, was built in the second quarter of the 19th century. Nos. 12 and 13 are double-fronted. The original front windows have cambered cement arches with key blocks. *No. 9 demolished.*

(202) GROVE TERRACE, Nos. 27–49 (odd), a row of twelve two-storey houses, four of which are double-fronted, has a datestone of 1824. The front elevation of

white brick was designed as one symmetrical architectural composition with centre and end features projecting forward. The entrances have fluted shafts supporting entablatures with geometrical patterns to the friezes, rectangular fanlights with marginal panes, and doors of six panels.

JEWBURY (Monuments 203–205)

The name Jewbury is derived from an ancient cemetery of the Jews first mentioned in 1230 (Raine, 281; *YAJ*, III (1875), 186).

(203) HOUSE, No. 13, built in the early 19th century, is of three very low storeys and very small dimensions. It has a horizontally-sliding sash window to the second floor and a pantiled roof.

(204) HOUSE, No. 14, built in the early 19th century, has a shallow bow window to the first floor and a modern pantiled roof. It is of two storeys.

(205) ROCKINGHAM HOUSE, No. 25, was erected by James Lamb in 1792 (YML, Subchanters Bk. IV, 77, No. 102, 268); it now forms part of a training school for nurses for the County Hospital. The house (Plate 94) is of three storeys and basement and has a symmetrical front with stone bands and two-storey bay windows on each side of the entrance, which has engaged Doric columns, reeded frieze blocks, and open pediment (Plate 107). The bay windows were reconstructed 1971.

The central entrance hall has arcaded walls; to the N.W. is a through room with bow windows to front and back; to the S.E. a smaller room with a staircase behind completes the plan of the original house, except for a small closet projection at the back. The staircase has stone steps and simple, hollow-sided, iron balusters. On the first floor in the N.W. room the plain arcading of the entrance hall is repeated in the two long walls. Doors throughout the house have six panels, fielded on one side and decorated with planted or attached mouldings on the other.

LAWRENCE STREET (Monuments 206–220)

Lawrence Street takes its name from the adjacent church of St. Lawrence and forms the beginning of the Hull road, the main road out of York to the S.E. from Walmgate Bar. The surrounding area is now largely industrial. Unless otherwise described Monuments 206–220 are of two storeys.

W. side:

(206) HOUSES, Nos. 4, 6, are a pair of small dwellings built c. 1800. *Demolished.*

(207) THE QUEEN, p.h., No. 12, is a three-storey building with a stucco façade of c. 1840. It was called the Queen Victoria on the 1852 OS map.

(208) HOUSES, NOS. 14–18 (even), of three storeys, were built in the early 19th century.

Front Elevation

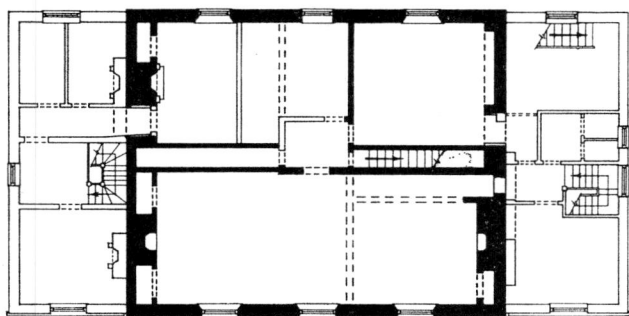

1st Floor Plan

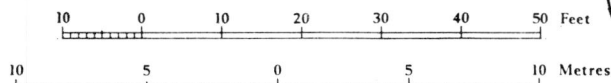

10 0 10 20 30 40 50 Feet

10 5 0 5 10 Metres

Fig. 70. (215) St. Lawrence Working Men's Club, Nos. 29, 31 Lawrence Street.

(209) Former FLAX MILL, No. 30, now a warehouse in a large industrial complex, was built in several stages in the first half of the 19th century. The earliest part may be that advertised for sale as a 'newly erected fire-proof flax mill . . . without Walmgate Bar', with engine and boiler houses and a reservoir, in *YCh* 20/11/1817. The main interest of the building lies in the extensive evidence of fire-proofing techniques which survives, despite drastic alterations.

The earliest part of the mill was of three storeys; a four-storey range was added to the S.W. later in the century. The windows have segmental heads of common bricks and the sills are of stone externally but of shaped bricks internally. The floors are solid and the staircases and landings of sandstone slabs. At the S.W. end cast-iron columns support a series of transverse iron joists between which shallow brick vaults carry the floor above. The original N.E. end is said to be of similar construction but the vaults are concealed by a later ceiling.

(210) HOUSES, NOS. 102, 104, are of the early 19th century. No. 102 has a symmetrical front to the street; No. 104 is an addition to S. Across the front are original cast-iron railings.

E. side:

(211) ROSE AND CROWN, p.h., was built probably as two houses in the early 18th century but has been completely modernised. Early 19th-century outbuildings at the back include a small cottage in which a reset pane of glass bears the former name of the house, The Friendly Inn.

(212) HOUSE, No. 17, has on a rainwater head the initials and date W.W. 1773, presumably the date of construction. The house is of two storeys and built on a normal square plan with front and back rooms each side of the entrance hall. To the E. is a slightly lower building of *c.* 1800 which, together with the E. side of the original house, now forms The Waggon and Horses public house. The original house has a timber doorcase with fluted pilasters and open pediment.

(213) HOUSE, NOS. 21, 23, is of two storeys and was built in the late 18th century. It has been much altered and the lower storey converted to two shops.

(214) HOUSE, NOS. 25, 27, is of two storeys and was built in the late 18th century. It has been much altered and partly rebuilt. On the S. front the original central entrance remains with fluted timber pilasters, entablature and open pediment. A shop-front and bay window on the ground floor and three windows above all represent later alterations.

(215) ST. LAWRENCE WORKING MEN'S CLUB, NOS. 29, 31, was built in the late 18th century, as a simple rectangular dwelling house of three storeys. In 1822 it was bought by Samuel Tuke, who died in 1853; during his ownership wings were added to each side but the W. wing was not complete at the beginning of 1849. The building has been extended at the back and drastically altered internally (Fig. 70).

(216) HOUSES, NOS. 45–59 (odd), form a terrace of three storeys and semi-basements dating from *c.* 1835. Nos. 51, 55, 57 and 59 have Doric doorcases with attached fluted columns. The window heads have segmental arches of cement with key blocks and simulated voussoirs.

(217) HOUSE, No. 61, built in the early 19th century,

is of three storeys and has first-floor balconies with scrolled cast-iron balustrades. The windows have rubbed brick arches.

(218) Houses, Nos. 83–91 (odd), were built c. 1830. Nos. 85 and 87 are a mirrored pair with shallow bow windows to the ground floor (Plate 105). The first-floor windows have cambered stuccoed arches with simulated voussoirs and key blocks.

(219) House, No. 93, was built c. 1830. Although more elaborate and roofed separately, it is very close in style to the terrace (218) which adjoins it on the W. and was probably part of the same development.

The front elevation (Plate 105) has a doorway with reeded attached three-quarter columns, flanked by shallow segmental bow windows with a simple raised geometrical pattern on the friezes. The three first-floor windows have a continuous sill band, which extends across the whole of the front elevation, and flat arches of stucco with simulated voussoirs and fluted key blocks. The roof is fully hipped to E. and W.

Regent Street:

Only two occupants were listed in the street in the Directory of 1830.

(220) Nos. 2–6 (even) are a pair of houses dating from c. 1830 with a third house of c. 1840–50 at the S. end. The entrances to Nos. 2 and 4 have reeded half-columns to the jambs and friezes with raised geometrical pattern.

LAYERTHORPE (Monuments 221–224)

Monuments 221–224 are all of two storeys.

(221) Nos. 9, 15, 29–31, 37, 39, 57–75, 83A, 85 and 114, are all small terraced houses of the first half of the 19th century, some in pairs or groups of three. No. 37 has a shallow segmental bay window to the first floor.

Bilton Street and Redeness Street:

(222) Streets of terraced houses constructed mainly between 1828 and 1833 on a site laid out after 1821 by

Fig. 71. (222) Nos. 35–40 Bilton Street and 43–49 Redeness Street.

Oswald Allen, Surgeon (Deeds). Bilton Street had thirteen occupants and Redeness Street fifteen in the 1830 Directory. Houses already occupied were for sale in Bilton Street in *YG* 11/10/1834. The houses, mostly of two storeys but some with attics, show a wide variety of plans and fittings (Fig. 71, p. 82). The roofs are tiled. *Demolished.*

Hallfield Place:

(223) TERRACE of four houses which appear on Robert Cooper's map of 1832 (surveyed in 1831) but not in the 1830 Directory. They have simple pilastered entrances with rectangular fanlights with geometrical glazing patterns. *Demolished.*

Hallfield Terrace:

(224) HOUSES, Nos. 21–31 (odd), were built between 1823 and 1830 (Directories). They have pantiled roofs. *Demolished.*

LORD MAYOR'S WALK (Monuments 225–237)

Lord Mayor's Walk, from Monkgate to Gillygate outside the City Wall, was formerly Goose Lane. In 1718 a broad walk was planted with elm trees. By 1818, however, some of the trees had been cut down and the street had 'lately become the site of several neat brick dwelling houses' (Drake, 254; Raine, 280; Hargrove, II, 560). Houses 'recently erected' were for sale in 1837 (*YG* 10/2/1837) and in 1849 (*YG* 3/11/1849) probably erected on the ground offered for sale in *YG* 26/6/1847. *See also* RCHM, *Monuments Threatened or Destroyed* (1961), 71.

Monuments 225–237 are of two storeys and date from the first half of the 19th century, unless otherwise described.

N.E. side:

(225) HOUSE, No. 18, was built *c.* 1840. The first-floor window has a stone lintel simulating an arch.

(226) HOUSE, No. 24, has a pantiled roof.

(227) HOUSES, Nos. 26–34 (even), are all of three storeys. No. 26 has an entrance with attached reeded shafts and a semicircular fanlight. There is a shallow bow window with reeded surround to the first floor; the upper windows have slightly segmental arches. The roof is pantiled, as is that of No. 28. No. 32 has a continuous sill-band to the first-floor windows and a platt-band at second-floor level. No. 34 has a main doorway and a smaller doorway on the ground floor, and a moulded and dentilled timber cornice.

(228) HOUSES, Nos. 36–42 (even), form a three-storey terrace. Nos. 36–40 retain their original entrances with fluted pilasters and doors of six fielded panels under semicircular fanlights.

(229) No. 50, dating from *c.* 1840, is a detached house of three storeys and semi-basements with a plastered E. wall suggesting that it was once one of a pair or the end of a terrace. There is a doorway with a pediment and a segmental fanlight, and a bay window to the ground floor. The windows of the upper floors have slightly segmental arches.

(230) HOUSES, Nos. 54–58 (even), are of three storeys with semi-basements, dating from *c.* 1845–50. The entrance porches have Ionic pilaster jambs; those to Nos. 56 and 58 are conjoined.

S.W. side:

(231) HOUSES, Nos. 17, 19, are a pair with pantiled roofs.

(232) OLD FARM, house, No. 27, was erected by George Darbyshire, coal dealer, between 1785 and 1794 on land which had been part of the garden of another house. The house was built on a rectangular plan with two front rooms but no passage between

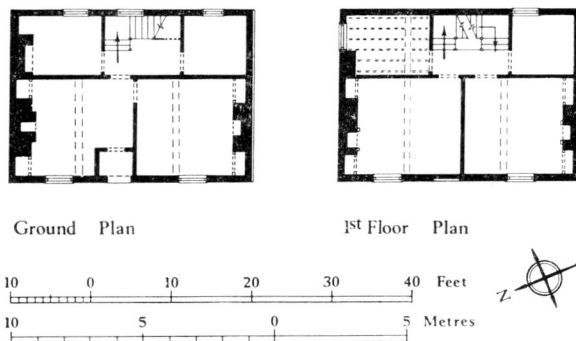

Ground Plan 1st Floor Plan

Fig. 72. (232) Old Farm, No. 27 Lord Mayor's Walk.

them, and two small rooms behind with the staircase between them, producing an unusually small version of the square four-room plan. The house retained most of its original simple fittings; they included doors with four fielded panels, fireplaces with iron hobgrates and simple pilastered timber surrounds, and a staircase with close string, square newels and slender square balusters. *Demolished in 1958.*

(233) HOUSE, No. 41, together with Nos. 43, 45, 49 and 51 (234–237), all built by 1808, was part of the development of the garden of No. 90 Gillygate by Mary Sickling. Nos. 41, 43 and 45 were sold shortly after her marriage to Edward Cowper which took place

in 1811 or 1812, but Nos. 49 and 51 and the piece of ground between them and No. 45 were not sold until 1828 after she and her husband had both died. No. 41 was sold in 1812 to Joseph Terry, linendraper, and inherited c. 1830 by his son Thomas who had already bought the site of No. 47 in 1829 (Deeds). It was of three storeys and had a symmetrical front elevation. The central doorway was similar to that of No. 47 of c. 1830; the windows had segmental heads. *Demolished.*

(234) HOUSES, Nos. 43, 45, were a part of the same development as No. 41. No. 43 was sold in 1812 to Christopher Newstead, gent., and No. 45 in 1813 to William Lockey, cheesemonger (Deeds). They were a three-storey pair with variations in plan and detail. No. 45 had a pantiled roof. *Demolished.*

(235) HOUSE, No. 47, of two storeys with attics and cellar, was built c. 1830. The site, together with Nos. 49 and 51, was sold in 1828 to Thomas Townend, yeoman, who in 1829 sold them to Thomas Terry, son of the owner of No. 41. In December 1831 Terry sold the properties, which by this time included the newly erected No. 47 already occupied by Ann Campbell, widow, to Miss Mary Brown of Clifton (Deeds). The entrance doorway was similar to that of No. 41 and to the W. of this doorway was a round-headed opening leading to the yard behind No. 51 and giving access to No. 49. *Demolished.*

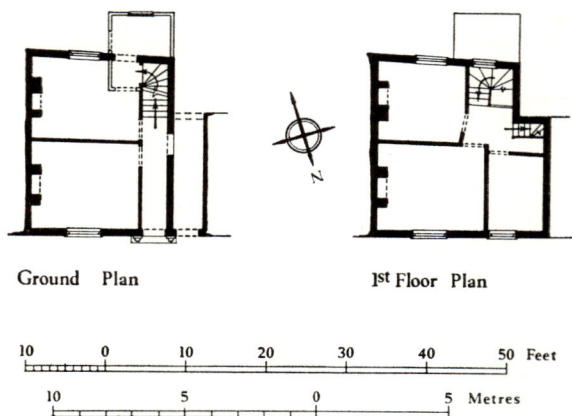

Ground Plan   1st Floor Plan

Fig. 73. (235) No. 47 Lord Mayor's Walk.

(236) HOUSE, No. 49, was, like Nos. 41, 43 and 45, referred to as 'lately erected' in a mortgage deed of 1808. It was one dwelling up to and including 1828 but when sold with No. 51 and the site of No. 47 in 1829 it was described as a 'dwelling house . . . now divided into two

tenements' (Deeds). It had plain doorways, probably of 1828–9, and a pantiled roof. *Demolished.*

(237) HOUSE, No. 51, was part of the same development as Nos. 41, 43, 45 and 49, but was probably slightly earlier for it was referred to as 'sometime since erected' in the mortgage deed of 1808 (Deeds). It had a symmetrical front elevation with a central doorway flanked by windows with shallow segmental heads, and a pantiled roof. *Demolished.*

LOWTHER STREET (Monuments 238, 239)

Lowther Street was built between 1830 and 1838 (Directories) and later. Building lots were for sale in YG 23/3/1844; five houses in Lowther Street and Eldon Street were in the course of erection and for sale in YG 17/1/1846. *See also* ELDON STREET. Monuments 238 and 239 are of two storeys.

(238) HOUSES, Nos. 66–78 (even), are a terrace dating from c. 1835. Nos. 66 and 68 are slightly larger than the others and have a more elaborate finish.

(239) HOUSES, Nos. 151–161 (odd), are a range of six small dwellings built c. 1849. The bay windows have carved decoration to the surrounds and the roofs are pantiled.

MALTON ROAD

(240) THE HERDSMAN'S COTTAGE, built c. 1820, is a single-storey rectangular dwelling with a central porch on the S. side roofed in one with the main building. Both the casement windows and the entrances are set beneath round-headed arches.

MARYGATE (Monuments 241–255)

S.E. side:

Only one house remains on this side of the road, all others having been cleared away to expose the precinct wall of St. Mary's Abbey.

(241) HOUSE, No. 29, of three storeys, basement and attic, was built in the late 18th century (Plate 95). It stands on a confined site against the wall of St. Mary's Abbey and the back wall, to S.E., probably incorporates parts of the abbey wall and another building of uncertain date.

The entrance doorway (Plate 108) has a timber surround with simple panelled pilasters and console brackets flanking the fanlight and rising to an open pediment over composition garlands in low relief; the pediment being raised clear above the fanlight gives exaggerated height to the design. On the upper floors the front windows to S.W. have been blocked, the S.W. rooms being lit by bay windows in the gable end; above the bays is a semicircular window to the attic.

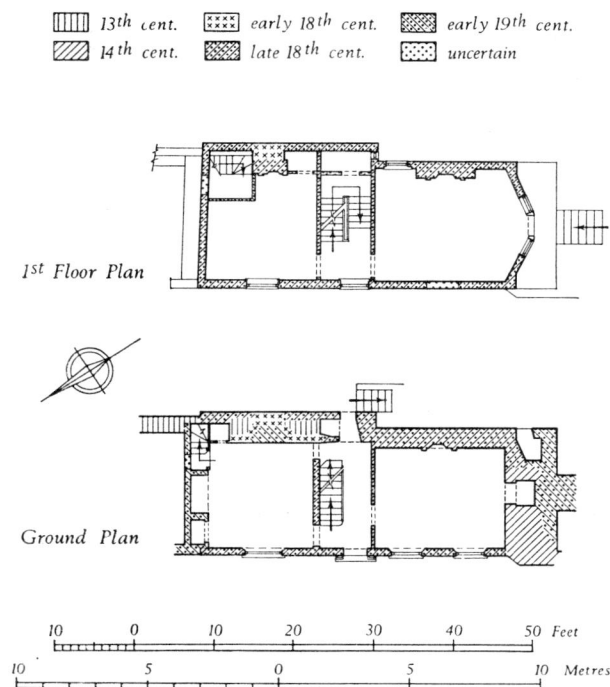

13th cent.   early 18th cent.   early 19th cent.
14th cent.   late 18th cent.   uncertain

1st Floor Plan

Ground Plan

10   0   10   20   30   40   50 *Feet*
10   5   0   5   10 *Metres*

Fig. 74. (241) No. 29 Marygate.

Many original fittings survive: the staircase has a close string and turned balusters; a fireplace on the first floor has decoration by Thomas Wolstenholme (Plate 115). A grate on the second floor is decorated with figures (Plate 132). On the ground floor at the S.W. end a cupboard has been formed in the doorway of the 14th-century abbey building adjoining.

N.W. side:

(242) HOUSE, No. 20, built *c.* 1800, has a narrow frontage. It is of two storeys and the roof is covered with pantiles.

(243) HOUSE, No. 22, of two storeys, was possibly built before 1850 but has been much altered. It has a pantiled roof.

(244) HOUSE, No. 28, of two storeys with brick walls and pantiled roof, was built in the late 17th century. The plan is L-shaped giving two front rooms on each floor and a back wing containing the chimney with the staircase beside it, and a third room beyond. The entrance, in a modern addition in the re-entrant angle, is now reached by a carriageway opened through one of the ground-floor rooms.

(245) HOUSE, No. 30, built in the first half of the 19th century, is of three storeys and attic. It has a bay window of the mid 19th century to the first floor; other windows have segmental arches. Inside, the fittings include a number of cast-iron grates of the early 19th century, some brought from elsewhere and some original to the house (Plate 132).

(246) HOUSES, Nos. 32, 34, comprise a four-storey L-shaped building of *c.* 1800 and a two-storey addition of the first half of the 19th century in the re-entrant angle. It is not clear whether the original building formed one house or two. The entrance doorway to No. 34 is flanked by engaged reeded shafts carrying scroll brackets which support an open pediment over the fanlight. The entrance to No. 32 has a simpler doorcase with a flat head. The internal fittings are simple; some of the fireplaces retain original iron grates, one signed Carron.

(247) HOUSES, Nos. 36, 38, built in the early 19th century, are a three-storey pair designed to present a unified three-bay façade to the street. The doorways are grouped in the centre on either side of a third door giving access to the rear and under the central blind windows of the first and second floors. The window openings have low segmental arches. The roof is pantiled.

(248) Nos. 40, 42, comprise a small house of *c.* 1760 with an early 19th-century extension at the S. end and a substantial back wing of the same date. The original house, of two storeys with cellar and attic, had a symmetrical front and contained, on plan, one room on each side of the entrance hall.

(249) ST. OLAVE'S HOUSE, No. 48, was built in the late 17th century on a U-shaped plan with unequal wings projecting N.W. In the late 18th century a building was erected behind the shorter wing, separate from the house and apparently not for domestic use, but in the early 19th century it was joined to the house and a chimney and bay windows were added to give one large room on each of two floors. At the same time the space between the original wings was filled in. The street front was rebuilt *c.* 1900.

The house is of two storeys with attics. Most of the original brick walls are concealed but at the back of the N. wing is an original gable with tumbled brick coping and a projecting band at eaves level. The interior fittings are mostly of the late 18th and early 19th centuries. The staircase to the first floor is of the early 19th century; the slender balusters are of square section with beaded angles and hollowed sides. Leading up to the attic is a reset staircase of the late 17th century which is

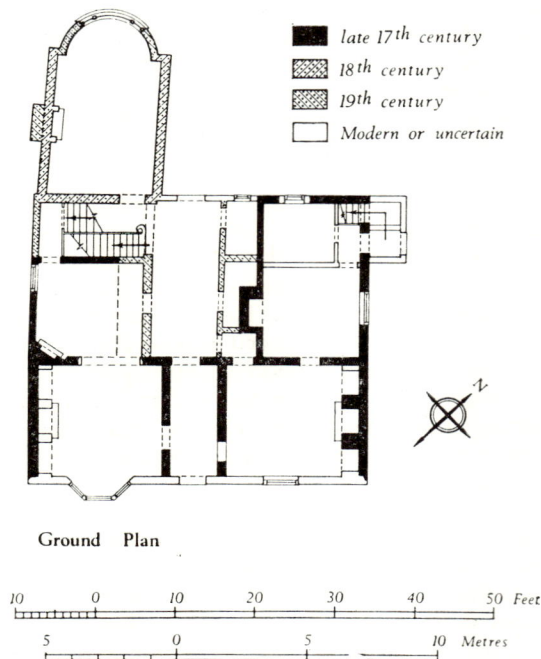

late 17th century
18th century
19th century
Modern or uncertain

Ground Plan

Fig. 75. (249) St. Olave's House, No. 48 Marygate.

probably the original staircase to the first floor; it has a close string and bulbous turned balusters.

Summerhouse, N.W. of the house, is an octagonal structure of two storeys built in the second quarter of the 19th century. The walls are of ashlar and rubble masonry. The ground floor provides a garden store entered by an arched doorway. The upper floor has three walls fully glazed with marginal panes to each window.

(250) HOUSE, No. 50, was built *c.* 1700; it is of three storeys with a pantiled roof. On plan it comprised two front rooms with an entrance passage between them and a small kitchen, staircase and store-room behind. The front wall was completely rebuilt *c.* 1905 with added bay windows, and at the back the kitchen has been enlarged. The staircase has original balusters cut out of flat planks to an undulating profile between a close string and a plain handrail. Some original three-panel doors remain on the first floor.

(251) HOUSE, Nos. 56, 58, of two storeys and now two tenements, was built *c.* 1700 on an **L**-shaped plan and enlarged soon after to give a **U**-shaped plan. The elevations have been drastically modernised. In No. 58 is a mid 18th-century staircase with alternately turned and twisted balusters.

(252) HOUSE, No. 60, formerly the Grey Coat School, was opened as a charity school for girls in 1705. Partial reconstruction and extensive refenestration were carried out at the end of the 19th century and later. Many of the large rooms in the house have been subdivided in conversion to flats and the N.W. end now forms part of Little Garth (*see* (254) and (27)).

The building is of three storeys **L**-shaped on plan, with frontages to Marygate and Marygate Lane (Plate 95). The storeys are marked by projecting brick bands; some original windows remain on the front to Marygate Lane with three-centred arches over two and three casements. Reset over a doorway to Marygate is a fragment of a 17th-century carved bargeboard said to have come from a house in High Ousegate dated 1635.

(253) ST. MARY'S COTTAGE, No. 62, is a small dwelling of the second half of the 18th century, facing S.W. In the N.W. gable is an eroded datestone, inscribed 1767 or 1787. The house is of two storeys and on plan originally comprised two rooms flanking a central entrance passage and a staircase behind the N.W. room. The house was refitted in the early 19th century and has absorbed a small part of No. 60 to which it is attached.

MARYGATE LANE

(254) ALMERY GARTH is a substantial house of *c.* 1745, standing on part of the garden ground formerly belonging to the Almonry of St. Mary's Abbey, from which it takes its name. A service wing to the S.E., together with some rooms of the adjoining No. 60 Marygate (252), now forms a separate residence known as Little Garth. The building is of three storeys; the walls are of pale brick with red brick dressings and the roofs are tiled. On plan it comprises a long straight range only one room thick, facing S.W. towards the river. The original S.W. elevation of the main block was of five bays with a pilaster at each end and with the central bay projecting slightly and stone bands separating the storeys; the windows had gauged flat arches with key stones. Early in the 19th century the ground floor was enlarged by the addition of a boldly projecting segmental bay and the extension of the hall, a second storey being added to these additions later in the same century. The arrangement of windows in the N.W. bays was also changed and what must have been an imposing symmetrical Georgian front has lost much of its original character.

The front of the service wing, Little Garth, now partly masked by additions, has projecting plat-bands and segmental arches over the windows. The elevation to Marygate Lane is very plain, with 19th-century surrounds to the doorways. The interior fittings are mostly of the early 19th century, including the lower part of the main staircase with open string and

Almery   Garth   .   Little   Garth

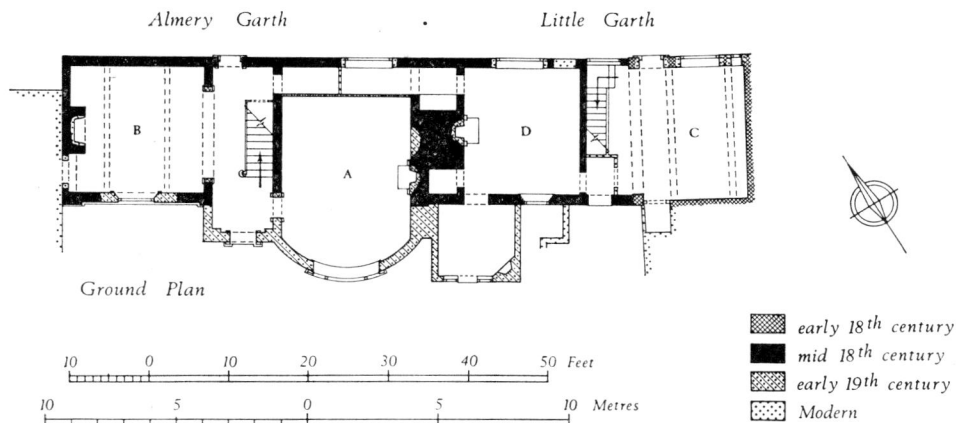

Ground   Plan

```
10      0      10     20     30     40     50 Feet
10           5           0           5          10 Metres
```

early 18th century
mid 18th century
early 19th century
Modern

Fig. 76. (254) Almery Garth, Marygate Lane, Marygate.

turned balusters (Plate 127), but the upper part of the staircase is of *c.* 1745 and has a close string and turned balusters.

(255) COTTAGES, range of four, 60 yds. W. of St. Olave's church, includes a small building, not originally domestic, which is probably of early 18th-century date though it does not appear on any map earlier than White's of 1785. The building has been divided into two to form tenements Nos. 1 and 2. Nos. 3 and 4 were built later in the same century and the whole range has been refronted.

The building is of two storeys with brick walls and tiled roof with stone slates at the eaves. The upper floor of Nos. 1 and 2 is carried on slightly chamfered oak beams and oak joists; the rafters are supported by purlins which overlap at the principal rafters to which they are fixed by tusk-tenons. *Demolished.*

MONKGATE (Monuments 256–288)

Monkgate is the main approach to the city from the N.E. The 12th-century church of St. Maurice, outside Monk Bar, which was pulled down in the 19th century, provides evidence that there was an extra-mural settlement here by the 12th century. None of the surviving domestic buildings antedate the destruction that occurred here in the Civil War.

Unless otherwise described monuments 256–288 are of two storeys.

N.W. side:

(256) HOUSES, Nos. 1, 3, were built by 1846 by James Bowman, coal dealer and wheelwright (YCL, Council Minutes III, 10 August 1846; Directories). They form a three-storey group of five bays comprising two houses and shops with a carriageway to the rear at the end. The front elevation has recessed vertical strips rising through the first and second floors.

(257) HOUSE, No. 11, is probably of late 18th-century date. The ground floor has been drastically altered in conversion to a shop but a drawing of *c.* 1840 shows that there was a central doorway with a window on each side. The interior arrangement included a staircase placed transversely between front and back rooms on the S.W. side. *Demolished.*

(258) HOUSE, No. 15, is L-shaped on plan: a back wing, of which part has been demolished, was built probably in the late 17th century and contains an original window with wooden mullion and bars; the front range was built in the late 18th century. The house is of two storeys. The street front is symmetrical with a central entrance, and the plan shows the entrance hall and staircase between two rooms. The staircase has a close string and turned balusters; some of the rooms retain moulded ceiling cornices, and doors with six fielded panels. *Demolished.*

(259) HOUSE, No. 17, was built in the late 17th century comprising two storeys with a single room on each floor. The house was increased in depth in the early 18th century to give a second room on each floor, and refronted. Further alteration to the front at the end of the 18th century raised the eaves line to match that of No. 15.

The front is of brown brick with red dressings and has a plat-band at the first floor. The entrance has a 19th-century doorcase. An original window, now blocked, has an ovolo-moulded wood frame and mullion and three vertical wood bar in each light. The balusters of the 18th-century staircase have been boxed in. *Demolished.*

(260) HOUSES, Nos. 19, 21, were built probably *c.*

1812, the date on a rainwater head. The houses are of three storeys, each with two front rooms to each floor, behind which the planning is very irregular, and a variety of projections containing back rooms, staircases and closets. The front door to No. 19 is recessed behind panelled reveals and has six moulded and fielded panels. One of the first-floor fireplaces has the sides enriched with composition decoration.

(261) HOUSE, No. 37, of three and four storeys, was built by Joseph Buckle just before 1848, the four-storey part to the N.E. being a remodelling of a three-storey house built by William Walker in 1794 (Deeds).

The street front has the upper storeys divided into two bays by giant pilasters with stone bases and capitals, and the sills of the upper windows are linked by stone bands. The entrance doorway is flanked by narrow windows to form a tripartite composition with Ionic columns (now missing) between pilasters carrying an entablature with central pediment; it is similar to the entrance to No. 61 Bootham (48) but the pilasters here are not fluted. The S.W. side is divided into three bays by brick pilaster strips without bases or capitals.

On plan the two parts of the house are very similar in idea though different in scale. Each has a staircase placed transversely between front and back compartments; in the earlier part to the N.E. the original staircase remains forming the secondary staircase to the whole; in the S.W. part the front door leads into a spacious entrance hall screened from the main staircase by Ionic columns but this grand design has been spoilt by the introduction of a modern partition. The handrails to the staircase are carried on elaborate cast-iron standards (Plate 130). On the first floor, over the entrance hall, a lofty reception room has a richly decorated plaster ceiling.

(262) HOUSE, No. 39, of three storeys, was built in 1794 by William Walker and at that time formed with No. 37 a symmetrical pair (Deeds). The semicircular arch over the entrance and flat arches over the front windows are of gauged rubbed brick. The house repeats the plan of the earlier part of No. 37; in both houses there are large cupboards between the staircases and the party wall.

(263) HOUSES, Nos. 45–51 (odd), dating from 1830–40, form a terrace and are of three storeys with basements and attics. The doorcases have reeded half-columns. The windows of Nos. 47–51 have decorated key blocks and continuous sill-bands on the first floor.

(264) HOUSE, No. 53, is of the early 19th century, but much altered. It has a pantiled roof.

(265) HOUSE, No. 55, (Plate 104) was probably built by John Mason between 1809 and 1815 (YCA, E96,

Front Elevation

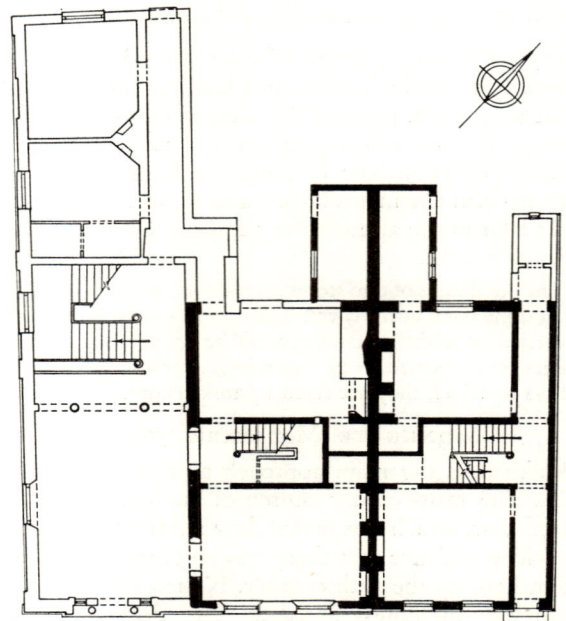

Ground Plan

10   0   10   20   30   40   50 Feet

5   0   5   10 Metres

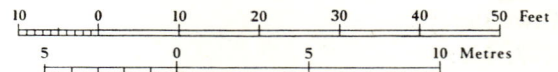

Fig. 77. (261, 262) Nos. 37, 39 Monkgate.

PLATE 101

(111) No. 32 Grange Garth, FULFORD ROAD. Ground-floor window. c. 1840.

(103) No. 33 FISHERGATE. Ground-floor window. Early 19th-century.

(110) No. 2 Grange Garth, FULFORD ROAD. c. 1835.

(90) No. 23 The Green, CLIFTON. 1839.

(89) No. 16 The Green, CLIFTON. c. 1840.

PLATE 102

(91–93) Nos. 2–12 Water End, CLIFTON. Late 17th to early 19th-century.

(167–170) Nos. 1 Mill Lane and 44–54 Heworth Green, HEWORTH. 1840–50.

PLATE 103

(100) Nos. 39–51 ELDON STREET and Nos. 92–96 LOWTHER STREET. Early 19th-century.

(297–299) Nos. 29–33 PENLEY'S GROVE STREET. Nos. 29, 31 by J. B. and W. Atkinson. 1843, 1847.

PLATE 104

(269) Nos. 67, 69. *c.* 1813

MONKGATE.

(265) No. 55. Early 19th-century.

Head of window.

Head of entrance porch.

From S.W.

(145) Belle Vue House, HESLINGTON ROAD. *c.* 1834.

PLATE 105

(218) Nos. 85, 87. *c.* 1830.

(219) No. 93. *c.* 1830.

LAWRENCE STREET.

PLATE 106

(42) No. 47. Rainwater head. *c.* 1753.

(38) No. 25. Rainwater head. 1768.

(49) Record House, No. 65. Railings. John Walker, mid 19th-century.

BOOTHAM.

(51) No. 77. Rainwater head. 1770.

(41) No. 45 BOOTHAM. 1748.

(128) No. 28 GILLYGATE. 1769.

(109) No. 37 Grange Garth, FULFORD ROAD.
Late 18th-century.

(205) Rockingham House, No. 25 JEWBURY. 1792.

PLATE 108   ENTRANCE DOORWAYS

(241) No. 29 MARYGATE. Late 18th-century.

(41) No. 39 BOOTHAM. *c.* 1800.

(279) No. 40 MONKGATE. *c.* 1800.

(121) No. 23 GILLYGATE. *c.* 1800.

No. 17.
(290) NEW WALK TERRACE. *c.* 1825.

No. 16.

(67) No. 2 CLIFTON. *c.* 1825.

(165) No. 26 Heworth Green, HEWORTH.
c. 1835.

(43) No. 49. Formerly on first floor. *c.* 1738.

BOOTHAM.

(42) No. 47. Ground floor. 1753.

(117) No. 5 GILLYGATE. First floor. Decoration by
Thomas Wolstenholme, 1797.

(44) No. 51 BOOTHAM. Ground floor. *c.* 1804.

(41) No. 43 BOOTHAM. Ground floor. Early 19th-century

(109) Fulford Grange, Grange Garth, FULFORD ROAD. First floor. c. 1835.

(147) Garrow Hill House, HESLINGTON ROAD. Ground floor. Decoration by the Wolstenholmes, early 19th-century.

PLATE 112   INTERNAL DOORWAYS

(117) No. 5 GILLYGATE. Second floor. 1797.

(117) No. 3 GILLYGATE. First floor. 1797.

(57) Bootham Lodge, No. 56 BOOTHAM. Ground floor. 1840–5.

DECORATION BY THE WOLSTENHOLMES.

PLATE 113

(147) Garrow Hill House, HESLINGTON ROAD. Ground floor. Early 19th-century.

(57) Bootham Lodge, No. 56 BOOTHAM. First floor. 1840–5.

(41) No. 43 BOOTHAM. First floor. Early 19th-century.

DECORATION BY THE WOLSTENHOLMES.

PLATE 114   FIREPLACES

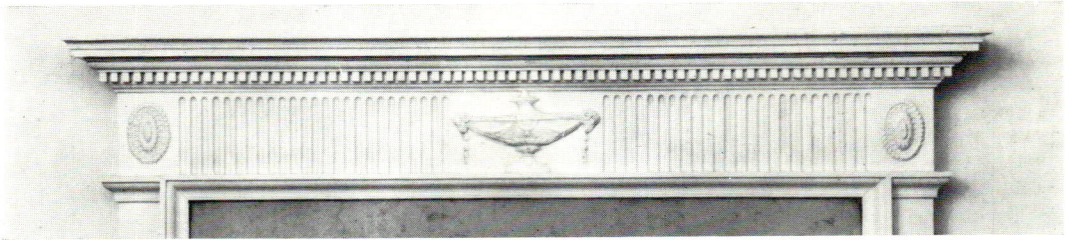

(68) No. 8 CLIFTON. Ground floor. 1782–4.

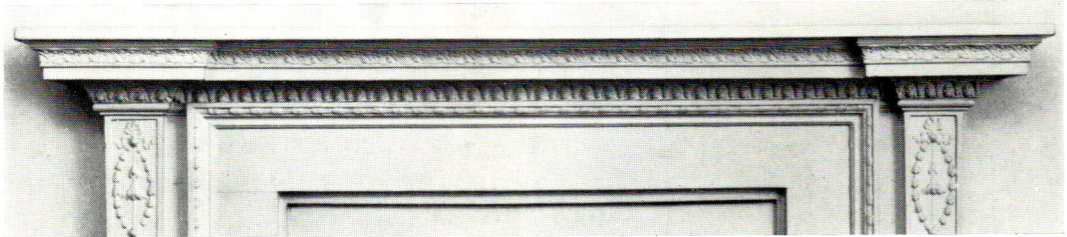

(68) No. 8 CLIFTON. First floor. 1782–4.

(44) No. 51 BOOTHAM. First floor. *c.* 1804.

(128) No. 28 GILLYGATE. Ground floor. *c.* 1769 with early 19th-century applied decoration.

(128) No. 28 GILLYGATE. First floor. Decoration by Thomas Wolstenholme, early 19th-century.

(117) No. 3 GILLYGATE. Second floor. 1797.

(117) No. 5 GILLYGATE. First floor. 1797.

(241) No. 29 MARYGATE. First floor. Late 18th-century.

(278) No. 38 MONKGATE. Ground floor. c. 1803.

DECORATION BY THE WOLSTENHOLMES.

(44) No. 51 BOOTHAM. First floor. c. 1804.

(57) Bootham Lodge, No. 56 BOOTHAM. First floor. 1840–5.

PLATE 116   FIREPLACE

(42) No. 47 BOOTHAM. First floor. 1753.

(42) No. 47. Ground-floor front room. 1753.

(44) No. 51. Hall and main stairs. *c.* 1804.

BOOTHAM.

PLATE 118   CEILING

(42) No. 47 BOOTHAM. Staircase ceiling. 1753.

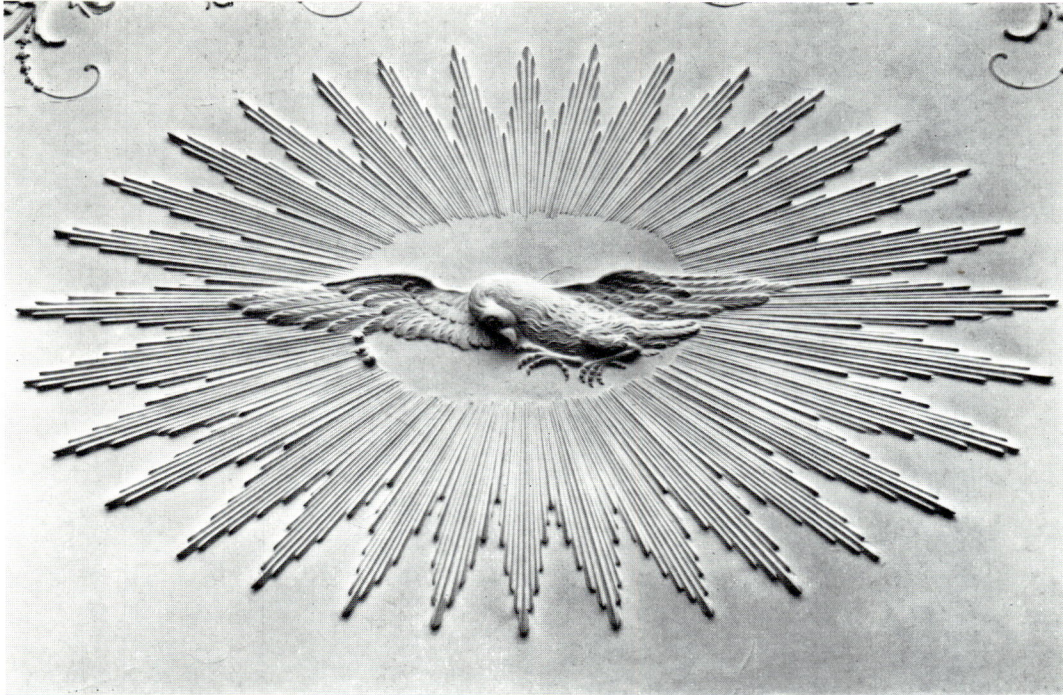

First-floor saloon.

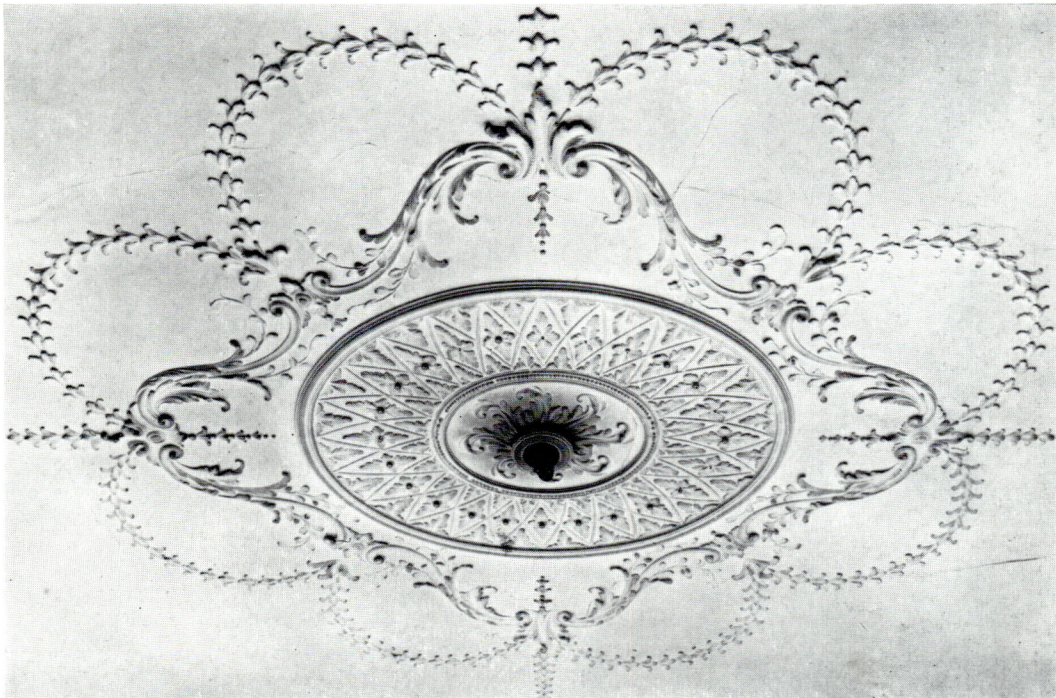

Ground floor. Front room to S.

(128) No. 28 GILLYGATE. Robert Clough junior, 1769.

PLATE 120   CEILINGS

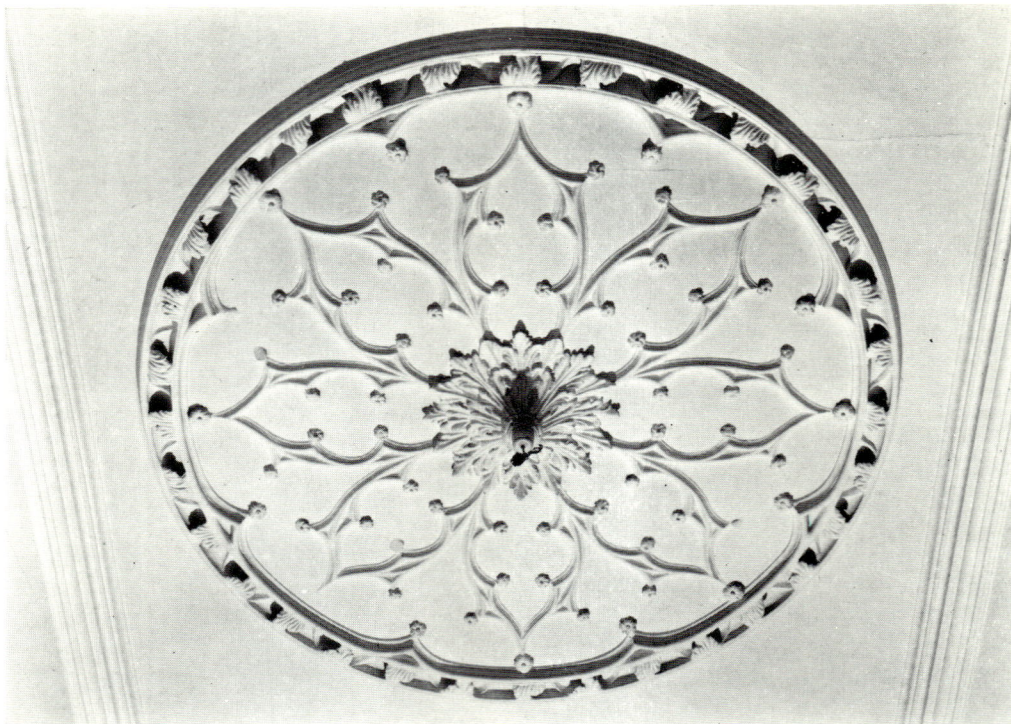

(128) No. 28 GILLYGATE. Staircase ceiling. Robert Clough junior, 1769.

(48) No. 61 BOOTHAM. First floor. Back room. *c.* 1840.

(82) Burton Cottage, No. 27 CLIFTON. Ground floor. Back rooms. *c.* 1835.

PLATE 122   PLASTER DETAILS

(128) No. 28 GILLYGATE. Ground floor. Ceiling cornice. Robert Clough junior, 1769.

(49) Record House, No. 65 BOOTHAM. Staircase hall. Frieze. *c.* 1827.

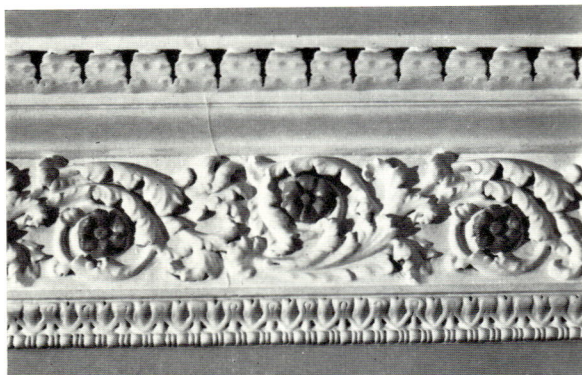

(48) No. 61 BOOTHAM. First floor. Frieze. *c.* 1840.

(48) No. 61 BOOTHAM. First floor. Frieze. *c.* 1840.

(48) No. 61 BOOTHAM. Frieze over staircase. *c.* 1840.

(57) Bootham Lodge, No. 56 BOOTHAM. Ground floor. Ceiling cornice. 1840–5.

(42) No. 47 BOOTHAM. Main staircase. 1753.

PLATE 124

STAIRCASES

(26) Wandesford House, No. 37 BOOTHAM. 1743.

(278) Middleton House, No. 38 MONKGATE.
c. 1700 and *c.* 1740.

(37) Nos. 21, 23 BOOTHAM. Late 17th-century.

(92) No. 8 Water End, CLIFTON. Late 17th-century.

PLATE 125

(128) No. 28 GILLYGATE. 1769.

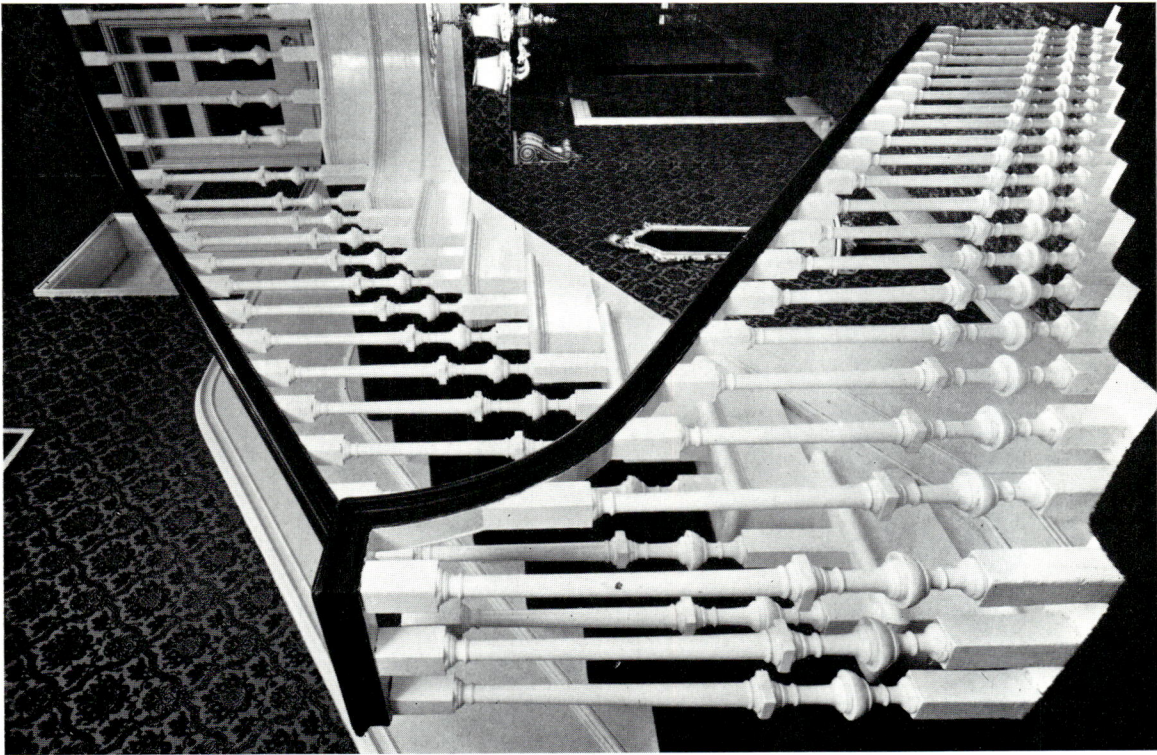

(38) No. 25 BOOTHAM. 1766.

PLATE 126

STAIRCASES

(40) No. 35 BOOTHAM. Mid 18th-century.

(39) No. 33 BOOTHAM. 1753–5.

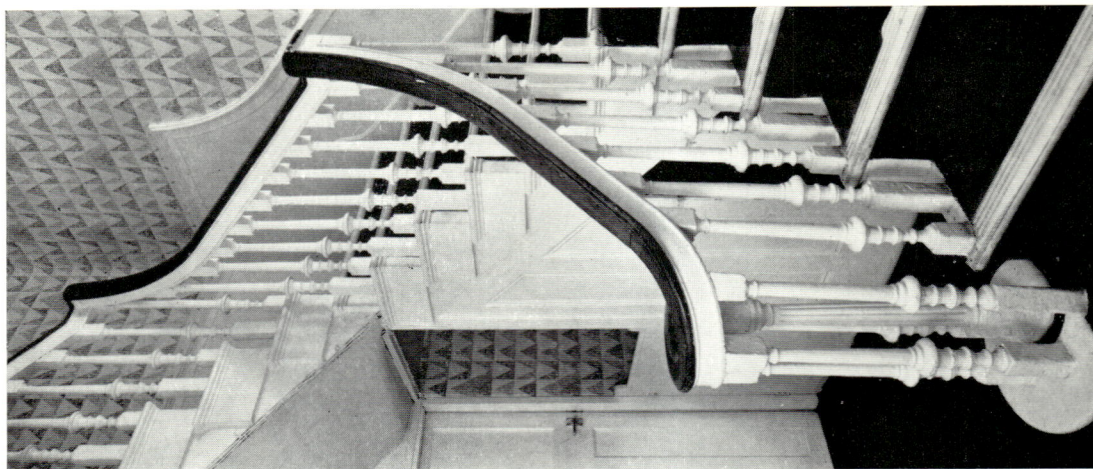

(254) Almery Garth, Marygate Lane, MARYGATE. c. 1745.

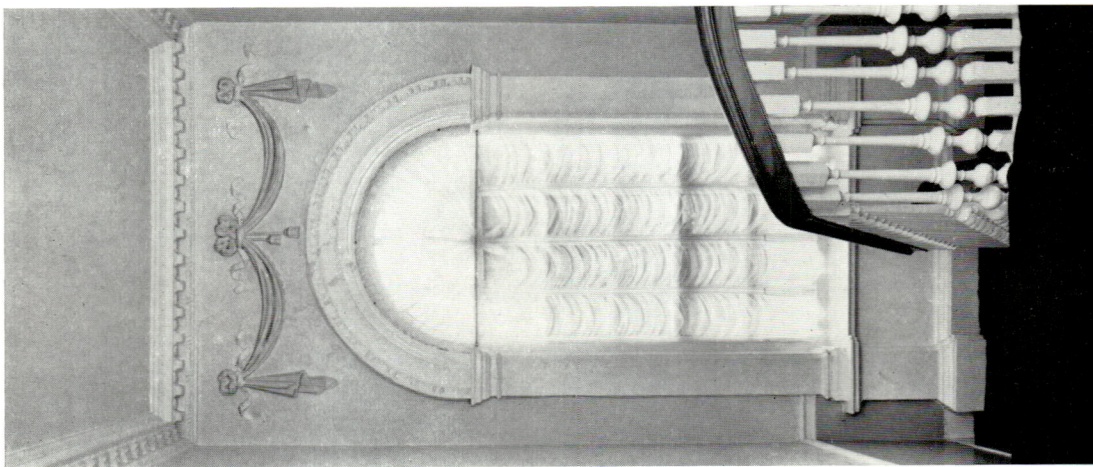

PLATE 127

STAIRCASES

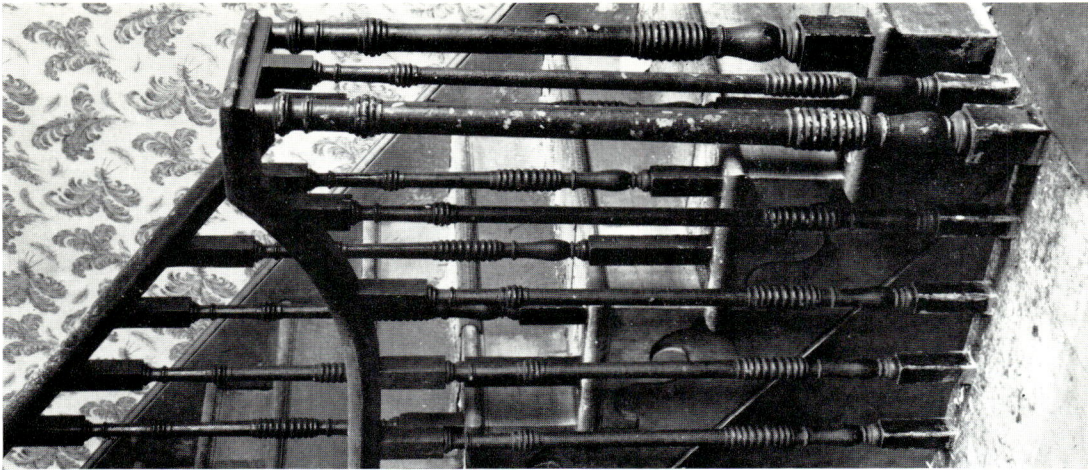

(127) No. 18 GILLYGATE. Early 19th-century.

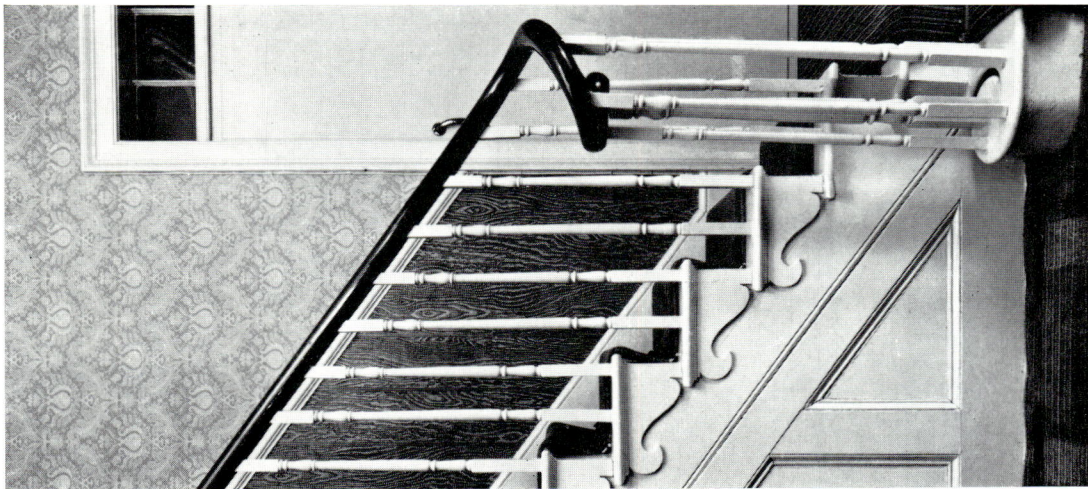

(294) No. 17 PENLEY'S GROVE STREET. 1846.

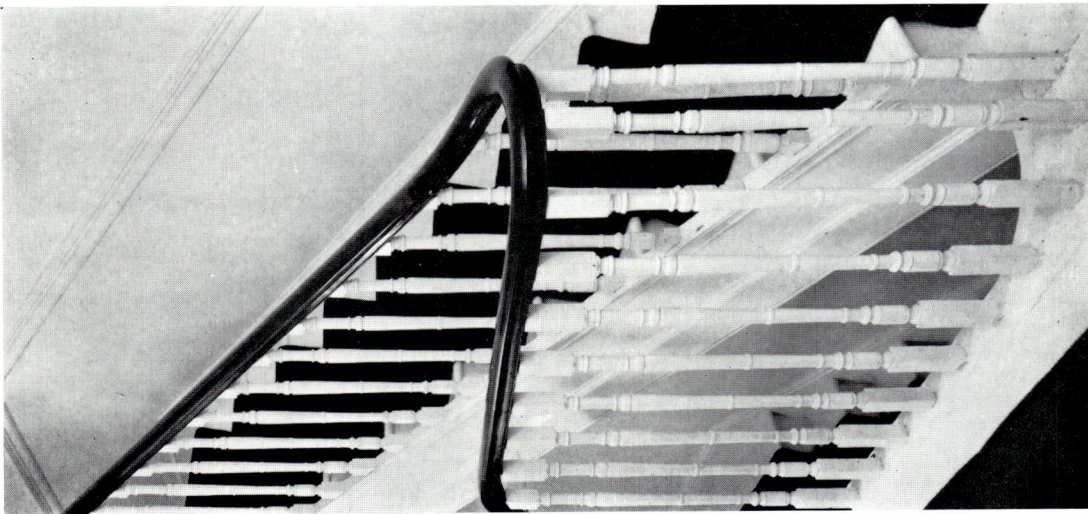

(254) Almery Garth, Marygate Lane, MARYGATE.
Early 19th-century.

PLATE 128

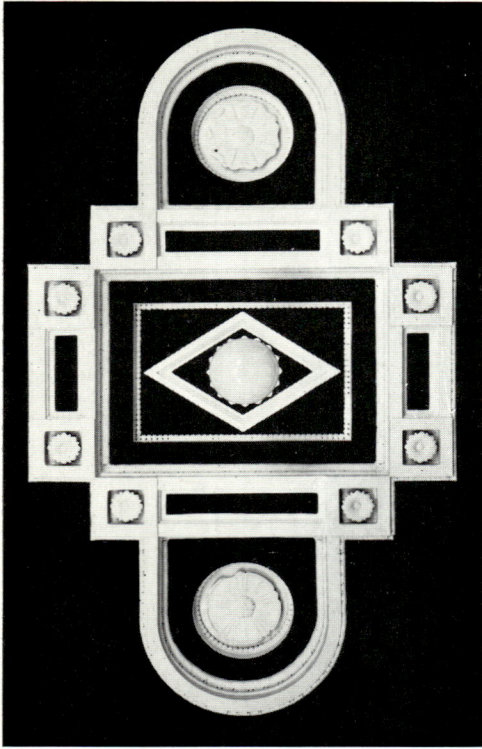

Entrance hall. Ceiling.

Central light well.

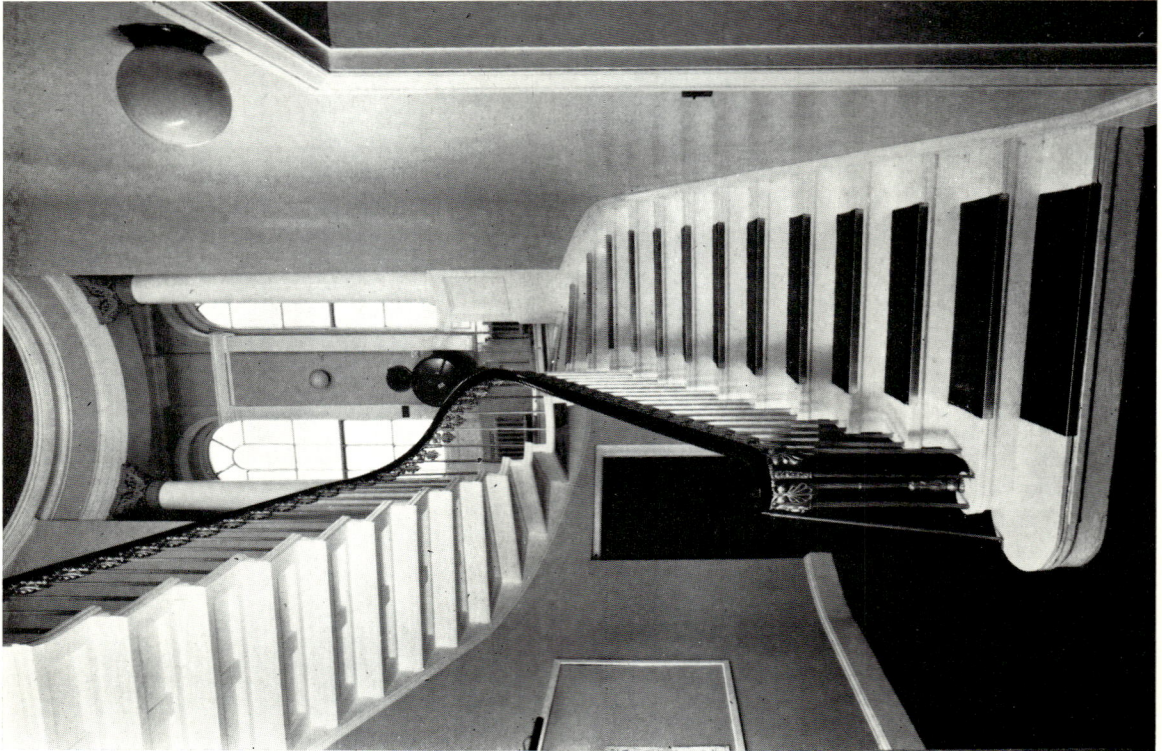

Main staircase.

(105) Fishergate House, FISHERGATE. J. B. and W. Atkinson, 1837.

PLATE 129

(147) Garrow Hill House, HESLINGTON ROAD. First floor. Hall. Early 19th-century.

(105) Fishergate House, FISHERGATE. Main staircase and first-floor landing. J.B. and W. Atkinson, 1837.

PLATE 130   STAIRCASES, CAST-IRON BALUSTRADES

(282) No. 44 MONKGATE. *c.* 1835.

(57) No. 56 BOOTHAM. Secondary staircase.
1840–5.

(147) Garrow Hill House, HESLINGTON
ROAD. Early 19th-century.

(261) No. 37 MONKGATE. *c.* 1847.

(109) Fulford Grange, Grange Garth,
FULFORD ROAD. *c.* 1835.

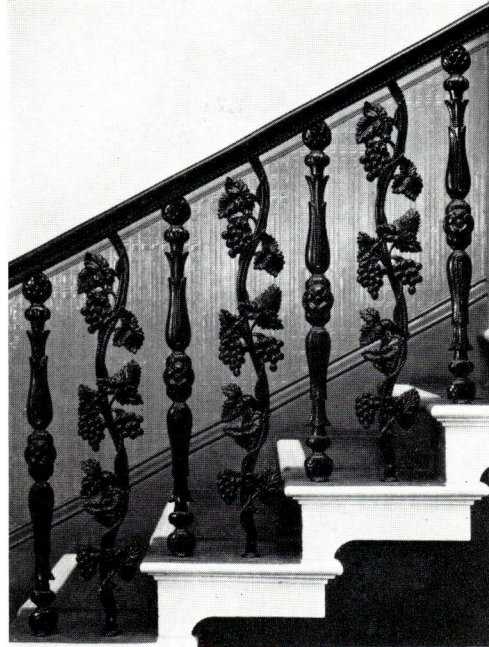

(48) No. 61 BOOTHAM. *c.* 1840.

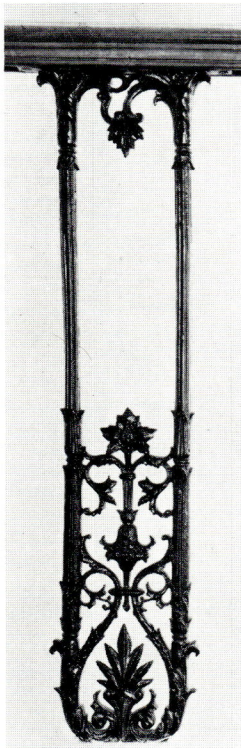

(57) Bootham Lodge,
No. 56 BOOTHAM.
1840–5.

(80) Bootham Grange, CLIFTON. *c.* 1835.

PLATE 132   CAST-IRON HOB GRATES

(92) No. 8 Water End, CLIFTON. Ground floor, reset. Late 18th-century.

(245) No. 30 MARYGATE. Ground floor, reset. Early 19th-century.

(245) No. 30 MARYGATE. First floor. Early 19th-century.

(245) No. 30 MARYGATE. First floor, reset. Early 19th-century.

(241) No. 29 MARYGATE. Second floor. Late 18th-century.

(245) No. 30 MARYGATE. Second floor. Early 19th-century

ff. 126–126v; YML, Vicars Choral Lease Book No. 4, 141). It has a symmetrical front elevation of three bays; the central doorway has recessed pilasters and a fanlight with geometrical glazing pattern.

(266) HOUSES, Nos. 57, 59, a three-storey pair now converted to a single shop on the ground floor, were probably built between 1830 and 1840 and were certainly there by 1843 when No. 57 was occupied by John Catton (Directory).

(267) HOUSE, No. 61, of three storeys, dates from c. 1845. It has a rendered plinth and the window arches have simulated voussoirs and decorated key blocks.

(268) HOUSES, Nos. 63, 65, are a pair, one or both of which were probably built between 1812 and 1815 by John Hart (YCA, E96, ff. 187–188v; 240v–241v).

(269) HOUSES, Nos. 67, 69, (Plate 104) are a pair built between 1812 and 1814 by James Woodburn and Joseph Dutton respectively (YCA, E96, ff. 187–188v; 225v–226). No. 69 was occupied by the architect George Fowler Jones in 1846 and 1851 (Directories). They have coupled doorways with reeded pilasters and semicircular fanlights.

(270) HOUSES, Nos. 71–75 (odd), were probably built between 1812 and 1830 (YCA, E96, ff. 187–188v; Directories). No. 71 retains its original doorcase with semicircular fanlight. Nos. 71 and 75 have pantiled roofs.

S.E. side:

(271) THE BAY HORSE, p.h., No. 4, was built probably soon after 1820 but the front was taken down and rebuilt in 1837 by order of the City Council 'so as to form a line with the rampart wall'. It is of three storeys and has a symmetrical, rather archaic, front elevation of three bays with continuous sill-bands to the first and second-floor windows, and rusticated quoins.

(272) ICEHOUSE, behind The Bay Horse, set into the rampart of the City Wall, was constructed probably at the beginning of the 19th century. It consists of a circular domed chamber 12½ ft. in diameter, entered by a passage 7 ft. long and formerly vaulted, all in brickwork except for a stone cornice over the entrance.

(273) HOUSES, Nos. 16, 18, are a three-storey pair, of differing heights but one build. They have pantiled roofs. Early or mid 19th century.

(274) HOUSE, No. 24, of mid 18th-century date, is of two storeys at the front and three at the back. On plan the transverse staircase between front and back rooms

rises to one side of a nearly central chimneystack with back-to-back fireplaces. The staircase is built around an open well, with turned balusters and moulded handrail. The lower part has a cut string but above the first floor there is a close string. *Demolished.*

(275) HOUSE, No. 28, of c. 1840, is of three storeys. It has a doorway with reeded elliptical shafts, and a tiled roof.

(276) HOUSE, No. 30, built probably between 1830 and 1840, is of three storeys. The doorcase has reeded pilasters and a radial fanlight within an open pediment.

(277) HOUSE, No. 36, was built in 1796–8 (Deeds; YCA, E95, ff. 206, 207, 225) partly over a carriageway to which the earlier arched entrance still remains. The owner was then Thomas Tate.

The house is of three storeys and, except for the carriage entrance, is symmetrical (Plate 97). The central doorway is recessed under a plain gauged brick semicircular arch; over the doorway are plain hung-sash windows but the windows to each side of the elevation are of three unequal lights and there are similar three-light windows at the back all under gauged brick flat arches. At ground floor the plan is that of a simple terrace house with staircase placed transversely between front and back rooms; on each of the upper floors there is an additional room over the carriageway. On the first floor is a fireplace surround enriched with rosettes and festoons modelled in plastic composition.

(278) MIDDLETON HOUSE, No. 38, (Plate 97) was built c. 1700, perhaps for Benjamin West, gentleman, who died in 1711. West also owned two adjoining tenements and all three were let (Deeds). The original house was of two storeys, L-shaped on plan, with a symmetrical five-bay front and finished with a Dutch gable at each end. Subsequent owners included Isaac Johnson, baker, and Joseph Beckett, silkweaver. A contract for the sale of the house and of No. 40 by the widow of John Preston in 1772 mentions two new-built chambers, indicating the building of rooms over the carriageway to the N.E., but these rooms, now part of No. 38, were originally an extension of No. 40; it was probably also Preston who added the third storey in the third quarter of the 18th century. In 1798 the house was bought by the Rev. Charles Wellbeloved, who in 1803 became principal of Manchester College, founded in Manchester in 1786 as a dissenting academy. The college was moved to York in 1803 and accommodated in Wellbeloved's house until 1811 (VCH, *York*, 449).

It was presumably for the college that the house was enlarged by additions at the back and the extension of the N. room on the ground floor. The present arch to the carriageway is of this date. The newly enlarged

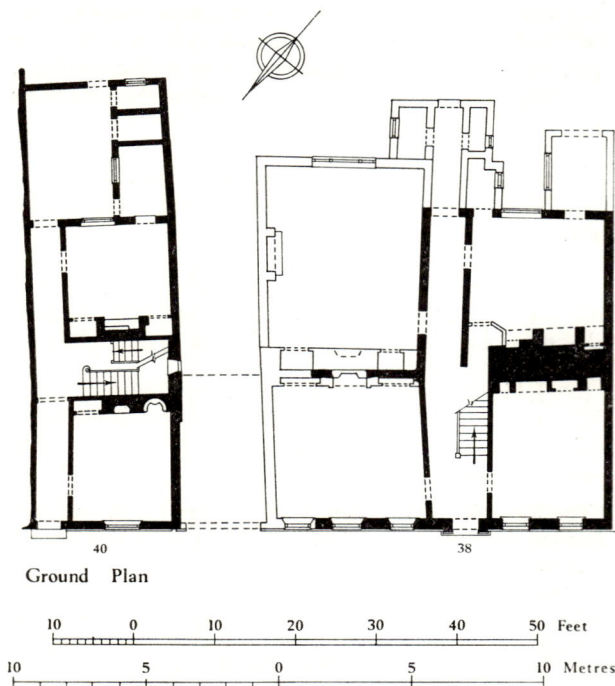

Ground Plan

Fig. 78. (278, 279) Nos. 38, 40 Monkgate.

N. room was entirely refitted, the fireplace surround (Plate 115) and flanking cupboards being decorated by Wolstenholme.

On the street front original sashes with thick glazing bars survive only in the ground-floor windows; the stucco dressing to the surround of the front door is not original. Inside, most of the original staircase remains (Plate 124) but the lowest flight was refitted with lighter, twisted balusters before the middle of the 18th century. An enriched fireplace surround of the mid 18th century remains in a room over the carriageway and in two other upper rooms are decorated iron firegrates by Carron, probably c. 1803, set in surrounds of the same date.

(279) HOUSE, No. 40, stands on the site of one of the three houses owned by Benjamin West (see No. 38 above) and was in the same ownership as No. 38 throughout most of the 18th century. The present house (Plate 97) appears to be of the second quarter of the 18th century but the present timber doorcase to the entrance (Plate 108) is of c. 1800 and the eaves cornice is of the 19th century.

The house, of three storeys, is undistinguished on plan; it has single front and back rooms with the staircase placed transversely between them and a small projecting wing containing a wash-house etc. at the back. The front is of mottled brown brick with red dressings; the first-floor windows have been lengthened. Inside, the staircase has open strings and turned balusters with square knops.

(280) HOUSE, No. 42, (Plate 97) was built by George Hudson by April 1828 (YCA, E98, ff. 61v–62v; Deeds of No. 44 Monkgate). It is of three storeys. The doorway has fluted Doric pilasters. See Fig. 79.

(281) MALTKILN, No. 42A, now used as a store, is a large single-storey building with brick walls, roofed in two spans. It was described as new-built in a conveyance of 1772. A second maltkiln behind Nos. 36 and 38 was demolished c. 1796.

(282) HOUSE, No. 44, (Plate 97) was built probably by Robert Edwards, yeoman, soon after he acquired the site in 1723 (Deeds). The deeds of the property show that there had been at least two earlier houses here, one probably destroyed in the siege of 1644. The house was leased between 1741 and 1747 to John Houghton, a gentleman of some standing, and later it became the property of Thomas Beckwith, painter and antiquary (1731–86). In 1827 it passed to George Hudson who drastically remodelled it; in 1828 he also acquired and rebuilt No. 42 adjoining. According to a conveyance of 1828 No. 42 was for the use of Richard Nicholson (YCA, E98, f. 61v, 62v), but the deeds of No. 44 speak of Hudson using the two properties as one mansion house. An advertisement for the sale of Hudson's house in YG 3/4/1847 describes it as containing the following rooms: on the ground floor, entrance hall, breakfast room, study, kitchen, and house-keeper's, butler's and servants' rooms; on the first floor, a suite of four drawing rooms, dining room with butler's room, two bedrooms and a dressing room; on the second floor,

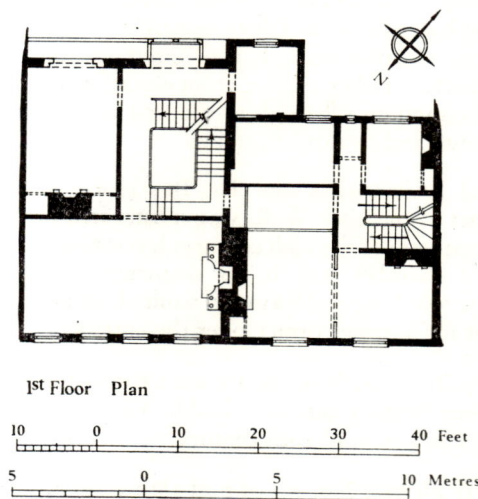

1st Floor Plan

Fig. 79. (280, 282) Nos. 42, 44 Monkgate

seven bedrooms, three dressing rooms and a bathroom with pipes for hot and cold water; accommodation for the servants was provided in the attics and over the servants' offices. The house was also provided with two water closets and, outside, three coach-houses with stabling for eight horses. George Hudson was described in a directory of 1830 as draper; he became well-known for his activities in local politics, serving as Lord Mayor three times, and in the promotion of railways. He became the first chairman of the York and North Midland Railway Company.

The house is of three storeys and attics. The front is of 18th-century brown brick with red dressings, and is of four bays. The entrance, with side windows between pilasters, was put in for Hudson. The windows have modern hung sashes in Georgian style replacing plate glass sashes of the 19th century. The back was rebuilt by Hudson but a projecting bay and balcony were added later in the 19th century. The interior contains a staircase built round an open well with iron balustrades (Plate 130) of c. 1835, and the interior woodwork of the same period is heavily moulded.

(283) HOUSES, Nos. 46, 48, (Plate 97) were built as a symmetrical pair soon after the site had been bought by Thomas Beckwith, painter, in 1768 (YCA, E94, f. 96v, 97). The upper parts of both houses have been rebuilt at different times and heightened to range with No. 44. The lower part of No. 46 has been converted to a showroom and the former entrance, which mirrored that of No. 48 with a 19th-century pilastered timber doorcase, has been destroyed. On the upper floors each house has two hung-sash windows to each floor. On plan the houses were generally similar to Nos. 36 and 40 (277, 279); in No. 46 the staircase has cast-iron balusters, of two fairly simple designs alternated.

(284) NOS. 54–58 (even) are a range of three houses of three storeys and semi-basements dating from c. 1840. There is a carriageway through No. 58. The windows have segmental arches.

(285) NOS. 62–66 (even) are a terrace of three-storey houses dating from the 1840s, built possibly by John Shaftoe, builder (Directories). The doorways have round-headed surrounds with rusticated blocks.

(286) HOUSES, Nos. 68–72 (even), are a terrace probably built between 1840 and 1850. There is a continuous sill-band to the first-floor windows; the stuccoed window heads have simulated voussoirs and patterned key blocks.

(287) HOUSES, Nos. 74–82 (even), are a terrace dating from c. 1845–50. There was originally a continuous sill-band at the first floor; the first-floor window heads are of stucco with simulated voussoirs.

(288) HOUSE, No. 84, built in the early 19th century, is of four bays and has a doorcase with Doric pilasters with sunk round-headed panels in the second bay from the S.W. end.

MUNCASTERGATE
(W. of the Malton Road)

(289) COTTAGE (61535345), adjoining Muncaster House which was built after 1850, appears on the OS map of 1852 and is probably of 18th-century date but it has been very much altered. It is of two storeys with rendered walls, and on plan originally comprised three rooms in line.

NEW WALK TERRACE

(290) HOUSES, Nos. 12a–18, form a terrace of eight superior dwellings with fairly spacious rooms, built c. 1825. Two houses, occupied but newly-erected, were advertised for sale in YG 12/11/1831.

The houses are of three storeys with semi-basements and are set behind long front gardens with smaller yards behind, with coach-houses reached by an accommodation road at the back. No. 12a now comprises two houses from which the top storey has been removed and which have been much altered. No. 18, the end house nearest the river, has its entrance in a symmetrically designed side elevation. The entrance doorways mostly have fluted side pilasters with palmette leaf capitals and rectangular fanlights with geometrical glazing patterns (Plate 109); that to No. 17 has fluted Roman Doric attached columns (Plate 109). The basements and ground floors have shallow segmental bow windows, those to the ground floors with recessed brick panels beneath. The windows of the upper floors have flat arches of gauged brick. (Fig. 80, p. 92.)

PENLEY'S GROVE STREET (Monuments 291–301)
The name of the street is derived from the Paynelathes Crofts of the Middle Ages (Drake, 598), enclosures on part of a larger area of land belonging to St. Mary's Abbey (Raine, 280; EPNS, XIV, 296). Houses in this street were first mentioned in Baines' Directory of 1823. Unless otherwise described monuments 291–301 are of two storeys.

(291) HOUSE, No. 1, dating from the 1840s, is of three storeys and attic. The windows have segmental arches of stucco with simulated voussoirs and decorated key blocks.

(292) HOUSES, Nos. 3, 5, 7, are a terrace of three storeys with basements, built c. 1840. The entrances of Nos. 3 and 5 are linked to form one composition with three fluted pilasters; that to No. 7 is similar. The window openings have segmental arches.

West Elevation

North Elevation

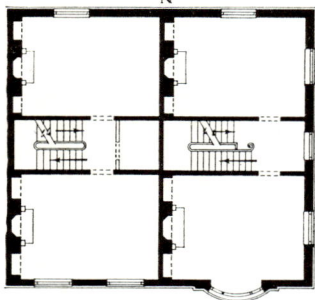

N

First Floor

Ground Floor

10 0 10 20 Feet

10 5 0 Metres

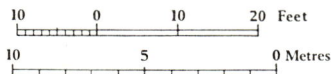

Fig. 80. (290) Nos. 17, 18 New Walk Terrace. Based on a photogrammetric survey by R. W. A. Dallas.

(293) No. 15 is a detached double-fronted house of c. 1845, L-shaped on plan, with low gables to each end elevation. The windows have stone lintels.

(294) No. 17, a detached double-fronted house, was designed in 1846 by J. B. and W. Atkinson (Brierley, Leckenby and Keighley). (Fig. 81, p. 93).

The central doorway has a pilastered surround and the windows stone lintels. It has a square plan with a small wing at the rear. The dining room and drawing room were on the ground floor at the front; at the back was a kitchen, with direct access by a secondary staircase to the servant's bedroom above it (stair now removed), also a water-closet, and a store-room. The main staircase remains and has very slender balusters (Plate 127). There were two principal bedrooms flanking a small 'plant room'; behind them were the servant's room and a 'lumber room'. Neither the store-room on the ground floor nor the lumber room above were designed to have any windows.

(295) HOUSE, No. 19, of three storeys, was built after No. 17.

(296) HOUSE, No. 21, of two storeys and attic, was built c. 1849. The entrance and bay window have a mixture of Gothic and Classical details. There is a passage to the rear. There is a brick corbel table at the eaves.

(297) HOUSE, No. 29, (Plate 103; Fig. 81, p. 93) was designed by J. B. and W. Atkinson for Miss Hazelwood in 1843 (Brierley, Leckenby and Keighley).

The doorway has a round-arched stone architrave and the windows hinged casements with marginal panes and stone lintels. On the first floor are balconies with anthemion pattern cast-iron railings. The plan shows a dining room and a 'small room' on the ground floor, and a very long wing at the rear containing kitchen, scullery, privy with outside entry, etc. On the first floor were two bedrooms with a servant's bedroom and lumber room in the back wing; this wing had a long lead-covered roof slope to the rear, and a staircase with turned balusters. The original cost was estimated at £500.

(298) HOUSE, No. 31, (Plate 103; Fig. 81, p. 93) which is similar to No. 29 but with differences in internal planning and details, was designed in 1847 by J. B. and W. Atkinson, also for Miss Hazelwood.

The front doorway leads to a through-passage, with access to the house itself by a door from the passage, directly opposite the staircase. The parlour and kitchen were on the ground floor, with scullery, pantry, privy, etc. in the back wing. There were four bedrooms on the first floor, and also attics above. At the bottom of the garden was a Tub House. The cost was £360.

(299) HOUSE, No. 33, was built c. 1847. The front has a continuous sill-band to the first-floor windows (Plate 103).

Front Elevation

Front Elevation

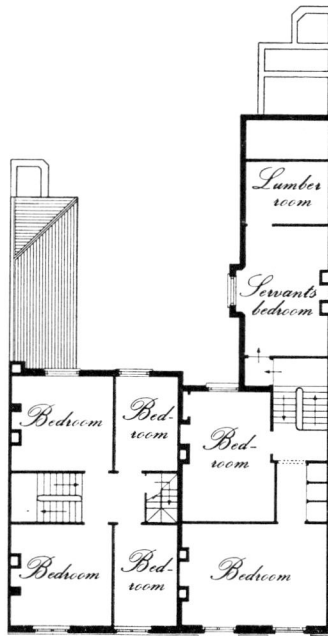

Chamber plan

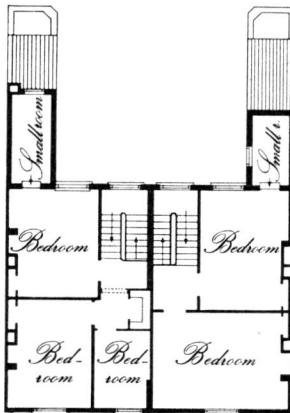

Chamber plan

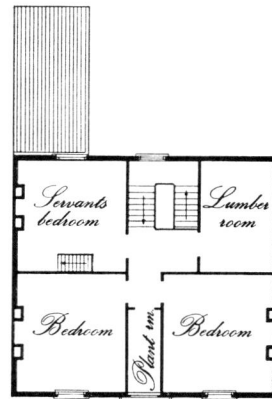

Chamber plan

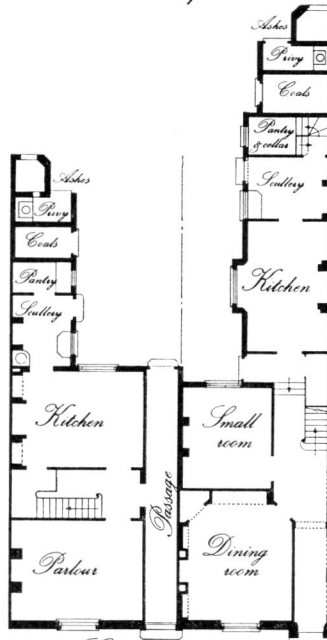

Ground floor plan
Nᵒ 31 | Nᵒ 29

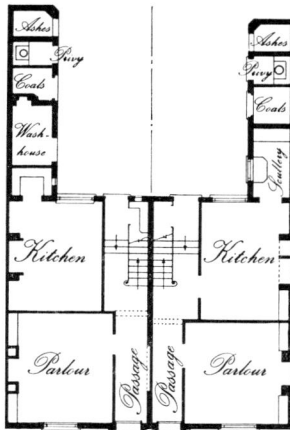

Ground floor plan
Nᵒ 37 | Nᵒ 35

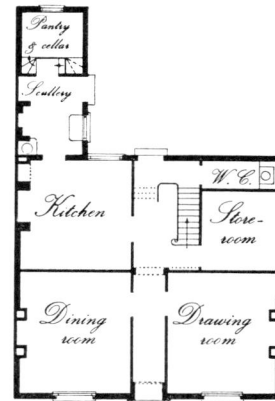

Ground floor plan
Nᵒ 17

Penley's Grove Street

10 0 10 20 30 40
Scale of Feet

Fig. 81.

(300) HOUSES, Nos. 35 and 37, were designed by J. B. and W. Atkinson for Mr. Benjamin Hopper in 1845 and 1847; although built separately they form a uniform mirror pair (Brierley, Leckenby and Keighley Fig. 81, p. 93).

No. 35 retains its original pilastered doorcase. The plan of No. 35 shows the parlour and kitchen on the ground floor, and also a scullery, privy, etc. in the rear wing. There were only two bedrooms on the first floor, though the frontage is only a little narrower than that of No. 31 (298). On plan No. 37 is similar to No. 35 (but reversed), though there are three bedrooms on the first floor. The cost was £220 each.

(301) HOUSES, Nos. 39–43 (odd), date from the second quarter of the 19th century. No. 39 has windows with cambered arches. No. 43, although ranging with the other two, was built separately. It has a carriageway with a segmental brick arch.

TANG HALL

(302) TANG HALL HOTEL, formerly Tang Hall, is a two-storey house of several dates between 1800 and 1850 and is built on the site of an earlier house, of which some stone walling remains in the cellars. It is a large detached house built to an irregular plan and with widely overhanging eaves; it has been much altered.

TOWNEND STREET

(303) HOUSES, Nos. 44–52 (even), were built c. 1840, forming a terrace of small single-fronted dwellings. Some have been converted to shops.

UNION TERRACE

(304) HOUSES, Nos. 1–5 (odd), 13–91 (odd), 2–70 (even), of two storeys, were built in the second quarter of the 19th century. Two occupied but newly erected houses were for sale there in YG 30/10/1847.

The houses are generally uniform in design, but some have basements and there are variations in the treatment of the eaves cornices. The simple doorcases have pilaster jambs and rectangular fanlights. Nos. 1–5 and 13–91 have small front gardens enclosed by dwarf brick walls and iron railings; some have gates hung to moulded circular standards, others to flat scrolled and foliated standards, all in cast iron (Fig. 82). Inside, the staircase usually occupies the back part of the entrance passage. *Nos. 1–68 and 70 demolished.*

ADDENDA

(305) HOUSE, Nos. 68, 70 Clifton, comprises a late 18th-century house of three storeys, now divided into two tenements. It was extensively damaged in the Second World War and has been largely rebuilt. A ground floor room in No. 68 contains a late 18th-century fireplace surround reset from No. 56 Skeldergate (*York* III (117), plate 73).

(306) BOUND STONES, in Clifton, of brown limestone with round heads (18th or 19th-century), inscribed ST. OL. P. (for St. Olave's Parish), two: (a) reset in a brick pier ajoining Green Tree Cottage (94); (b) built into the E. wall of No. 16 Clifton Green (89).

Fig. 82.
From (304) Union Terrace.

GLOSSARY

OF THE MEANING ATTACHED TO THE TECHNICAL TERMS USED IN THE INVENTORY

ABACUS—The uppermost member of a capital.

ABUTMENT—The solid lateral support of an arch.

ACANTHUS—A plant represented in stylised form in Classical and Renaissance ornament, in particular in the capitals of the Corinthian and Composite Orders.

ACHIEVEMENT—In heraldry, the shield accompanied by the appropriate external ornaments, helm, crest, mantling, supporters, etc. In the plural the term is also applied to the insignia of honour carried at the funerals and suspended over the monuments of important personages, comprising helmet and crest, shield, tabard, sword, gauntlets and spurs, banners and pennons. (*See also* HATCHMENT.)

ACROTERIA—In Classical architecture, blocks on the apex and lower ends of a pediment, often carved with honeysuckle or palmette ornament.

ALTAR TOMB—A modern term for a tomb of stone or marble resembling, but not used as, an altar.

ANNULET—In architecture, a small flat fillet encircling a column or shaft.

ANTHEMION—Honeysuckle or palmette ornament in Classical architecture.

APSE—A semicircular or polygonal recess, semi-domed or vaulted, in or projecting from a building.

ARABESQUE—A highly stylised fret-ornament in low relief, common in Moorish architecture, found in 16th and 17th-century work in England.

ARCADE—A range of arches carried on piers or columns. *Blind arcade*, a series of arches, frequently interlaced, carried on shafts or pilasters against a solid wall.

ARCH—The following are some of the most usual forms:

Flat or straight—Having the soffit horizontal.

Four-centred, depressed, Tudor—A pointed arch of four arcs, the two outer and lower arcs struck from centres on the springing line and the two inner and upper arcs from centres below the springing line. Sometimes the two upper arcs are replaced by straight lines.

Lancet—A pointed arch struck with radii greater than the span.

Ogee—A pointed arch of four or more arcs, the two uppermost being reversed, *i.e.*, convex instead of concave to the base line.

Pointed or two-centred—Two arcs struck from centres on the springing line, and meeting at the apex with a point.

Relieving—An arch, generally of rough construction, placed in the wall above the true arch or head of an opening, to relieve it of most of the superincumbent weight.

Segmental—A single arc struck from a centre below the springing line.

Segmental-pointed—A pointed arch, struck from two centres below the springing line.

Three-centred, elliptical—Formed with three arcs, the middle or uppermost struck from a centre below the springing line.

ARCHITRAVE—The lowest member of an entablature (*q.v.*); often adapted as a moulded enrichment returned round the jambs and head of a doorway or window opening.

ARCHIVOLT—In Classical architecture, the moulding round an arch.

ARRIS—The sharp edge formed by the meeting of two surfaces.

ASHLAR—Masonry wrought to an even face and square edges.

ASTRAGAL—A small semicircular moulding or bead.

ATTIC—A low storey above an entablature or cornice; also, a storey wholly or partly within the roof.

ATTIC BASE—A moulded column-base with a profile comprising two *torus* (convex) mouldings divided by a *scotia* (concave) between two fillets. In Romano-British examples the fillets are often omitted.

BAND or PLAT BAND—A flat projecting horizontal strip of masonry or brickwork across the face of a building, as distinct from a moulded string.

BARGE BOARD—A timber plank, often carved, fixed to the edge of a gabled roof at a short distance from the face of the wall, to protect projecting timbers.

BARREL VAULTING—*See* VAULTING.

BATTLEMENT—In fortification, the alternating merlons and embrasures on the parapet or breastwork of a rampart walk; hence *Battlemented* (or *Embattled*, a usage for the decorative adaptation of the feature).

BAY—The main divisions of a building or feature, on plan or in elevation, defined by recurring structural members, as in an arcade, a fenestrated façade or a timber frame.

BEADING—Small round moulding.

BILLET—In architecture, an ornament used in the 11th and 12th centuries consisting of short attached cylinders or rectangles with spaces between.

BOLECTION MOULDING—A bold moulding of double curvature raised above the general plane of the framework of a door, fireplace or panelling.

BOND—*See* BRICKWORK.

BOSS—A projecting square or round ornament, covering the intersections of the ribs in a vault, panelled ceiling or roof, etc.

BOTTOM RAIL—The lowest horizontal timber of a door, partition, window sash, etc.

BRACE—In timber framing and timber roof construction, subsidiary timber rising obliquely from a major vertical member to support a major horizontal member (in contradistinction to a STRUT, *q.v.*). *Arch-brace*, when curved. *Wind-brace*, a subsidiary timber placed diagonally between the principals and purlins of a roof to increase resistance to wind-pressure.

BRESSUMER—A spanning beam forming the direct support of a wall or timber framing above it.

BRICKS and BRICKWORK

Cutter—A brick of very fine quality used for arches, quoins, etc. and capable of being cut.

Header—A brick laid so that the end appears on the wall face.

Stretcher—A brick laid so that the side appears on the wall face.

English Bond—A method of laying bricks so that alternate courses appear as all headers and all stretchers on the wall face.

Flemish Bond—In which alternate headers and stretchers in each course appear on the wall face.

BUTTRESSES—Projecting masonry or brickwork support to a wall.

Angle—Two meeting, or nearly meeting, at right-angles at a corner.

Clasping—Returned to encase an angle.

Diagonal—Projecting diagonally at a corner.

Lateral—At a corner of a building and axial with one wall.

CABLE MOULDING—A moulding carved in the form of a rope or cable.

CANOPY—A projection or hood over a door, window, etc.; the covering over a tomb or niche.

CARTOUCHE—In Renaissance ornament, a tablet imitating a scroll with ends rolled up, used ornamentally or bearing an inscription or arms.

CASEMENT—1. A wide hollow moulding in a window jamb, etc.; 2. the hinged part of a window which opens sideways.

CAVETTO—A hollow moulding, in profile a quarter-circle.

CENTRING—Temporary wooden framework used to support an arch or vault during construction.

CHALICE—The name used in the Inventory to distinguish the pre-Reformation type of Communion cup with a shallow bowl from the post-Reformation cup with a deeper bowl.

CHAMFER—The small plane formed when a sharp edge or arris is cut away, usually at an angle of 45°; *hollow chamfer*, when the plane is concave; *sunk chamfer*, when it is recessed.

CHEVRON—In architecture, a decorative form resembling an inverted V and often used in a consecutive series.

CINQUEFOIL—*See* FOIL.

CLERESTOREY—An upper stage, pierced by windows, in the main walls of a church or domestic building.

COFFERS—Sunk panels in ceilings, vaults, domes and arch-soffits.

COLLAR BEAM—In a roof, a horizontal beam framed to and serving to tie together a pair of rafters at some distance above wall-plate level.

COLLAR PURLIN—*See* PURLIN.

CONSOLE—A bracket with a compound-curved outline.

COPED SLAB—A slab of which the upper face is ridged down the middle, and sometimes hipped at each end.

CORBEL—A projecting stone or piece of timber for the support of a super-incumbent weight. *Corbel-table*—A row of corbels, usually carved, and supporting a projection.

CORNICE—A crowning projection. In Classical architecture, the crowning or upper portion of the entablature.

COVER PATEN—A cover to a Communion cup, used as a paten.

CREST—A device worn upon a helm or helmet. *Cresting*, an ornamental finish along the top of a screen, etc.

CROCKETS—Carved projections spaced, usually at regular intervals, along the vertical or sloping sides of spires, canopies, pinnacles, hood-moulds, etc.

CROP-MARK—Trace of a levelled or buried feature revealed on the land surface by differential growth of crops, especially after drought.

CROWN POST—A vertical post standing centrally on a tie-beam to give direct support to a collar and collar purlin, and additionally to the collar purlin through two-way braces.

CUSHION CAPITAL—A capital cut from a cube with its lower angles rounded off to adapt it to a circular shaft.

CUSP—A pointed projection from the soffit of an arch, formed by two arcs of smaller radius. The foils in Gothic windows, arches, panels, etc., are formed by *cusping* and *sub-cusping*, often ornamented at the ends (*cusp-points*) with carving.

DADO—The separate protective or decorative treatment applied to the lower parts of wall-surfaces to a height, normally, of 3 ft. to 4 ft. *Dado-rail*, the moulding or capping at the top of the dado.

DENTILS—The small rectangular tooth-like blocks used decoratively in Classical cornices.

DIAPER—All-over decoration of surfaces with squares, diamonds, or other patterns.

DIE—The part of a pedestal between the base and the cornice.

DOG-LEG STAIRCASE—*See* STAIRCASE.

DOG-TOOTH ORNAMENT—A typical 13th-century carved enrichment consisting of a series of pyramidal flowers of four petals; often used to enrich hollow mouldings.

DORMER WINDOW—A vertical window on the slope of a roof and having a roof of its own.

DOUBLE-OGEE MOULDING—*See* OGEE.

DRESSINGS—The building materials, specially chosen or treated, defining or emphasising the architectural features of an elevation.

DRIP STONE—*See* HOOD MOULD.

EMBATTLED—*See* BATTLEMENT.

EMBRASURES—The openings in an embattled parapet, or the recesses for windows, doorways, etc.

ENTABLATURE—In Classical and Renaissance architecture, the part of an order above the column, the full entablature comprising *architrave*, *frieze*, and *cornice*; often used alone, in whole or in part, as a horizontal feature.

EXTRADOS—The outer curve of the voussoirs of an arch.

FANLIGHT—Glazed opening immediately over, and integrated within the framing of, a doorway.

FASCIA—A plain or moulded facing board.

FILLET—In mouldings, a plain narrow band between, or adjacent to, more complex mouldings.

FINIAL—A stylised ornament at the top of a pinnacle, gable, canopy, etc.

FOIL (*trefoil, quatrefoil, cinquefoil, multifoil,* etc.)—The shape defined by the curves formed by cusping.

FOLIATED (of a capital, corbel, etc.)—Carved with leaf ornament.

FOUR-CENTRED ARCH—*See* ARCH.

FRIEZE—The middle zone in an *entablature*, between the *architrave* and the *cornice*; generally any band of ornament or colour immediately below a cornice.

GADROONING—Decorative enrichment comprising a series of convex ridges, the converse of fluting, forming an ornamental edge or band.

GARDEROBE—Wardrobe. Antiquarian usage applies it to a latrine or privy chamber.

GAUGING—In brickwork, cutting and rubbing bricks to a particular shape. Specially made soft bricks are used for the purpose.

GESSO—A mixture of whiting and size, spread on stone or wood as a ground for painting.

GRAFFITO—Scratched inscription or design.

GRISAILLE—Painting in shades of grey.

GROINING, GROINED VAULT—*See* VAULTING.

GUILLOCHE—A geometrical ornament consisting of two or more undulating bands intertwining to form a series of circles.

GUTTAE—Small stud-like projections under the triglyphs and mutules of the Doric entablature.

HALL—In a mediaeval house, the principal room, often open to the roof.

HATCHMENT—In modern usage, the large square or lozenge-shaped framed painting displaying the armorial bearings of a deceased person. It was first hung outside his house and then laid up in the church.

HIPPED ROOF—A roof with sloping instead of vertical ends. *Half-hipped*, a roof with ends partly vertical, and partly sloping.

HOG-BACK—A type of late Saxon stone grave-cover shaped with a curved ridge forming a 'hog-back'.

HOOD MOULD or LABEL—A projecting moulding on the wall face above an opening or feature; it may follow the form of the arch or head of the same or be square in outline.

IMPOST—The projection, often moulded, at the springing of an arch, upon which the arch appears to rest.

INDENT—The sinking in a slab for a monumental brass.

INTRADOS—The inner curve of an arch.

JAMBS—The sides of an archway, doorway, window, or other opening.

JETTY—The projection of an upper storey of a building beyond the plane of a lower storey.

JOGGLING—The method of cutting the adjoining faces of the voussoirs of an arch with rebated, zigzagged or wavy surfaces to provide a better key or lodgement.

KEYSTONE—The middle voussoir in an arch.

KING-POST—A vertical post extending from a tie-beam or a collar-beam to the apex of a roof, and supporting a ridge-piece.

KNEELER—In a parapeted gable, the stone or block built well into the wall to resist the sliding tendency of the coping.

LACING COURSE—In masonry or brickwork, a bonding course binding the wall-facing together or to the wall core.

LANCET—A tall, narrow window with a pointed head, typical of the 13th century. *See also* ARCH.

LINENFOLD PANELLING—Wainscot ornamented with stylised representation of folded linen.

LINTEL—The horizontal beam or stone bridging an opening.

LOOP—A small narrow light, often unglazed.

LOUVRE—A lantern–like structure surmounting the roof of a hall or other building, with openings for ventilation or the escape of smoke; the openings are usually crossed by sloping slats (*louvre boards*) to exclude rain. Louvre boards are also used in the windows of church belfries, instead of glazing, to allow the bells to be heard.

MANSARD—*See under* ROOFS.

MASK STOP—*See under* STOPS.

METOPES—The panels, often carved, filling the spaces between the triglyphs in the Doric entablature.

MIDDLE RAIL—A horizontal rail between ground sill and wall-plate in a timber–framed wall.

MITRE—The junction of mouldings or strips meeting at an angle; in joinery the joint is commonly on the line of the mitre. A junction in which the joint is straight and not at the angle of the mitre is termed a *mason's mitre*.

MODILLIONS—Brackets under the cornice in a Classical entablature.

MORTICE—A socket cut in a piece of wood, usually to receive the end, the *tenon*, of another piece.

MULLION—An upright of timber, stone or brick dividing an opening into lights.

MUNTIN—In panelling, an intermediate vertical timber between panels and butting into or stopping against the rails.

NAIL HEAD—Small architectural enrichment of pyramidal form, used extensively in 12th–century work.

NECKING or NECK MOULDING—The narrow moulding round the lower extremity of a capital.

NEWEL—The central post in a circular or winding staircase; also the principal post at each angle of a dog–legged or well staircase.

OFFSET—A ledge formed by the set–back of a wall.

OGEE—A compound curve of two parts, one convex, the other concave. A *double–ogee* moulding is formed by two ogee mouldings meeting at their convex ends.

ORDERS (of arches)—Receding concentric rings of voussoirs.

ORDERS OF ARCHITECTURE—In Classical or Renaissance architecture, the five systems of columnar architecture, known as Tuscan, Doric, Ionic, Corinthian, and Composite. *Colossal Order*, one in which the columns or pilasters embrace more than one storey of the building.

ORIEL WINDOW—A projecting bay–window carried upon corbels or brackets.

OVERSAILING COURSE—A brick or stone course projecting beyond the course below it.

OVOLO MOULDING—A Classical moulding forming a quarter round or semi–ellipse in section.

PALLADIAN or VENETIAN WINDOW—A three–light window, with a tall round–headed middle light and shorter lights on either side, the side lights with flanking pilasters and small entablatures forming the imposts to the arch over the centre light.

PATEN—A dish for holding the Bread at the celebration of Holy Communion.

PATERA, –AE—A square or circular flat ornament applied to a frieze, moulding or cornice; in Gothic work it commonly takes the form of a four–lobed leaf or flower.

PEDIMENT—A low–pitched gable used in Classical and Renaissance architecture above a portico, at the end of a building, or above doorways, windows, niches, etc.; sometimes the gable angle is omitted, forming a *broken pediment*, or the horizontal members are omitted, forming an *open pediment*. A curved gable form is sometimes used in this way.

PELLET ORNAMENT—An enrichment consisting of balls or flat discs.

PILASTER—A shallow pier of rectangular section attached to a wall.

PISCINA—A basin for washing the sacred vessels and provided with a drain, generally set in or against the wall to the S. of the altar, but sometimes sunk in the pavement.

PLAT BAND—*See* BAND.

PLINTH—The projecting base of a wall, generally chamfered or moulded at the top.

PODIUM—In Classical architecture, a basis, usually solid, supporting a temple or other superstructure.

PORTICO—A covered entrance to a building, colonnaded, either constituting the whole front of the building or forming an important feature.

PRINCIPALS—In a roof of double–frame construction, the main as opposed to the common rafters.

PURLIN—*Collar purlin*, a beam running longitudinally immediately beneath the collars joining pairs of common rafters. *Side purlin*, a horizontal longitudinal member resting on or tenoned into the principal rafters of a truss and giving intermediate support to the common rafters.

QUARRY—In glazing, small panes of glass, generally diamond–shaped or square set diagonally.

QUEEN-POSTS—In a roof truss, a pair of vertical posts equidistant from the centre line of the roof. *See also under* ROOFS.

QUOINS—The dressed stones at the angle of a building, or distinctive brickwork in this position.

RAFTERS—Inclined timbers supporting a roof–covering. *See also under* ROOFS.

RAIL—A horizontal member in the framing of a door, screen, or panel.

REAR ARCH—The arch on the inside of a wall spanning a doorway or window–opening.

REBATE—A continuous rectangular notch.

REEDING—The converse of fluting, *i.e.* with convex not concave moulding.

REREDOS—A screen of stone or wood at the back of an altar, usually enriched.

RESPONDS—The half–columns or piers at the ends of an arcade or at each side of a single arch.

REVEAL—The internal side surface of a recess, especially of a doorway or window opening.

RIDGE (or RIG) AND FURROW—Remains of old cultivations.

RISER—The vertical piece connecting two treads in a flight of stairs.

ROOD (*Rood beam, Rood screen, Rood loft*)—A cross or crucifix. The *Great Rood* was set up in the E. end of the nave with accompanying figures of St. Mary and St. John; it was generally carved in wood, and fixed on the loft or head of the rood screen, or on a special beam (the *Rood beam*).

ROOFS

Collar–beam—a principal–rafter roof (*q.v.*) with collar–beams connecting the principals.

Crown–post—a trussed–rafter roof with a central post (crown-post) standing on a tie–beam and carrying a central purlin supporting the collars.

King–post—in which a central post (king-post) standing on the tie–beam or collar–beam of a truss directly supports the ridge.

Mansard—characterised in exterior appearance by two pitches, the lower steeper than the upper.

Principal–rafter—with rafters of greater scantling than the common rafters framed to form trusses at regular intervals along the roof; normally called by the name of the connecting member used in the truss, tie–beam or collar–beam. Post-mediaeval roofs of this kind often have queen-posts.

Queen–post—with two vertical or nearly vertical posts (queen-posts) standing towards either end of the tie–beam of a truss and supporting the collar–beam or the principal–rafters.

Tie–beam—a principal–rafter roof with a simple triangulation of a horizontal beam linking the lower ends of the pairs of principals to prevent their spread.

RUBBLE—Walling of rough unsquared stones or flints. *Coursed Rubble*, rubble walling with the stones or flints very roughly dressed and levelled up in courses; in *Regular Coursed Rubble* the stones or flints are laid in distinct courses, being kept to a uniform height in each course.

RUSTICATION—Primarily, masonry in which only the margins of the stones are worked; also used for any masonry where the joints are emphasised by mouldings, grooves, etc.; rusticated columns are those in which the shafts are interrupted by square blocks of stone or broad projecting bands.

SACRISTY—A room generally in immediate connection with a church, in which the holy vessels and other valuables are kept.

SARCOPHAGUS—A stone coffin, usually inscribed and often embellished with sculptures, intended to be viewed above ground or in a tomb chamber.

SCALLOPED CAPITAL—A development of the cushion capital (q.v.) in which the single cushion is elaborated into a series of truncated cones.

SHAFT—A slender column. *Shafted jambs*, reveals of a wall opening elaborated with one or more shafts, either engaged or detached.

SIDE-PURLIN—*See* PURLIN.

SILL—The lower horizontal member of a window or door-frame; the stone, tile or wood base below a window or door-frame, usually with a weathered surface projecting beyond the wall face to throw off water. In timber-framed walls, the lower horizontal member into which the studs are tenoned.

SOFFIT—The under side of an arch, staircase, lintel, cornice, canopy, etc.

SOFFIT CUSPS—Cusps springing from the flat soffit of an arch, and not from its chamfered sides or edges.

SPANDREL—The more or less triangular space between an angle and a contained curve.

SPLAY—A sloping face making an angle of more than a right angle with another face, as in internal window jambs, etc.

SPRINGING LINE—The level at which an arch springs from its supports.

SPUR—Carved tongue, foliage or grotesque filling each spandrel between a circular base and a square or polygonal plinth.

SQUINCH—An arch thrown across the angle between two walls to support a superstructure, such as the base of a stone spire.

SQUINT—A piercing through a wall to allow a view of an altar from places whence it could otherwise not be seen.

STAGES—The divisions (*e.g.* of a tower) marked by horizontal string-courses.

STAIRCASE—A *close-string* staircase is one having a raking member into which the treads and risers are housed. An *open-string* staircase has the raking member cut to the shape of the treads and risers. A *dog-leg* staircase has adjoining flights running in opposite directions with a common newel. A *well staircase* has stairs rising round a central opening more or less as wide as it is long.

STILE—The vertical members of a frame into which are tenoned the ends of the rails or horizontal pieces.

STOPS—Blocks terminating mouldings or chamfers in stone or wood; stones at the ends of labels, string-courses, etc., against which the mouldings finish, frequently carved to represent shields, foliage, human or grotesque masks; also, plain or decorative, used at the ends of a moulding or a chamfer to form the transition thence to the square.

STOUP—A receptacle, normally by the doorway of a church, to contain holy water; those remaining are usually in the form of a deeply-dished stone set in a niche or on a pillar.

STRAIGHT JOINT—A vertical joint in a wall usually signifying different phases of building.

STRING, STRING COURSE—A projecting moulded band across a wall. *See also* STAIRCASE.

STRUT—In timber framing and roof-construction a subsidiary oblique timber rising from a horizontal member to give support to a vertical post or to a rafter (in contradistinction to BRACE, *q.v.*).

STUDS—The common posts or uprights in timber-framed walls.

SWAG—Decorative representation of a festoon of cloth or flowers and fruit suspended from both ends.

TIE-BEAM—The horizontal transverse beam in a roof, tying together the feet of pairs of rafters to counteract thrust.

TIMBER-FRAMED BUILDING—A building of which the walls are built of open timbers and the interstices filled in with brickwork or lath and plaster ('wattle and daub') etc., the whole often covered with plaster or boarding.

TRABEATION—The use of horizontal beams in building construction; descriptive in the Inventory of conspicuous cased ceiling-beams.

TRACERY—The ornamental work in the head of a window, screen, panel, etc., formed by the curving and interlacing of bars of stone or wood, grouped together, generally over two or more lights or bays.

TRANSOM—An intermediate horizontal bar of stone or wood across a window-opening. The horizontal member of a door-frame beneath a fanlight.

TREAD—The horizontal platform of a step or stair.

TRELLIS—Latticework of light wood or metal bars.

TRIGLYPHS—Blocks, with vertical channels, placed at intervals along the frieze of the Doric entablature.

TRUSS—A number of timbers framed together to bridge a space, to be self-supporting, and to carry other timbers. The trusses of a roof are generally named after a particular feature in their construction, e.g. *King-post, Queen-post*; see under ROOFS.

TYMPANUM—The triangle in the face of a pediment or the semicircle in the head of an arch.

VAULTING—An arched ceiling or roof of stone or brick, sometimes imitated in wood or plaster. *Barrel vault*, a tunnel vault unbroken in its length by cross vaults. *Groined vault* (or *cross vault*), resulting from the intersection of simple vaulting surfaces. *Ribbed vault*, with a framework of arches carrying the covering of the spaces between them. One bay of vaulting divided into four quarters or compartments is termed *quadripartite*.

VENETIAN WINDOW—*See* PALLADIAN WINDOW.

VERGE—The slightly projecting edge of a roof-covering along the sloping gable-end of a roof.

VOUSSOIRS—The wedge-shaped stones forming an arch.

WAINSCOT—Wood panelling. Oak imported for this purpose from the Baltic was also so called.

WALL-PLATE—A timber laid lengthwise at the wall top to receive the ends of the roof rafters and other joists. In timber-framing, the studs are also tenoned into it.

WAVE MOULDING—A compound moulding formed by a convex curve between two concave curves.

WEATHERING (to sills, tops of buttresses, etc.)— A sloping surface for casting off water.

WELL STAIRCASE—*See* STAIRCASE.

WIMPLE—Scarf covering chin and throat.

WIND BRACE—*See* BRACE.

WINDOW—*Casement*, with hinged glazed panels. *Sash* (or *hung sash*), glazed panels sliding vertically. *Sliding-sash*, with glazed panels sliding horizontally (also *Yorkshire sash*).

YORKSHIRE SASH—*See* WINDOW—*Sliding-sash*.

INDEX

Numbers in brackets refer to the serial numbers of the monuments.
The letters 'a' and 'b' denote left- and right-hand columns respectively.

Park Place, (201) 80a.

Parliament Street: burial ground in, xxxv; grave-slabs from, xlv, Pl. 25.

Pavilions, 19th-century: (13) 45b; (29) 53b.

Paynelathes Crofts: xxxviii; name of street derived from, 91b.

Pearson, Dr., rector of South Kilworth, Leics., roof and instruments given by, (12) 45a.

Peckitt, William, see Craftsmen and Tradesmen, Glass-Painters and Glaziers.

Penley's Grove Street, 91–4.

Penty, W. G., door-hood added by, (86) 68a.

Percy, William de, monks quarrelled with, (4) 3a.

Phillips, John, keeper of Yorkshire Museum, (4) 7b.

Pickard, Leonard and the Rev., house owned by, (42) 58a.

Pikeing Well, New Walk, (30) 53b, Fig. 48, p. 54.

Pinnacle, on merlon, (4) 20b.

Piper Lane Close, gasworks on part of, (20) 47b.

Piscina, 14th-century, (4) 23a, Pl. 42.

Place, Francis, see Artists and Surveyors.

Plaque, bronze, (4) 22a.

Plaster, see Building Materials.

Plasterwork, l, liv.
  16th-century, (11) 32a, 40b, Pl. 65.
  18th-century: (39) 56b, Pl. 126; (128) 74a, Pl. 122.
  19th-century: (48) (49) 62a, Pl. 122; (57) 63a, Pl. 122; (155) 77a.
  See also Ceilings, Plaster Decorated.

Plate, Church
  Brass
    Almsdishes
      16th-century (German), (7) 25a.
      18th-century, (9) 29a.
    Silver
    Almsdish
      19th-century, (6) 24b.
    Cups
      15th/16th-century, (9) 29a (stem and foot).
      16th-century, (8) 25b.
      17th-century: (7) 25a; (9) 29a.
      18th and 19th-century, (6) 24b.
    Flagon
      18th-century: (8) 25b (whereabouts unknown); (9) 29a.
    Patens
      17th-century, (8) 25b.
      18th-century: (6) 24b; (7) 25a; (9) 29a.
      19th-century: (6) 24b; (8) 25b.
    Salver
      18th-century, (6) 24b.

Plate tracery, (8) 25b, Pl. 48.

Plows, William Abbey, see Sculptors and Monumental Masons.

Plumpton, Roger, bequest of, (9) 27a.

Poole
  David 1830, and two daughters, monument, (9) 29a.
  James, drawing by, (9) 27b.

Porches
  16th-century, (11) 39a.
  19th-century: (44) (45), 60a, Pl. 87; (49) 62a; (57) 63a, Pl. 90; (95) 68b; (97) (98) 69a; (105) 71a, Pl. 99; (109) 71b; (145) 75b, Pl. 104; (147) 76a; (164) 78a, Pl. 99; (171) (174) 78b; (200) 80a; (230) 83b.

Portcullis, slot for, (4) 17b.

Postern, 15th-century, (4) 6a, 21b, Pl. 23.
  „ Tower, see St. Mary's Abbey, Precinct Wall and Towers.

Poulter, John, public house kept by, (177) 78b.

Prescott, Col. Robert, house occupied by, (128) 74a.

Presidents, of the Council of the North
  Hastings, Henry, Earl of Huntingdon, (11) 32a, 40b.
  Manners, Henry, Earl of Rutland, (11) 31a, b, 41a.
  Radcliffe, Thomas, Earl of Sussex, xlviii, (11) 31b, 41a.
  Sheffield, Edmund., Lord, (4) 34a, 36b, 43b.
  Wentworth, Thomas, Earl of Strafford: (11) 34b, 41b; arms of (11) 35a, 42a.
  Young, Archbishop Thomas, (11) 31b, 41a.

Preston, John, widow of, contract for sale of houses by, (278) 89b.

Price, view by, (4) 21a.

Pritchett, J. P., see Architects.

Proprietary School, buildings erected for, (29) 53b.

Punch Bowl Hotel, (141) 75a.

Queen, The, public house, (207) 80b.
  „ Margaret's Arch, see St. Mary's Abbey, Precinct Wall and Towers.
  „ Victoria, The, public house formerly called, (207) 80b.

Queen's Villa, house formerly known as, (164) 77b.

Radcliffe, Thomas, Earl of Sussex, President of the Council of the North, xlviii, (11) 31b, 41a.

Railings, see Ironwork.

Rainwater Heads and Pipes, see Leadwork.

Ranges, see Ironwork.

Rawcliffe Lane, stone formerly standing at the corner of, (32) 54a.

Record House, (49) 62a, Pl. 98.

Redeness Street, 82b, Fig. 71.

Regent Street, 82b.

Religious Foundations
  Augustinian
    St. Andrew, priory of: xxxviii; (5) 24b; 69a.
  Benedictine
    St. Mary's Abbey: xl; (4) 3a.
  Carmelite
    Friary, sites of, xxxvii, xlvi.
  Hospitals, sites of
    St. Anne, xxxvii.
    St. Anthony, xxxvii.
    St. Helen, Fishergate, xxxviii.
    St. Loy, xxxviii.
    St. Mary, xxxvii.
    St. Mary Magdalene, xxxvii.

Rennie, John, report of, (19) 47b.

Reresby, Sir John, Governor of York, (11) 36a.

Retreat, The: (24) 51a, Pl. 82; earthwork in grounds of, (1) 1a; (3) 2a; property acquired by, (145) (147) 75b.

Richards, Jacob, drawing by, (11) 35b.

Richmond, Constable of, xxxvii.

Riddle, R. W., house owned by, (49) 62a.

Ridge-and-furrow, (3) 2a.

Rigg, John and Ann, monument to sons and daughters of, 1830, (7) 25a.

Roberts, William, mercer, property bought by, (69) 65a.

Robinson
  Mrs., house occupied by, (45) 60a.
  Sir Tancred, Bart., (11) 36a, b.

Rockingham House, (205) 80b, Pl. 94.

Roman
  Building stone, reused, xli.
  Coffins, reused, xxvii, n. 1.
  Fortifications, xxviii.
  Settlement, xxv.
  Tooling, marks of, (4) 7b.
  Wharves, xxvi.

Roofs and Roof-trusses
  Collar-Beam
    17th-century: (11) 41b; (23) 51a; (43) 59b.
    Undated, (11) 41a.
  King-Post
    15th-century, xlviii, (11) 40a, b, Pl. 63.
  Mansard
    18th-century, (47) 61b.
  Queen-Post
    18th-century, (42) 59a.
  Queen-Strut
    17th-century, (11) 40a.
  Tie-Beam
    16th-century, (11) 41a.

Printed in England for Her Majesty's Stationery Office
by Ebenezer Baylis & Son Ltd, The Trinity Press
Worcester, and London
Dd. 500613 K24